Biocolours

Ravi Kant Bhatia • Abhishek Walia
Editors

Biocolours

Revolutionizing the Palette of a Sustainable World

Springer

Editors
Ravi Kant Bhatia
Department of Biotechnology
Himachal Pradesh University
Summer Hill
Shimla, Himachal Pradesh, India

Abhishek Walia
Department of Microbiology
College of Basic Sciences, CSKHPKV
Palampur, Himachal Pradesh, India

ISBN 978-981-95-1737-4 ISBN 978-981-95-1738-1 (eBook)
https://doi.org/10.1007/978-981-95-1738-1

This Springer imprint is published by the registered company Springer Nature Singapore Pte Ltd.
The registered company address is: 152 Beach Road, #21-01/04 Gateway East, Singapore 189721, Singapore

Foreword

In an era where sustainability has become a global imperative, the quest for eco-friendly alternatives to synthetic products has gained unprecedented momentum. Among these alternatives, biocolours stand out as a transformative innovation, redefining the palette of a sustainable world. This book, *Biocolours: Revolutionizing the Palette of a Sustainable World*, delves into the fascinating realm of natural dyes and pigments derived from biological sources, offering a comprehensive exploration of their potential to replace conventional, often harmful, synthetic colourants.

The journey of biocolours is not just a scientific or industrial endeavour; it is a cultural and environmental revolution. From ancient civilizations that utilized plant-based dyes for textiles and art to modern advancements in microbial and algal pigments, biocolours have always been intertwined with human creativity and ingenuity. Today, as we face the pressing challenges of environmental degradation and resource depletion, biocolours emerge as a beacon of hope, offering a sustainable and renewable solution to meet the growing demand for colour in industries ranging from textiles and food to cosmetics and pharmaceuticals.

This book brings together the collective expertise of researchers and sustainability stakeholders to provide a holistic understanding of biocolours. It explores their sources, extraction methods, applications, and the profound environmental benefits they offer. Moreover, it highlights the challenges and opportunities in scaling up their production and adoption, emphasizing the need for interdisciplinary collaboration and innovative thinking. As we stand at the crossroads of tradition and innovation, biocolours represent a harmonious blend of both. They remind us that sustainability is not just about reducing harm but also about rediscovering and reimagining the natural world's boundless potential. This book is a testament to that vision, inspiring readers to embrace biocolours as a vibrant and sustainable alternative for a brighter, greener future. We invite you to embark on this enlightening journey and discover how biocolours are revolutionizing the palette of a sustainable world.

Naveen Kumar
Hon'ble Vice-Chancellor, CSK Himachal Pradesh
Agricultural University
Palampur, Himachal Pradesh, INDIA

Preface

The world is at a pivotal moment in its history, where the need for sustainable solutions has never been more urgent. As industries grapple with the environmental consequences of synthetic products, the search for eco-friendly alternatives has become a global priority. Among these alternatives, biocolour-natural dyes and pigments derived from biological sources have emerged as a groundbreaking innovation, offering a sustainable and vibrant solution to replace conventional synthetic colours.

The book entitled *Biocolours: Revolutionizing the Palette of a Sustainable World* is a culmination of extensive research, innovation, and collaboration aimed at exploring the potential of biocolours to transform industries and contribute to a greener future. This book is designed to serve as a comprehensive resource for researchers, industry professionals, students, and sustainability enthusiasts who are keen to understand the science, applications, and impact of biocolours to create a sustainable world. The journey of biocolours is deeply rooted in history, with ancient civilizations harnessing the power of natural sources like plants, insects, and minerals to create dyes for textiles, art, and cultural practices. Today, advancements in biotechnology and sustainable practices have expanded the scope of biocolours, enabling the production of pigments from microorganisms, algae, and even waste materials. These innovations not only reduce reliance on petrochemical-based synthetic dyes but also minimize environmental pollution, conserve resources, and promote circular economy principles.

This book is structured to provide a holistic understanding of biocolours, beginning with their historical significance and progressing to modern extraction techniques, applications across industries, and their role in achieving sustainability goals. Each chapter is meticulously crafted to highlight the latest research, case studies, and future directions, offering readers a balanced perspective on the opportunities and challenges in this field. One of the key objectives of this book is to inspire interdisciplinary collaboration. The development and adoption of biocolours require the collective efforts of scientists, engineers, policymakers, and industry leaders. By fostering a deeper understanding of biocolours, we hope to encourage innovation and drive the transition toward sustainable practices in colour production and application.

As editors of this amazing book we are deeply grateful to the countless authors, institutions, and organizations whose work has paved the way for this book. Their dedication to sustainability and innovation has been a constant source of inspiration. We also extend our heartfelt thanks to our readers, whose curiosity and commitment to a sustainable future motivate us to explore and share knowledge. We invite you to delve into the pages of this book and discover the transformative potential of biocolours. Together, let us embrace this vibrant revolution and contribute to a sustainable world where colour is not just a visual delight but also a testament to our respect for nature and its resources to create a sustainable world.

Shimla, Himachal Pradesh, India — Ravi Kant Bhatia
Palampur, Himachal Pradesh, India — Abhishek Walia

Contents

Editors and Contributors

About the Editor

Ravi Kant Bhatia is an Assistant Professor in the Department of Biotechnology at Himachal Pradesh University, Shimla, Himachal Pradesh, India and recognized in the World's Top 2% Scientists List published by Stanford University and Elsevier. He has cleared UGC-NET, ICAR-NET, and GATE, and completed a UGC-PDF-funded postdoctoral fellowship. With extensive experience in both industry and academia, he is actively engaged in teaching and research, focusing on environmental biotechnology for sustainable growth and development. He has authored various research publications and, completed multiple funded projects. A recipient of the SERB-ITS fellowship for presenting research in Scotland, he also oversees the biotechnology incubation centre (BIC-HPU) to promote bio-entrepreneurship.

Abhishek Walia is an Assistant Professor in the Department of Microbiology, College of Basic Sciences at Chaudhary Sarwan Kumar Himachal Pradesh Agricultural University, Palampur, Himachal Pradesh, India. He holds a PhD in microbiology, recognized with a gold medal by HIMOTKARSH Parishad Una (HP) for securing the top position in the college of forestry. With 9 years of teaching and research experience, he has previously headed the Department of Microbiology at DAV University, Jalandhar, Punjab, India.He specializes in agricultural and environmental microbiology, focusing on sustainable technologies with industrial applications. He has an impressive publication record, including various research papers, a few review articles, and some book chapters and conceptual articles. He has received prestigious accolades, including the DST-Inspire Fellowship by the Government of India, and serves as an editorial board member and review editor for reputed national and international journals.

Contributors

Sayasta Ahmed Environmental Molecular Microbiology and Applications Laboratory, Department of Molecular Biology & Bioinformatics, Tripura University (A Central University), Suryamaninagar, Tripura, India

Jyoti Bhardwaj Department of Biotechnology, CSIR-IHBT, Palampur, India

Sakshi Bhardwaj Department of Food Science and Technology, Dr. Y. S. Parmar University of Horticulture and Forestry, Nauni, Solan, HP, India

Ravi Kant Bhatia Department of Biotechnology, Himachal Pradesh University, Summer Hill, Shimla, Himachal Pradesh, India

Tanvi Bisht Department of Biotechnology, Himachal Pradesh University, Summer Hill, Shimla, Himachal Pradesh, India

Priyanjana Bose Amity Institute of Biotechnology, Amity University, Noida, Uttar Pradesh, India

Basanti Brar Department of Biotechnology, Om Sterling Global University, Hisar, Haryana, India

Anu Chaudhary Department of Biotechnology, Himachal Pradesh University, Summer Hill, Shimla, Himachal Pradesh, India

Maithilee Chaudhary GN Ramachandran Protein Centre, CSIR-IMTECH, Chandigarh, India

Pritam Chhetri Environmental Molecular Microbiology and Applications Laboratory, Department of Molecular Biology & Bioinformatics, Tripura University (A Central University), Suryamaninagar, Tripura, India

Barsha Datta Environmental Molecular Microbiology and Applications Laboratory, Department of Molecular Biology & Bioinformatics, Tripura University (A Central University), Suryamaninagar, Tripura, India

Laurent Dufossé Laboratoire de Chimie et Biotechnologie des Produits Naturels (CHEMBIOPRO), Université de La Réunion, ESIROI Agroalimentaire, Saint-Denis Cedex, France

Neha Gautam Department of Food Science and Technology, Dr. Y. S. Parmar University of Horticulture and Forestry, Nauni, Solan, HP, India

Charu Gupta Amity Institute of Herbal Research and Studies, Amity University, Noida, UP, India

Seema Joshi Department of Environmental Science, Om Sterling Global University, Hisar, Haryana, India

Poonam Jyoti Department of Biotechnology, CSIR-IHBT, Palampur, India

Preeti Mehta Kakkar Amity Institute of Biotechnology, Amity University, Noida, Uttar Pradesh, India

Kamakshi Amity Institute of Biotechnology, Amity University, Noida, Uttar Pradesh, India

Neelam Kaushik Department of Microbiology, Himachal Pradesh University, Shimla, India

Neha Kesharwani Faculty of Chemical Sciences, INSH, Shri Ramswaroop Memorial University, Barabanki, Uttar Pradesh, India

Nitesh Kumar Department of Biosciences, Himachal Pradesh University, Shimla, India

Rakshak Kumar Environmental Molecular Microbiology and Applications Laboratory, Department of Molecular Biology & Bioinformatics, Tripura University (A Central University), Suryamaninagar, Tripura, India

Dolly Kumari Faculty of Chemical Sciences, INSH, Shri Ramswaroop Memorial University, Barabanki, Uttar Pradesh, India

Neelam Kumari Department of Biosciences, Himachal Pradesh University, Shimla, India

Gulab Singh Maurya Faculty of Physical Sciences, INSH, Shri Ramswaroop Memorial University, Barabanki, Uttar Pradesh, India

Bharti Minhas Department of Microbiology, Himachal Pradesh University, Shimla, India

Loveena Pathak Department of Microbiology, Punjab Agricultural University, Ludhiana, Punjab, India

Chayanika Putatunda Department of Biotechnology, Om Sterling Global University, Hisar, Haryana, India

Department of Microbiology, Om Sterling Global University, Haryana, Hisar, India

Suruchi Rai Environmental Molecular Microbiology and Applications Laboratory, Department of Molecular Biology & Bioinformatics, Tripura University (A Central University), Suryamaninagar, Tripura, India

Rekha Rani Amity Institute of Biotechnology, Amity University, Noida, Uttar Pradesh, India

Aditya Sharma Department of Biotechnology, Himachal Pradesh University, Summer Hill, Shimla, Himachal Pradesh, India

Rashmi Sharma VRDL, Department of Microbiology, Shri Lal Bahadur Shastri Govt. Medical College, Mandi, Himachal Pradesh, India

Shivani Department of Microbiology, Himachal Pradesh University, Shimla, India

Gurjoban Singh Amity Institute of Biotechnology, Amity University, Noida, Uttar Pradesh, India

Rohit Kumar Singh Department of Biotechnology, Himachal Pradesh University, Summer Hill, Shimla, Himachal Pradesh, India

Preeti Solanki University Institute of Biotechnology, Chandigarh University, Mohali, Punjab, India

Tamanna Department of Biotechnology, Himachal Pradesh University, Shimla, India

Divya Tandon Department of Microbiology, Shoolini Institute of Life Sciences and Business Management, Solan, HP, India

Monika Thakur Department of Food Science and Technology, Dr. Y. S. Parmar University of Horticulture and Forestry, Nauni, Solan, HP, India

Aayushi Verma Department of Biotechnology, Himachal Pradesh University, Shimla, India

Pratibha Vyas Department of Microbiology, Punjab Agricultural University, Ludhiana, Punjab, India

Abhishek Walia Department of Microbiology, College of Basic Sciences, CSKHPKV, Palampur, Himachal Pradesh, India

Sudesh Kumar Yadav Department of Biotechnology, CSIR-IHBT, Palampur, India

The Science of Biocolour: An Overview

1

Loveena Pathak and Pratibha Vyas

Abstract

Biocolours or natural pigments are derived from microorganisms, plants along with animals and have turned into sustainable substitutes in place of synthetic dyes within industries such as food, textile, cosmetics, and pharmaceuticals. Ancient civilizations throughout India, China and Egypt as well as Europe historically used natural dyes. Biocolours are attractive because they are biodegradable and eco-friendly, with functional benefits also. Microbial pigments display a quick change because of better yields, faster cycles, and bigger scalability, compared to plant versions lately. Yet challenges like poor extraction rates, higher production expenses and sensitivity for surrounding climate restrict wider market acceptance. For overcoming these hurdles, a variety of advanced extraction technologies has been created including ultrasound-assisted extraction (UAE), microwave-assisted extraction (MAE), pressurized liquid extraction (PLE), supercritical fluid extraction (SFE), and enzyme-assisted extraction (EAE). These methods are able to offer more improved pigment recovery along with a minimized solvent use in order to achieve a shorter processing time with improved pigment stability. These biocolours offer antioxidant as well as UV-protective effects that further expand health-related and commercial applications. This chapter gives an overview of the sources, extraction strategies, wide-ranging applications of biocolours, recent technological advancements, and future prospects for pigment production. It also shows that as the demand for health-supporting products increases, the acceptance of biocolours as replacements for chemical dyes steadily increases.

L. Pathak · P. Vyas (✉)
Department of Microbiology, Punjab Agricultural University, Ludhiana, Punjab, India
e-mail: vyasp2000@pau.edu

R. K. Bhatia, A. Walia (eds.), *Biocolours*,
https://doi.org/10.1007/978-981-95-1738-1_1

Keywords

Biocolour · Natural pigments · Sustainable dyes · Eco-friendly colourants · Natural dye application · Bio-based colourants · Green chemistry

1.1 Introduction

The application of colour is a significant component of people's lives throughout history. Archival documents indicate that dyeing methods were developed in India and China around 2600 BC with the use of madder dye which was found in fabrics as well as during excavations in Mohenjo Daro and Harappa civilizations. Europe also practiced similar dyeing techniques during the Bronze Age (3500 BC). Further evidences were recovered from Egypt where mummy wrappings contained colour residues which had been dyed using natural dyes (Monishdeep et al. 2022; Chittaragi and Menon 2023). Over time the search for sustainable effective alternatives for colouring has shifted from plant-based dyes to biocolours, which is an unconventional approach of obtaining pigments from plants, animals and microorganisms like fungi, bacteria or microalgae. While plant-based dyes are eco-friendly, yet their large-scale production has several issues such as low yield of active compounds, complex purification and high costs of manufacturing.

Additionally, the intricate details of their chemical makeup frequently obstruct the extraction and synthesis processes. Despite this, microbial bioprocesses are capable of producing biocolours which serve as a sustainable alternative to traditional dyes, which results in easier processing, efficient production, better downstream handling and a more robust yield. There is a gradual transition from dyes of plant origin to pigments of microbial origin, which shows that biocolours can be seen as a novel area in science and industrial applications (Ruhil and Nagpal 2024). The eco-friendly nature of biocolours, along with the advances in biotechnology, generally increases their suitability for large-scale industrial applications (Fig. 1.1). Genetic engineering in combination with fermentation technology has made tremendous strides toward the production and stability of biocolours increasing their commercial viability. Biocolours also exhibit antioxidant, antimicrobial and UV protective characteristics, enhancing their commercial value (Orlandi et al. 2022).

In order to address the problems posed by traditional techniques, some modern methods have been developed. For example, microwave-assisted extraction (MAE) uses microwaves to improve solvent penetration; ultrasound-assisted extraction (UAE) utilizes ultrasonic waves to aid in pigment release by rupturing the plant cell structures; supercritical fluid extraction (SFE) uses supercritical fluids CO_2 for selective compound removal; pressurized liquid extraction (PLE) increases efficiency through high temperature and pressure combination, along with enzyme-assisted extraction (EAE), which employs specific enzymes that help disrupt cell walls to access pigments more easily. These techniques not only improve the efficiency and effectiveness of the extraction process but also enhance retention of pigments and reduce solvent requirements and processing time while increasing

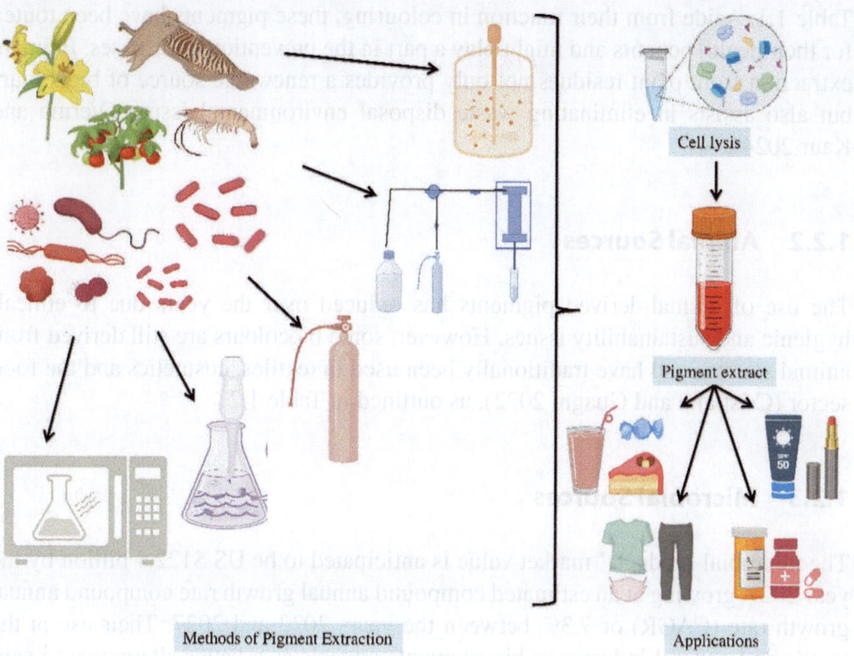

Fig. 1.1 An overview of biocolours sources, extraction and applications

stability of the pigments making these methods excellent choices for industrial use (Mazzocchi et al. 2023; Bürck et al. 2024). Biocolours have become increasingly important due to the awareness regarding their eco-friendly attributes while still sustaining their traditional significance, which continues to propel their usage. In the food sector, natural pigments find widespread application as colouring agents to increase the aesthetic value of the products while providing health and safety advantages (Patil and Hajoori 2024). Biocolourants have also been of cultural and artistic significance for centuries, with artists having used them in works of art and decoration to maintain the elegance and symbolism of the natural colours (De Mol and Vandamme 2024). In the cosmetics industry, these pigments are appreciated for their gentle characteristics and the absence of toxic chemicals and therefore qualify for application in skincare and makeup products (Bashir et al. 2024).

1.2 Sources of Biocolours

1.2.1 Plant Sources

Plants contain pigments that account for their characteristic natural colours. Common pigments of fruits and vegetables are red-yellow betalains, red-purple anthocyanins, green chlorophylls and yellow-orange carotenoids as outlined in the

Table 1.1. Aside from their function in colouring, these pigments have been touted for their health benefits and might play a part in the prevention of diseases. Pigment extraction from plant residues not only provides a renewable source of biocolours but also assists in eliminating waste disposal environmental issues (Verma and Kaur 2024).

1.2.2 Animal Sources

The use of animal-derived pigments has reduced over the years due to ethical, hygienic and sustainability issues. However, some biocolours are still derived from animal sources and have traditionally been used in textiles, cosmetics and the food sector (Cristiano and Guagni 2022), as outlined in Table 1.2.

1.2.3 Microbial Sources

The microbial products' market value is anticipated to be US $122.4 billion by the year 2027, growing at an estimated compound annual growth rate compound annual growth rate (CAGR) of 7.3% between the years 2022 and 2027. Their use in the textile and apparel industry as bio pigments appeals as a better alternative to synthetic dyes. These foster tremendous opportunities towards growth for the biodegradable textile printing and dyeing technology. The fashion industry depends greatly on fossil fuel resources such as fertilizers and chemicals, using over 98 million tons for producing textiles (Source: IndustryARC). Microorganisms are widely recognized to possess unique ability to synthesize natural pigments that differ in chemical composition, functionality and solubility behaviour. Such pigments produced by microorganisms are considered as secondary metabolites which are of high commercial value. Table 1.3 exemplifies various types of microbial pigments.

Table 1.1 Key biocolours, their plant sources and the colours obtained

Biocolours	Plant sources	Colours obtained	References
Curcumin	Turmeric	Bright yellow	Bora et al. (2019)
Anthocyanins	Grapes, elderberries, Blood orange and red cabbage	Red, blue	Sharma (2014), Ghareaghajlou et al. (2021)
Paprika	*Capsicum annum*	Orange–red	Bora et al. (2019), Poorniammal et al. (2022)
Betalains	Beetroot	Violet	Hernández-Aguirre et al. (2021)
Chlorophyll	Pineapple waste	Olive green	Roskan et al. (2022)
Lutein	Marigold flowers	Yellow	Manzoor et al. (2022)
Carotenoids	Carrot	Light yellow to orange	Poorniammal et al. (2022), Upadhyay et al. (2024)

Table 1.2 Key biocolours, their animal sources and the colours obtained

Biocolours	Animal sources	Colours obtained	References
Carminic acid (carmine)	*Dactylopius coccus* (Cochineal insect)	Bright red	Galappaththi and Patabendige (2021)
Melanin	*Sepia esculenta* (Cuttle fish)	Black	Xie et al. (2021)
Astaxanthin	*Trachypenaeus curvirostris* (Shrimp)	Red, orange-red	Zhu et al. (2022)
Tyrian purple	*Murex* species	Purple	Stubbs (2022)
Haemocyanin	Octopus	Blue	Ji et al. (2024)

Table 1.3 Key biocolours, their microbial sources and the colours obtained

Biocolours	Microbial sources	Colours obtained	References
Violacein	*Chromobacterium violaceum*	Violet	Aruldass et al. (2018), Naziya (2021)
Monascins, monapurfluore and ankaflavins	*Monascus* spp.	Yellow	Wu et al. (2021)
Phycoerythrin	*Rhodomonas* sp., *Colaconema* sp.	Red	Lee et al. (2021b), Derbel et al. (2022)
Monasfluore B	*Monascus purpureus*	Red	Kaur et al. (2023)
Melanin	*Aspergillus niger, Aspergillus nidulans*	Black	Salomi et al. (2023), Campanhol et al. (2023)
Phycocyanin	*Spirulina* sp., *Spirulina platensis*	Blue	Lee et al. (2021a), Thevarajah et al. (2024)
Carotenoids	*Rhodotorula* sp.	Bright red colour	Mahajan et al. (2024)
Prodigiosin	*Serratia marcescens, Serratia rubidaea*	Red	Paul et al. (2024)

1.3 Methods of Pigment Extraction

1.3.1 Microwave-Assisted Extraction

Microwave-assisted extraction (MAE) is a novel method that facilitates the extraction of bioactive compounds such as natural pigments using microwave energy coupled with solvents. In contrast to traditional methods, it operates using dipole rotation and ionic conduction which destroy the plant matrix and facilitate the release of desired compounds. Ionic conduction occurs when ions migrate under the electric field of the microwave, generating friction and heat with the surrounding material. Concurrently, dipole rotation occurs as polar molecules continuously orient and reorient to the fluctuating microwave field and thus collide with one another, producing heat (Bürck et al. 2024). In the process, solvent transfers microwave-induced energy to the material, making the solid absorb more microwaves and

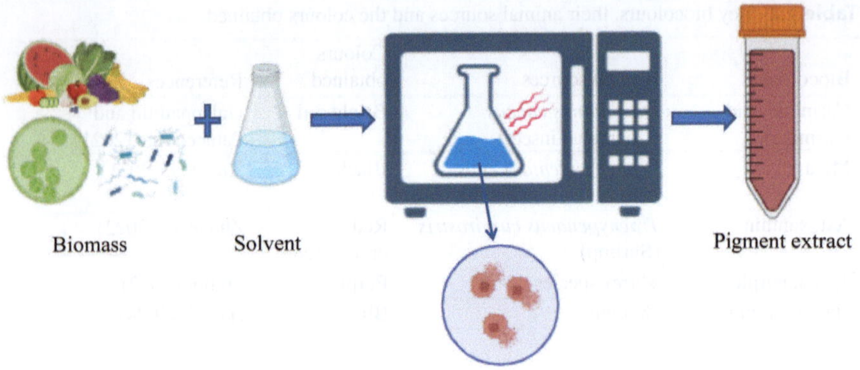

Fig. 1.2 Schematic representation of microwave-assisted extraction of bio pigments. Microwaves generate energy that facilitates the disruption of biomass cells, releasing bio pigments into the solvent which are collected as an extract

quickly heat up. This results in evaporation of moisture and rise in vapour pressure rise, which eventually causes cell wall rupture and release of pigments into the solvent (Fig. 1.2). Through this, MAE achieves efficient solvent penetration, pigment dissolution and transfer into the solution via natural or forced convection and thus is quite faster and efficient than traditional extraction techniques (Rafiq et al. 2022; Nonglait and Gokhale 2024).

This technique has been used to increase the pigment yield from coat of *Bixa orellana* L. by testing different seed-to-water ratios (up to 1:3 w/v), temperatures (up to 70 °C) and extraction times (up to 30 min) at 700 W. The study optimized microwave extraction at a 1:1 ratio for 30 min for the recovery of bixin and norbixin (Sudha et al. 2024). For microwave extraction of pigments from red sorghum, varying power (50–70 W), times (90–120 s) and particle sizes were evaluated. Optimal conditions (63 W, 101 s and 100 μm) yielded 10.5 mg/g anthocyanins, 606 mg/g phenolic, 195 mg/g flavonoids and antioxidant activity of 79.5% (DPPH) and 82.2% (ABTS) and redness of 13.6 (Shukla et al. 2022).

1.3.2 Ultrasound-Assisted Extraction

Ultrasound-assisted extraction (UAE) is one of the fastest methods for extracting heat-sensitive bioactive compounds. This is a two-step process where the solvent penetrates the plant cells and the cell contents are released. It involves acoustic cavitation in which ultrasound waves produce tiny bubbles in the liquid (Fig. 1.3). As these bubbles expand and contract quickly, they exert pressure thereby breaking them apart, which boosts mass transfer and release of target compounds (Lefebvre et al. 2021a, b; Manzoor et al. 2021). A study was conducted to optimize UAE to recover compounds namely betalains, betacyanins, betaxanthins, polyphenols,

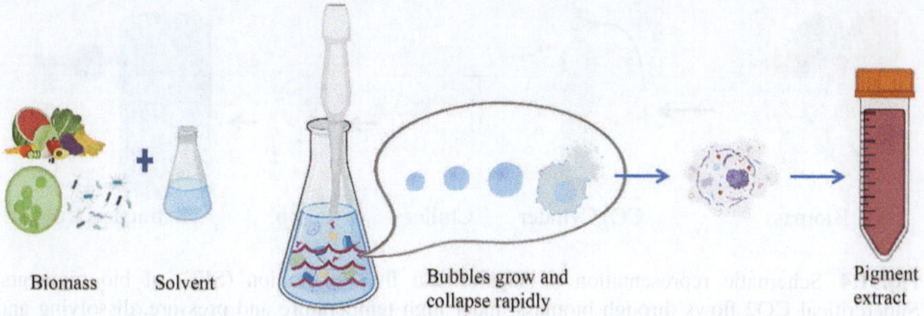

Fig. 1.3 Schematic representation of ultrasound-assisted extraction of bio pigments. High-frequency ultrasound waves generate cavitation within the solvent, disrupting biomass cells and facilitating bio pigment release into the solvent for extraction

antioxidant activity, colour and two major pigments, amaranthine and isoamaranthine from *Amaranthus hypochondriacus* var. *Nutrisol*. In comparison to the conventional techniques, UAE enhanced pigment and antioxidant extraction significantly with ~1.4–1.7 fold increase in yields (Tabio-García et al. 2021). In another research, UAE was optimized to extract valuable pigments and antioxidants were recovered from *Spirulina platensis* (Martins et al. 2023). It was observed that with a solvent ratio of 70 ml/mg at 60 °C for 20 min, high yields of pigments (165.19 mg/g), phenolics (36.50 mg/g) and antioxidant activity (37.98 mg/g) are obtained. These studies indicated that UAE is an efficient method to extract different kinds of pigments, phenolic compounds and antioxidants.

1.3.3 Supercritical Fluid Extraction

The history of supercritical fluids is traced back to 1822 when Baron Charles Cagniard de la Tour noticed that materials have varying behaviour when they come under certain temperature and pressure conditions. The supercritical fluids may be defined as a substance that is kept over its critical temperature as well as pressure, beyond which it no longer separates into phases of liquid and gas. Thus, supercritical fluid extraction is a method that utilizes fluids under supercritical conditions to dissolve and differentiate natural compounds. Utilizing the tuneable solvating ability of supercritical CO_2 (critical point: 31.1 °C, 7.38 MPa), SFE allows the selective pigmentation with no destruction of their stability and bioactivity (Abou Elmaaty et al. 2022). The extraction can be carried out in two modes: static and dynamic. In static extraction, the sample matrix is exposed to a specific quantity of supercritical CO_2 for a predetermined period, so that the fluid can penetrate the material and leach out the target analytes. In dynamic extraction, a stream of fresh CO_2 is continuously passed through the matrix, driving the transport of dissolved compounds.

Many SFE protocols utilize both modes together—first, static conditions to maximize analyte dissolution and then dynamic flow to sweep the dissolved compounds through a restrictor into the collection system. Upon reaching the separators, the

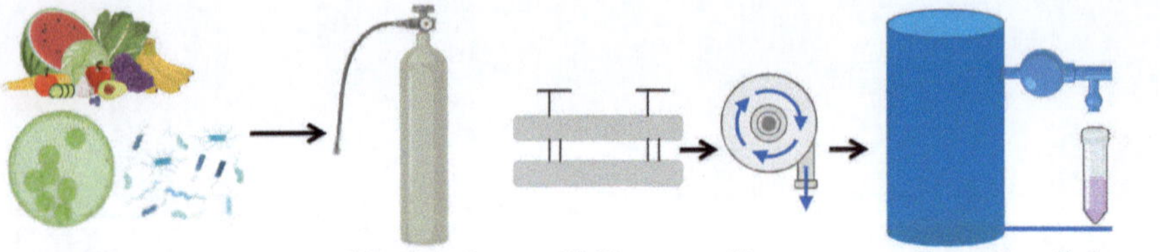

Fig. 1.4 Schematic representation of supercritical fluid extraction (SFE) of bio pigments. Supercritical CO2 flows through biomass under high temperature and pressure, dissolving and extracting bio pigments, which are finally collected in the extract container

products are collected through a tap at the bottom. Depressurization is followed by recycling or venting of the CO_2 and recovery of the extract for redistillation (Fig. 1.4). Yet, the optimal parameters differ depending on the kind of pigment to be extracted to attain a maximum yield and purity (Uwineza and Waśkiewicz 2020; Tzima et al. 2023). For instance, anthocyanins extracted from haskap berry pulp by $scCO_2$ with water as the co-solvent showed the optimal conditions of 45 MPa and 65 °C with 15 min static and 20 min dynamic extraction. The process provided higher anthocyanin yield (52.7% compared to 38.3%) and antioxidant activity (89.8% compared to 72.2%) than conventional methods (Jiao 2018).

This method was also employed to selectively separate bioactive compounds and pigments from rosemary. Four separate fractions were obtained: carotenoids with pure CO_2 (25 °C, 20 MPa), carnosic acid with 3% ethanol–water, rosmarinic acid with 10% modifier and chlorophylls with 30% ethanol. The technique, combined online using supercritical fluid chromatography (SFE-SFC), facilitated real-time separation every 5 min and later ultra high performance liquid chromatography (UHPLC)-MS analysis proved selective extraction of target compounds (Lefebvre et al. 2021a, b). Similarly, microbial pigment such as astaxanthin was isolated from *Paracoccus carotinifaciens*, in which high-temperature extraction (120 °C) greatly improved carotenoid recovery by 56.4% and Z-isomerization by 54.2% compared to 11.1% and 28.4% at 60 °C, respectively (Honda et al. 2022).

1.3.4 Pressurized Liquid Extraction

Accelerated solvent extraction or pressurized liquid extraction employs higher temperature, pressure and solvent to extract bioactive compounds. The heat helps the solvent penetrate better, breaking apart cells and releasing pigments, while the pressure drives the solvent into the material further (Rajendran et al. 2023). One of the advantages of this technique is that the solvent remains in liquid form under pressure, which enables faster and more selective removal even at elevated temperatures (Fig. 1.5). However, this method is not suited for temperature-sensitive compounds as high temperature may cause them to denature (Manzoor et al. 2021).

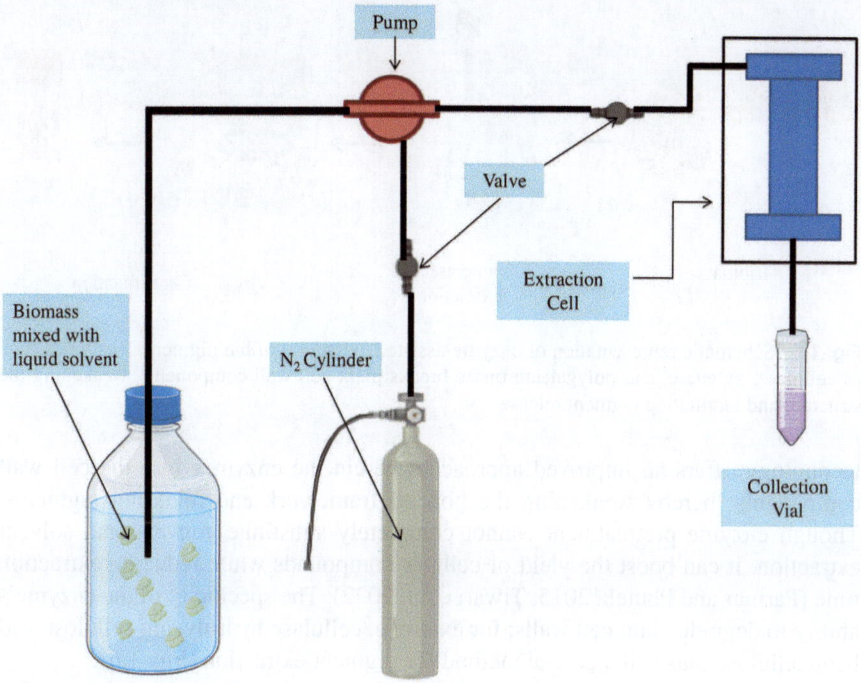

Fig. 1.5 Schematic representation of pressurized liquid extraction of bio pigments. High temperature enhances solvent penetration and cell disruption, while pressure drives the solvent through the biomass, efficiently extracting pigments. Depending on the solvent used, both polar as well as non-polar compounds can be extracted

A new PLE system was developed with in-line solid-phase extraction (SPE) and a deep eutectic solvent (DES) to isolate natural pigments from Brazilian berry waste (*Plinia cauliflora*). Eco-friendly "green solvents" were used to extract anthocyanins with minimal environmental impact, which are important for food and cosmetics. This approach enhanced not only extraction purity and yield but also thermal stability of anthocyanins resulting in a longer shelf life (de Souza Mesquita et al. 2023). Depending on the solvent used, both polar and non-polar compounds can be extracted by this technique. An eco-friendly solvent such as ethanol is commonly used to extract lipids, carotenoids and chlorophyll at 20–200 °C temperature. However, a key drawback is that high-pressure equipment requires costly setting and running costs (Bürck et al. 2024).

1.3.5 Enzyme-Assisted Extraction

Plant cell walls are composed of polysaccharides such as hemicellulose, pectin and cellulose, which act as a major barrier to the release of bioactive compounds. Therefore, to improve the extraction and recovery of these biocolours, enzyme

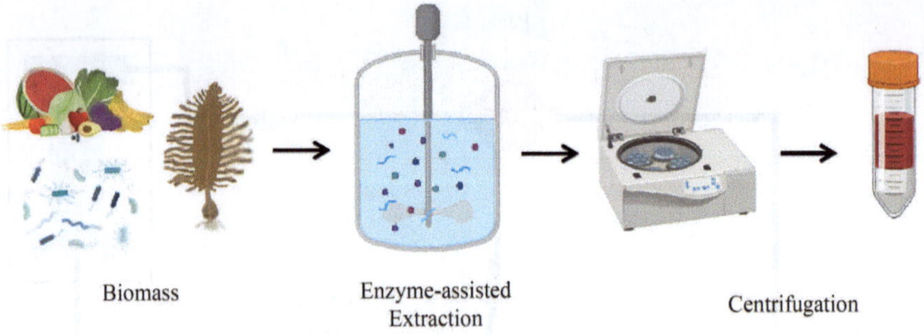

Fig. 1.6 Schematic representation of enzyme-assisted extraction of bio pigments. Enzymes such as cellulase, xylanase, and polygalacturonase breaks plant cell wall components, weakening the structure and facilitating pigment release

technology offers an improved approach wherein the enzymes lyse the cell wall components thereby weakening the bonded framework and releasing pigments. Though enzyme pretreatment cannot completely substitute conventional solvent extraction, it can boost the yield of cellular compounds while reducing extraction time (Parmar and Phutela 2015; Tiwari et al. 2022). The specificity of the enzyme's ability to degrade plant cell walls; for example, cellulase hydrolysing cellulose and hemicellulose makes it a crucial method for pigment extraction (Fig. 1.6).

Other benefits include mild operating conditions (pH and temperature), no risk of corrosion, faster extraction, and bypassing drying steps. However, the high cost of enzymes and the need to retain stable conditions during extraction limits its application (Pagels et al. 2021). EAE has been optimized for extraction of anthocyanin from *Prunus nepalensis* using cellulase enzyme. The optimal conditions (45.6 °C, 12.9% E/S and 3.98 h) resulted in a 62.9% yield and 9.89 mg/g anthocyanins (Swer et al. 2018). In another research study conducted by Mazzocchi et al. 2023, chlorophyll pigment was extracted from spinach for use as a natural food colourant. The cell wall makeup was analyzed, and a specific enzyme mix was designed with cellulase, xylanase and polygalacturonase. The optimal conditions (25 °C, 150 ppm zinc, B/S ratio 17.5, under 2 h and 12–45 U/g enzyme dose) were identified to maximize both chlorophyll yield and colour quality.

1.4 Purification of Extracted Pigments

The purification of biocolours remains vital for quality, chemical stability and numerous commercial applications. Since, biocolours originate from complex biological substrates like plant tissues, microorganisms or even animal products isolation is necessary to separate the desired pigment molecules from lipids, proteins and solvents that may hinder the proposed use of biocolours in any application. Once extraction is done, the mixture containing solid residues such as plant remnants or microbial biomass is either filtered or centrifuged. The suspended particles are discarded to obtain a solution consisting only the dissolved pigment (Nemer et al. 2021). Other methods such as ammonium sulphate precipitation help separate phycobiliprotein-rich extracts to get pure phycocyanin (PC) in two steps with the first step starting off at 20% saturation to remove impurities (increasing purity from 1.12 to 1.44) and the second step at 70% saturation for approximately two-fold increase in purity (~2.32) (Krakauskaitė et al. 2025). To achieve full removal of the uncompounded impurities different methods of chromatography can also be applied to further separate and refine biocolours from other closely related compounds. For example, in a study, researchers aimed to purify C-phycocyanin (C-PC) extracted from *Spirulina* sp. wherein they filtered the crude extract through 0.2 μm membranes followed by two-step ammonium sulphate precipitation at 20% and then 50% saturation with subsequent centrifugation. The sedimented C-PC pellet was resuspended in a phosphate buffer and dialysed using a membrane of 12–14 kDa to remove salts. The final purification was performed using gel filtration chromatography on a Superdex 200 column and the purified fractions were concentrated using protein concentrators (Katari et al. 2023). Likewise, Nemer et al. (2023) utilized column chromatography to separate and purify violacein and deoxyviolacein produced by genetically modified yeast *Yarrowia lipolytica* employing a cyclohexane-saturated silica column to effectively separate biocolours whilst preventing lipid and debris interference.

In another research, green pigment of *Metarhizium flavoviride* was purified successfully with Sephadex LH-20, which used both size exclusion and adsorption mechanisms to separate the extract into four fractions. The abundant dark green fraction (fraction 1) was then further subjected to two cycles of Sephadex LH-20 purification to obtain green pigment soluble in dimethyl sulphoxide (Wei et al. 2025). Over the past decades, green and eco-friendly purification methods have gained prominence to reduce the use of harmful chemicals and impact of purification process. There are more sustainable techniques like supercritical fluid extraction wherein using carbon dioxide as solvent and aqueous two-phase systems are becoming more prevalent in biocolour purification. This technique has been employed to obtain selective extraction of specific bioactive compounds and pigments from rosemary in high concentrations by adjusting pressure, temperature and adding modifiers (Abou Elmaaty et al. 2022). The process was optimized using SFE-SFC through the use of two columns with differing polarities to distinguish target compounds every 5 min. This method yielded four different fractions: carotenoids (with pure CO_2 at 25 °C, 20 MPa), carnosic acid (with 3% ethanol–water modifier), rosmarinic acid (with 10% modifier) and chlorophylls (with 30% ethanol). All the fractions were then profiled using UHPLC-DAD-ESI-QTOF-HRMS (Lefebvre et al. 2021a, b).

1.5 Applications of Biocolours

1.5.1 Food Products

The growing global trend towards healthier diets has intensified the interest in substituting artificial food colours with bio-based substitutes. Recent regulatory measures, including the plan of the U.S. FDA to phase out petroleum-derived colouring agents such as Red No. 3, Citrus Red No. 2 and Orange B indicate rising concerns with human health threats and the environmental effect of synthetic dyes. Currently, the initiatives are being made to encourage the food industry to phase out the frequently used artificial dyes like Red No. 40, Yellow No. 5 and Blue No. 1, as well as to expedite the testing of natural colourants like calcium phosphate, butterfly pea flower extract and gardenia blue. This change meets the consumer-need for cleaner labels and also presents newer opportunities for formulating natural bio-based pigments in a variety of food products. In an investigation it was revealed that the addition of microalgal astaxanthin extract from *Haematococcus pluvialis* to shrimp feed enhanced shrimp growth and muscle nutritional content with higher levels of free amino acids (Zhang et al. 2022). Another blue-coloured pigment, phycocyanin (PC), isolated from spirulina (*Arthrospira* spp.) is used in food products like ice cream, jellies, beverages, candies, dairy products etc. The microalgal culture conditions such as agitation rate (180 rpm), culture volume (250 ml) and growth time (14 days) have been optimized to obtain higher phycocyanin yield and colour quality. Ice cream tinted with this PC exhibited evident concentration-dependent colour intensity and stability (Damaciano et al. 2024). Similarly, other microorganisms have also been studied for natural pigment production in food industry. For instance, *Monascus* species synthesize different pigments called ankaflavine, monascine and monascorubrine, which also find application as colourants and flavouring agents. Correspondingly, *Rhodotorula glutinis* yields carotenoids which yield intense orange-red colours for food colouring, and *Penicillium oxalicum* yields arpink red used to improvise the visual appearance of food. Furthermore, advances in microbial fermentation, such as growth optimization and incorporating stress factors, have increased the efficiency of carotenoid production thereby making these microbes valuable as natural food colourants (Charalampia et al. 2017; Adin et al. 2023; de Lima et al. 2025).

1.5.2 Pharmaceuticals

Biocolourants are increasingly adopted in pharmaceutical industries to colour tablets, capsules and ointments and even topical formulations not only to improve their visual appeal but also to help differentiate products (Chen et al. 2024). These biopigments including anthocyanins, betalains, curcumin, chlorophyll and carotenoids also possess additional therapeutic effects including antioxidant, anti-inflammatory and antimicrobial activities that can enhance the therapeutic values of formulations. Additionally, they alleviate concerns over allergic reactions and toxicity often associated with chemical-based dyes (Manzoor et al. 2021; Krishnakripa and Thoppil 2025). Recent research has revealed bacterial pigments like prodigiosin from *Serratia marcescens* for their pharmaceutical potential. Prodigiosin has shown strong redox-balancing, anticancer and neuroprotective activities, with no toxicity to healthy cells (Koyun et al. 2022). Another example is violacein, a purple pigment produced by *Chromobacterium violaceum*, which has demonstrated potent antitumor effects by inducing the death of osteosarcoma and rhabdomyosarcoma cells. This broadens the prospects of microbial pigments like prodigiosin and violacein as natural sources for developing novel anticancer agents alongside their applications as biocolourants in pharmaceutical formulations (Milosevic et al. 2023).

1.5.3 Cosmetics

Cosmetic products are designed to protect and improve the condition of the skin by using various ingredients suitable for different skin types. Early formulations focused on efficacy rather than safety considerations but today biologically sourced pigments are appreciated not only for their unique colours but also their beneficial properties. Many biopigments demonstrate protective traits against oxidation, hyperpigmentation and ultraviolet damage, which may prove useful in manufacturing products that combine beauty with health (Aman Mohammadi et al. 2022; Kiki 2023). A biobased lip ointment was developed by combining melanin pigment extracted from *Streptomyces bellus* MSA1 with beeswax, lanolin and coconut oil (Srinivasan et al. 2017). Another study found that out of nineteen isolates of bacteria that produced pigments, seventeen exhibited sunscreen activity between 0.4 and 8.34, while six displayed significant antibacterial and antioxidant properties with *Staphylococcus xylosus* exhibiting consistent pigment production and multifunctional bioactive properties (Choksi et al. 2020). Other fungi such as *Talaromyces australis* and *Penicillium murcianum* along with some *Pseudomonas* species have also been studied for their use in sunscreen formulations thus providing natural photo-protective alternatives (Contreras-Machuca et al. 2022). Overall, these studies suggest that the use of microbial pigments in skin-care products addresses both consumer demand for natural ingredients and the growing focus on improving skin health.

1.5.4 Textile Industry

Textile wastewater is hazardous due to sulphur compounds, chromium, naphthols, heavy metals and other chemical additives. Additional pollutants often contain fixatives such as formaldehyde, hydrocarbon softeners, non-biodegradable dyes or chlorinated stain removers. When combined with chlorine disinfectants, these organic compounds can produce carcinogenic by-products. Industries are changing their focus to more natural product lines that are in harmony with the eco-friendly demands of shifting towards alternatives. Although the use of natural pigments in industries is still limited, there is continuing research about their application on different fabrics (Oginawati et al. 2022; Barreto et al. 2023). *Talaromyces* sp., grown in milk whey, yielded 0.523 ± 0.231 mg/g of pigment from a pineapple peel-based medium. The further analysis pursued evaluating its efficacy as an ecological textile dye through fabric dyeing experiments using different mordents (Umesh et al. 2023).

The violacein pigment obtained from halophilic bacterium *Chromobacterium violaceum* PDF23 demonstrated strong antimicrobial activity and free radical scavenging with an IC50 value of 14.40 μg/ml. The pigment was also assessed for its application in dyeing silk satin and cotton fabrics utilizing different mordants (Anahas et al. 2022). Likewise, *Chromobacterium piscinae* isolated from a soil sample was tested in a minimal medium. The ethanol extraction and experimental analyses, including HPLC, FTIR, and ^{1}H-NMR, confirmed the pigment as violacein which was successfully applied for dyeing cotton and polyester fabrics, highlighting its promising role for textile applications (Deeppal et al. 2024).

1.6 Conclusions and Future Perspectives

The use of biocolours has grown from their historical origins towards innovations in industry, entrepreneurship and eco-friendliness. There is an increased attention on these natural pigments due to all the benefits they offer besides their alluring colours. Even though bio pigments are appearing as a new potential solution to synthetic dyes, issues with extraction efficiency and cost-effectiveness still need to be addressed for them to be commercially adopted. More recent technologies that have been developed such as microwave-assisted, ultrasound-assisted, supercritical fluid, pressurized liquid and enzyme-assisted extractions seem to focus on the recovery of pigments while improving selectivity and quality, which aids in eliminating solvent-dependent methods. These changes can greatly improve the economic aspects of biocolour production making them much more beneficial for different industrial sectors. The future research should be directed towards the integration of biotechnological tools including synthetic biology, metabolic engineering and fermentation optimization since it may contribute even more to higher pigment yields and the engineering of pigment traits as per requirement. Besides that, the engineering of renewable solvents, circular extraction systems and waste valorization strategies may also help resolve other environmental and cost issues. With increased consumer knowledge about sustainable and healthy products, the market for biocolours will expand exponentially in the coming years, providing immense opportunities for research, innovation and commercialization. Cross-disciplinary collaborations among microbiologists, chemists, engineers and business entities will play a decisive role in maximizing the potential of biocolours in the coming years to place them as stalwart drivers of a more sustainable and health-focused global marketplace.

References

Abou Elmaaty T, Sayed-Ahmed K, Elsisi H, Magdi M (2022) Optimization of extraction of natural antimicrobial pigments using supercritical fluids: a review. PRO 10(10):2111

Adin SN, Gupta I, Panda BP, Mujeeb M (2023) Monascin and ankaflavin – biosynthesis from *Monascus purpureus*, production methods, pharmacological properties: a review. Biotechnol Appl Biochem 70(1):137–147

Aman Mohammadi M, Ahangari H, Mousazadeh S, Hosseini SM, Dufossé L (2022) Microbial pigments as an alternative to synthetic dyes and food additives: a brief review of recent studies. Bioprocess Biosyst Eng 45(1):1–12

Anahas AMP, Kumaran S, Kandeel M, Muralitharan G, Silviya J, Adhimoolam GL et al (2022) Applications of natural violet pigments from halophilic *Chromobacterium violaceum* PDF23 for textile dyeing with antimicrobial and antioxidant potentials. J Nanomater 2022(1):3885396

Aruldass CA, Masalamany SRL, Venil CK, Ahmad WA (2018) Antibacterial mode of action of violacein from *Chromobacterium violaceum* UTM5 against *Staphylococcus aureus* and methicillin-resistant *Staphylococcus aureus* (MRSA). Environ Sci Pollut Res 25:5164–5180

Barreto JVDO, Casanova LM, Junior AN, Reis-Mansur MCPP, Vermelho AB (2023) Microbial pigments: major groups and industrial applications. Microorganisms 11(12):2920

Bashir I, Pandey VK, Dar AH, Dash KK, Shams R, Mir SA et al (2024) Exploring sources, extraction techniques and food applications: a review on biocolours as next-generation colorants. Phytochem Rev 23(4):1–26

Bora P, Das P, Bhattacharyya R, Barooah MS, Saikia Barooah M (2019) Biocolour: the natural way of colouring food. J Pharmacogn Phytochem 8(3):3663–3668

Bürck M, Ramos SDP, Braga ARC (2024) Enhancing the biological effects of bioactive compounds from microalgae through advanced processing techniques: pioneering ingredients for next-generation food production. Foods 13(12):1811

Campanhol BS, Ribeiro BD, Casellato F, Medina KJD, Sponchiado SRP (2023) Improvement of DOPA-melanin production by *aspergillus nidulans* using eco-friendly and inexpensive substrates. J Fungi 9(7):714

Charalampia D, Antonios KE, Constantina N, Haralabos KC (2017) Current trends and emerging technologies in biopigment production processes: industrial food and health applications. Int J Hortic Agric Food Sci 1:33–46

Chen F, Zhong H, Chan G, Ouyang D (2024) A comprehensive analysis of biopharmaceutical products listed in the FDA'S purple book. AAPS PharmSciTech 25(5):88

Chittaragi D, Menon JS (2023) Bio-colours from spices. In: Herbs and spices-new advances. IntechOpen. https://doi.org/10.5772/intechopen.109830

Choksi J, Vora J, Shrivastava N (2020) Bioactive pigments from isolated bacteria and its antibacterial, antioxidant and sun protective application useful for cosmetic products. Indian J Microbiol 60:379–382

Contreras-Machuca PI, Avello M, Pastene E, Machuca Á, Aranda M, Hernández V, Fernández M (2022) Chemical characterization and microencapsulation of extracellular fungal pigments. Appl Microbiol Biotechnol 106(24):8021–8034

Cristiano L, Guagni M (2022) Zooceuticals and cosmetic ingredients derived from animals. Cosmetics 9(1):13

Damaciano SF, Kurpan D, Ribeiro RDO, Torres IB, França JVF, Barbarino E et al (2024) Technological and scientific prospection of phycocyanin production from spirulina (*Arthrospira* spp.): optimization and application in ice cream. Br J Food Technol 27:e2024037

de Lima JGO, Veríssimo NVP, de Azevedo Lima C, Picheli FP, de Paula AV, Santos-Ebinuma VDC (2025) Improvement of torularhodin production by *Rhodotorula glutinis* through the stimulation of physicochemical stress and application of the bioproduct as an additive in the food industry. Bioprocess Biosyst Eng:1–21

De Mol ML, Vandamme EJ (2024) Arts, cultural heritage, sciences and micro-/bio-/techno–/logy: impact of biomaterials and biocolorants from antiquity till today! J Ind Microbiol Biotechnol 51:kuae049
de Souza Mesquita LM, Contieri LS, Sosa FH, Pizani RS, Chaves J, Viganó J et al (2023) Combining eutectic solvents and pressurized liquid extraction coupled in-line with solid-phase extraction to recover, purify and stabilize anthocyanins from Brazilian berry waste. Green Chem 25(5):1884–1897
Deeppal, Raj KK, Chowdhury T, Umarajashekhar A, Soni R (2024) Screening and characterization of *Chromobacterium piscinae* AMA-5 for enhanced production of violacein and its ability of textile dyeing. Biotechnol Appl Biochem 71(6):1293–1303
Derbel H, Elleuch J, Tounsi L, Nicolo MS, Rizzo MG, Michaud P et al (2022) Improvement of biomass and phycoerythrin production by a strain of *Rhodomonas* sp. isolated from the Tunisian coast of Sidi Mansour. Biomolecules 12(7):885
Galappaththi M, Patabendige N (2021) Cochineal chemistry, related applications and problems: a mini review. Academia Letters 1792:1–12
Ghareaghajlou N, Hallaj-Nezhadi S, Ghasempour Z (2021) Red cabbage anthocyanins: stability, extraction, biological activities and applications in food systems. Food Chem 365:130482
Hernández-Aguirre OA, Muro C, Hernández-Acosta E, Alvarado Y, Díaz-Nava MDC (2021) Extraction and stabilization of betalains from beetroot (*Beta vulgaris*) wastes using deep eutectic solvents. Molecules 26(21):6342
Honda M, Murakami K, Osawa Y, Kawashima Y, Wasai M, Hirasawa K, Kuroda I (2022) Supercritical CO_2 extraction of carotenoids (astaxanthin, adonirubin, and adonixanthin) from *Paracoccus carotinifaciens*: improved Z-isomer ratio and recovery of carotenoids via high-temperature extraction. Eur J Lipid Sci Technol 124(7):2200021
https://www.fda.gov/news-events/press-announcements/hhs-fda-phase-out-petroleum-based-synthetic-dyes-nations-food-supply. https://www.industryarc.com. Chemicals and Materials
Ji R, Guan L, Hu Z, Cheng Y, Cai M, Zhao G, Zang J (2024) A comprehensive review on hemocyanin from marine products: structure, functions, its implications for the food industry and beyond. Int J Biol Macromol:132041
Jiao G (2018) Extraction of anthocyanins from haskap berry pulp using supercritical carbon dioxide: influence of co-solvent composition and pretreatment. LWT 98:237–244
Katari JK, Khan MRUZ, Trivedi V, Das D (2023) Extraction, purification, characterization and bioactivity evaluation of high purity C-phycocyanin from *Spirulina* sp. NCIM 5143. Process Biochem 130:322–333
Kaur M, Goel M, Mishra RC, Lahane V, Yadav AK, Barrow CJ (2023) Characterization of the red biochromes produced by the endophytic fungus *Monascus purpureus* CPEF02 with antimicrobial and antioxidant activities. Fermentation 9(4):328
Kiki MJ (2023) Biopigments of microbial origin and their application in the cosmetic industry. Cosmetics 10(2):47
Koyun MT, Sirin S, Aslim B, Taner G, Dolanbay SN (2022) Characterization of prodigiosin pigment by *Serratia marcescens* and the evaluation of its bioactivities. Toxicol In Vitro 82:105368
Krakauskaitė U, Aboobacker S, Kitrytė-Syrpa V, Syrpas M (2025) Optimised extraction and purification of dual-function cosmetic-grade Phycocyanin and Allophycocyanin from dried *Arthrospira platensis* biomass using conventional methods. Appl Sci 15(2):532
Krishnakripa PK, Thoppil JE (2025) Beyond a pigment-a review on the beneficial aspects of anthocyanins, chlorophylls, carotenoids, and betalains. Vegetos:1–11
Lee CW, Bae GY, Bae SH, Suh HJ, Jo K (2021a) Increased thermal stability of phycocyanin extracted from *Spirulina platensis* by cysteine addition during enzyme extraction. Food Sci Technol 42:e15021
Lee MC, Yeh HY, Jhang FJ, Lee PT, Lin YK, Nan FH (2021b) Enhancing growth, phycoerythrin production, and pigment composition in the red alga *Colaconema* sp. Through optimal environmental conditions in an indoor system. Bioresour Technol 333:125199
Lefebvre T, Destandau E, Lesellier E (2021a) Selective extraction of bioactive compounds from plants using recent extraction techniques: a review. J Chromatogr A 1635:461770

Lefebvre T, Destandau E, Lesellier E (2021b) Sequential extraction of carnosic acid, rosmarinic acid and pigments (carotenoids and chlorophylls) from Rosemary by online supercritical fluid extraction-supercritical fluid chromatography. J Chromatogr A 1639:461709

Mahajan S, Nikhanj P, Singh A (2024) Microbial biopigment production by *Rhodotorula* sp. using horticultural waste; its functionality, stability and applications. Indian J Microbiol:1–10

Manzoor M, Singh J, Gani A, Noor N (2021) Valorization of natural colors as health- promoting bioactive compounds: phytochemical profile, extraction techniques, and pharmacological perspectives. Food Chem 362:130141

Manzoor S, Rashid R, Panda BP, Sharma V, Azhar M (2022) Green extraction of lutein from marigold flower petals, process optimization and its potential to improve the oxidative stability of sunflower oil. Ultrason Sonochem 85:105994

Martins R, Mouro C, Pontes R, Nunes J, Gouveia I (2023) Ultrasound-assisted extraction of bioactive pigments from *Spirulina platensis* in natural deep eutectic solvents. Bioresour Bioprocess 10(1):88

Mazzocchi C, Benucci I, Lombardelli C, Esti M (2023) Enzyme-assisted extraction for the recovery of food-grade chlorophyll-based green colorant. Foods 12(18):3440

Milosevic E, Stanisavljevic N, Boskovic S, Stamenkovic N, Novkovic M, Bavelloni A et al (2023) Antitumor activity of natural pigment violacein against osteosarcoma and rhabdomyosarcoma cell lines. J Cancer Res Clin Oncol 149(13):10975–10987

Monishdeep V, Deshpande H, Poshadri A, Kumar B (2022) A review on potential microbial sourced bio-colours from fungi. Pharma Innov J 11:3200–3205

Naziya BS (2021) Exploring the biomedical applications of violacein extracted from chromobacterium violoceum. Doctoral dissertation, Thiruvalluvar University

Nemer G, Louka N, Vorobiev E, Salameh D, Nicaud JM, Maroun RG, Koubaa M (2021) Mechanical cell disruption technologies for the extraction of dyes and pigments from microorganisms: a review. Fermentation 7(1):36

Nemer G, Louka N, Rabiller Blandin P, Maroun RG, Vorobiev E, Rossignol T et al (2023) Purification of natural pigments violacein and deoxyviolacein produced by fermentation using *Yarrowia lipolytica*. Molecules 28(11):4292

Nonglait DL, Gokhale JS (2024) Review insights on the demand for natural pigments and their recovery by emerging microwave-assisted extraction (MAE). Food Bioprocess Technol 17(7):1681–1705

Oginawati K, Susetyo SH, Sulung G, Chazanah N, Kusumah SWD, Fahimah N (2022) Investigation of dermal exposure to heavy metals (Cu, Zn, Ni, Al, Fe and Pb) in traditional batik industry workers. Heliyon 8(2)

Orlandi VT, Martegani E, Giaroni C, Baj A, Bolognese F (2022) Bacterial pigments: A colorful palette reservoir forbiotechnological applications. Biotechnol Appl Biochem 69(3):981-1001

Pagels F, Pereira RN, Vicente AA, Guedes AC (2021) Extraction of pigments from microalgae and cyanobacteria—A review on current methodologies. Appl Sci 11(11):5187

Parmar M, Phutela UG (2015) Biocolours: the new generation additives. Int J Curr Microbiol App Sci 4(7):688–694

Patil D, Hajoori M (2024) An overview of microbial pigment: a reliable source of biocolor. Int Pharm Res Appl 9:1758–1765

Paul T, Mondal A, Bandyopadhyay TK, Bhunia B (2024) Prodigiosin production and recovery from *Serratia marcescens*: process development and cost–benefit analysis. Biomass Convers Biorefinery 14(3):4091–4110

Poorniammal R, Prabhu S, Sakthi AR (2022) Biocolourants and their applications in food industry. Biotica Res Today 4:524–528

Rafiq N, Gupta N, Bhat A, Singh J, Bandral J, Sood M et al (2022) Extraction techniques and utilization of natural pigments. Chem Sci Rev Lett 11:239–245

Rajendran P, Somasundaram P, Dufossé L (2023) Microbial pigments: eco-friendly extraction techniques and some industrial applications. J Mol Struct 1290:135958

Roskan NFM, Zainol N, Samad KA (2022) Chlorophyll extraction from pineapple plantation waste through mechanical extraction. Biocatal Agric Biotechnol 39:102264

Ruhil S, Nagpal K (2024) Microbial pigments as vegan Colors for food and pharmaceuticals: a sustainable approach. In: Harnessing microbial potential for multifarious applications. Singapore, Springer Nature Singapore, pp 419–438

Salomi V, Alfaris AA, Saranraj P, Swetha MS, Gayathri K (2023) Extraction of fungal pigment melanin from Aspergillus Niger and analysis of its antimicrobial activity. Eur Chem Bull 12(11):431–438

Shukla S, Lohani UC, Shahi NC, Dubey A (2022) Extraction of natural pigments from red sorghum (*Sorghum bicolor*) husk by ultrasound and microwave assisted extraction: a comparative study through response surface analysis. J Food Process Eng 45(10):e14130

Srinivasan M, Keziah SM, Hemalatha M, Devi CS (2017, November) Pigment from *Streptomyces bellus* MSA1 isolated from marine sediments. In: IOP conference series: materials science and engineering, vol 263, No. 2. IOP Publishing, p 022049

Stubbs D (2022) The purple tide: murex dye and the formation of the Minoan State. Master's thesis, The University of Arizona

Sudha P, Manoja V, Deepa J, Jayakumar J, Kishore SG, Pandiselvam R (2024) Optimization of microwave-assisted aqueous extraction of pigments from annatto seeds using Box-Behnken design. Biomass Convers Biorefinery 14(16):18775–18788

Swer TL, Mukhim C, Bashir K, Chauhan K (2018) Optimization of enzyme aided extraction of anthocyanins from *Prunus nepalensis* L. LWT 91:382–390

Tabio-García D, Paraguay-Delgado F, Sánchez-Madrigal MÁ, Quintero-Ramos A, Espinoza-Hicks JC, Meléndez-Pizarro CO et al (2021) Optimisation of the ultrasound-assisted extraction of betalains and polyphenols from *Amaranthus hypochondriacus* var. Nutrisol Ultrason Sonochem 77:105680

Thevarajah B, Piyatilleke S, Nimarshana PHV, Koushalya S, Malik A, Ariyadasa TU (2024) Exploring effective light spectral conversion techniques for enhanced production of spirulina-derived blue pigment protein, c-phycocyanin. Bioresour Technol:130612

Tiwari S, Yawale P, Upadhyay N (2022) Carotenoids: extraction strategies and potential applications for valorization of under-utilized waste biomass. Food Biosci 48:101812

Tzima S, Georgiopoulou I, Louli V, Magoulas K (2023) Recent advances in supercritical CO_2 extraction of pigments, lipids and bioactive compounds from microalgae. Molecules 28(3):1410

Umesh M, Suresh S, Santosh AS, Prasad S, Chinnathambi A, Al Obaid S et al (2023) Valorization of pineapple peel waste for fungal pigment production using *Talaromyces albobiverticillius*: insights into antibacterial, antioxidant and textile dyeing properties. Environ Res 229:115973

Upadhyay N, Kamble K, Tiwari S, Meena SK, Malhotra R, Singh AK, Kaur C (2024) Novel avenue of carotenoids & omega-3 rich carrot waste extract in innovative table spread: optimization (based on texture and sensory) and characterization. Discov Food 4(1):67

Uwineza PA, Waśkiewicz A (2020) Recent advances in supercritical fluid extraction of natural bioactive compounds from natural plant materials. Molecules 25(17):3847

Verma S, Kaur K (2024) Biocolorants from plant sources: a novel approach to enhance the aesthetics and health profile of foods. Bioscene 21(2):330–370

Wei J, Zhao X, Yang X, Jia W, Qin J, Li W et al (2025) Extraction, purification, and structural analysis of green pigments from *Metarhizium flavoviride*. J Mol Struct 1334:141913

Wu L, Zhou K, Chen F, Chen G, Yu Y, Lv X et al (2021) Comparative study on the antioxidant activity of Monascus yellow pigments from two different types of Hongqu—functional Qu and colouring Qu. Front Microbiol 12:715295

Xie J, Li H, Che H, Dong X, Yang X, Xie W (2021) Extraction, physicochemical characterisation, and bioactive properties of ink melanin from cuttlefish (*Sepia esculenta*). Int J Food Sci Technol 56(7):3627–3640

Zhang S, Chang Z, Wang W, Zheng Y, Li J (2022) Dietary supplement of microalgal astaxanthin extraction improved shell pigmentation and nutritional value of *Litopenaeus vannamei* in an indoor industrial aquaculture system. Aquac Nutr 2022(1):3981246

Zhu K, Yan W, Dai Z, Zhang Y (2022) Astaxanthin extract from shrimp (*Trachypenaeus curvirostris*) by-products improves quality of ready-to-cook shrimp surimi products during frozen storage at– 18 C. Foods 11(14):2122

2 Biochemistry of Biocolours

Dolly Kumari, Gulab Singh Maurya, and Neha Kesharwani

Abstract

Colour plays a significant role in every aspect of our lives from the clothes we wear and the furniture in our homes to religious rituals and even the food we consume. A chemical compound is perceived as a colour when it absorbs certain wavelengths of visible light while reflecting or transmitting others. Colours are generally classified into four categories: natural colours, nature- identical colours, synthetic colours and inorganic colours. Biocolour refers to a dye or pigment derived from biological sources such as plants, insects, animals or microorganisms. When applied to food, medicines, fabrics or cosmetics, biocolours impart colour through eco-friendly means. Natural colouring agents obtained from plant sources such as chlorophyll, carotenoids, and betacyanins are widely used in medicine, food preservation, textiles and molecular biology, offering environmentally safe alternatives to synthetic dyes. Colours derived from natural sources are collectively termed 'biocolours'. These are typically extracted from various plant parts, including fruits, vegetables, seeds and roots, and also from microbial sources. Among all sensory attributes, colour is often the most critical factor in food, as it significantly influences consumer perception of quality. This chapter discusses the types, biosynthetic pathways, molecular structures and biochemical properties of various biocolours.

Author Neha Kesharwani has been co-corresponding author.

D. Kumari (✉) · N. Kesharwani
Faculty of Chemical Sciences, INSH, Shri Ramswaroop Memorial University, Barabanki, Uttar Pradesh, India

G. S. Maurya
Faculty of Physical Sciences, INSH, Shri Ramswaroop Memorial University, Barabanki, Uttar Pradesh, India

R. K. Bhatia, A. Walia (eds.), *Biocolours*,
https://doi.org/10.1007/978-981-95-1738-1_2

Keywords

Biocolour synthesis · Molecular structure · Biochemical pathways · Biological systems · Pigments

2.1 Introduction to Biocolours

The desire for sustainable and natural alternatives has increased recently across a number of industries, including the food industry. The use of biocolourants made from plant sources as a cutting-edge strategy to improve the appearance and nutritional value of foods is one important area of focus. The food industry's move to plant-based ingredients is in line with customer demands for better nutritional values, clean labels and environmental sustainability. Because of their natural bioactive ingredients, these pigments not only give food goods brilliant colours, but they may also have health advantages (Verma and Kaur 2024). Dyes or pigments, when added to products, are considered natural colourants and commonly known as 'Biocolour' (Chaitanya 2014). The word 'Biocolour' combines the words 'bio,' which indicates its natural source, and 'colour,' which indicates its ability to provide colour. Biocolourants are colouring compounds that can give food products vivid colours. They are derived from biological elements such as plants, algae, insects, fungi and animals. The majority of these biocolourants come from plants, and they are all made from sustainable resources. In addition to their use in food colouring, biocolourants have been used as medicinal agents due to their bioactive qualities. Biocolours are a good substitute for artificial food colouring that will not harm people or the environment. Certain plants have a tendency to be dominated by particular pigments, such as capsanthin in red peppers and lycopene in tomatoes (Arunkumar et al. 2024).

Food colour is a fundamental aspect of our culture and way of life. They have been applied to improve dishes' visual appeal. Making food more aesthetically pleasing and identifiable is the main goal of adding colour to it. Because food colour can either increase or decrease hunger, everyone is sensitive to it (Sharma 2014). Given the negative effects of artificial food colourings, which can include attention deficit hyperactivity disorder (ADHD), behavioral issues, depression, food allergies, headaches and migraines, natural colours are preferred over artificial ones. According to the European Food Safety Authority (EFSA 2011), artificial food colouring is commonly used in desserts, doughnuts, fruit preserves, ice creams, colas and confectionery to improve their look. Biocolourants give food products a range of colours, from vivid yellows and purples to deep reds and greens. Food producers may produce aesthetically pleasing items without the need for artificial additives because of this natural diversity. Natural colours are frequently linked by consumers to authenticity and freshness, which increases the commercial appeal of bio colourants. Because the source plants contain bioactive substances, plant-based colourants frequently provide extra health benefits. For example, the anthocyanins in berries have antioxidant qualities that may improve health and wellness in addition to giving them their red and purple colours. Food products may be able to provide both nutritional value and aesthetic appeal by adding biocolourants (Verma and Kaur 2024).

A big part of today's lifestyle is cosmetics. In addition, the current trend of adopting a more natural lifestyle is becoming green practically in every industry, including cosmetics (Kole et al. 2005). Natural food products, herbal medicines and biological therapeutic methods are the superior alternatives for leading a wholesome life. Organically cultivated vegetables and fruits are also in high demand. Colour, in any form, has been a part of our diet for millennia. Traditional colouring agents include paprika, turmeric, saffron and others. Early Egyptian texts mention the use of colourants in cosmetics. Additionally natural colouring compounds came from flowers, spinach, beetroot, carrots, grapes and pomegranates. Food, cosmetics and medications all use colours made from beets, spinach, turmeric and other plants (Joshi and Shinde 2017). Numerous biological qualities that are beneficial to the body's health are boosted by biocolours. Polyphenols are antioxidants found in the majority of pigments. It has been demonstrated that microbial pigments such as naphthoquinone, carotenoids, anthocyanins and violacein are strong antioxidants. Numerous pigments exhibited antimicrobial properties. Sen et al. (2019) demonstrated that an endophytic fungal pigment was more effective than the commercial antibiotic streptomycin agents. This pigment can be used to combat bacteria, including *Vibrio cholerae*, *Salmonella typhi*, *Staphylococcus aureus* and *Klebsiella pneumoniae*. Additionally, vitexin has demonstrated antimicrobial properties, including the ability to inhibit and even kill germs.

The textile industry produces textile effluent which is challenging to handle and dispose of, and uses vast quantities of synthetic dyes. The negative environmental effects of synthetic dyes used in textiles have prompted concerns about natural alternatives that provide eco-friendly products with fewer negative effects (Juailova et al. 1997). A much-needed substitute for the complicated term 'chemical dye' is natural dye. When added to textiles, biocolourants can be applied directly or in combination with particular mordants that are typically employed to maintain the stability of natural pigments. These pigments also have the ability to absorb UV light and exhibit antimicrobial properties which raise the value of dyed textiles (Mehta and Shah 2016). The present chapter explains different types of biocolours obtained from various sources with special reference to plant-based, derived microbial species and animal-origin biocolours. Biochemical methods for synthesis of biocolours along with their molecular structures, characteristics and roles in biological systems are also detailed.

2.2 Types of Biocolours

Anthocyanin, betalain, carotenoid, chlorophyll and caramel are the five categories into which bio-colours are divided. Typically, spices are used to flavor and season food. Additionally, several spices can be utilized as bio-colourants. The following spices are frequently employed as colourants: tamarind (leucoanthocyanidine), paprika (capsanthin), saffron (crocin), turmeric (curcumin) and kokum (anthocyanin) (Chittaragi and Menon 2023).

2.2.1 Plant-Derived Biocolours

2.2.1.1 Chlorophylls

The green pigment chlorophyll is present in plants and algae. The term 'E140' refers to the usage of chlorophyll and chlorophyllins as a food additive. In pasta, absinthe, cheeses, preserved vegetables, jams, jellies and marmalade, it is typically produced from nettles, grass and alfalfa. Even though there are five distinct kinds of chlorophylls, only two, a and b, are used as colourants in the food business. The primary obstacle to employing chlorophylls is their complex structure, which makes stability challenging. Usually, commercial chlorophyll colourants are made from alfalfa. They are found in many food items, including dairy products, soups, drinks and sweets (Anonymous 2023).

2.2.1.2 Carotenoids

Orange-yellow pigments called carotenoids are fat-soluble substances that can be found in a variety of fruits and vegetables, including oranges, pumpkins, mangoes and carrots. Carotene was the name given to the carotenoid after it was first discovered in carrots. The amount of carotenes present in a vegetable may frequently be determined by looking at how vibrant its colour is. The three most prevalent carotenoids—lutein, violaxanthin and neoxanthin—are mostly found in green leaves. Carotenoids are used as natural food colouring agents and are frequently derived from foods including tomatoes, paprika, annatto, saffron and more (Marmion 1991).

2.2.1.3 Anthocyanins

The Greek words 'anthos,' which means flower, and 'kyanos,' which indicates dark blue, are the roots of the word 'anthocyanin'. Anthocyanin is a polyphenolic molecule and water-soluble plant pigment that is widely utilized as a natural food colouring in the global food and beverage industry. A range of colours, from orange, red, pink, purple and blue, are imparted. In drinks, food items, confections and fruit preparations, anthocyanins are necessary food colouring agents. The pH of the surrounding environment determines the colour that anthocyanins provide; they might be red, violet or blue. Many anthocyanins have red colour under acidic conditions, but they tend to turn blue under less acidic ones (Bora et al. 2019). These pigments can vary in concentration and are found in a variety of plant components, such as fruits, stems and leaves. One important class of secondary metabolites that gives fruits their varied colours is called anthocyanins. They are essential for shielding plants from pests and UV damage.

Anthocyanins are thought to have a number of health advantages for people, such as anti- inflammatory, anti-ulcer, antioxidant and antimutagenic qualities (Arunkumar et al. 2024).

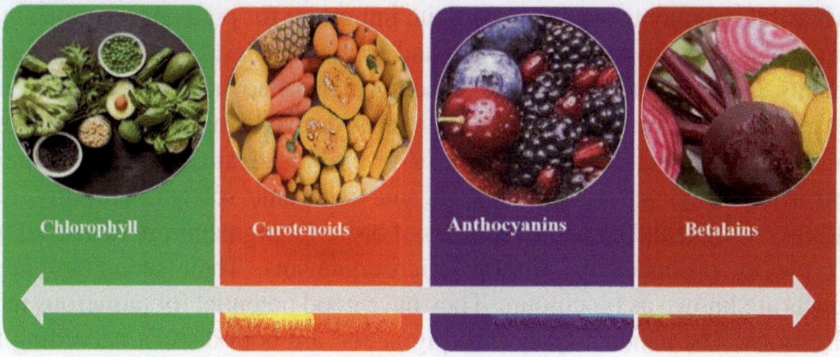

Fig. 2.1 Classification of plant based biocolours

2.2.1.4 Betalains

Water-soluble pigments called betalains give red beets, cactus fruits and some flowers their red and yellow hues. Out of all the sources, beetroot has the most potential for production. The pH range of 4–6 keeps these pigments stability; however, heat processing can cause them to degrade. The use of betalains as food colouring is restricted because of their sensitivity (Delgado-Vargas et al. 2000). Figure 2.1 shows all four categories of plant-based biocolours.

2.2.2 Microbial-Derived Biocolours

Pigments produced by bacteria, fungi, algae and other organisms can be utilized to create biocolours. Carotenoids, melanins, flavins, phenazines and quinones are some of the compounds that microbes create (Pankaj and Kumar 2016).

2.2.2.1 Bacterial Pigments

The utilization of bacteria is increased by the health benefits of the microbial pigments, which include antioxidant, non-carcinogenic, and antimicrobial properties. Examples of microbes that produce pigments include *Flavobacterium* sp. (yellow pigment: zeaxanthin), *Agrobacterium aurantiacum* (pink-red pigment: astaxanthin), *Micrococcus* sp. (carotenoids, various coloured pigments), *Serratia marcescens* (red pigment), *Chromobacterium* sp. (violet pigment) and *Rheinheimera* sp. (blue pigment) (Venil et al. 2013). These days, canthaxanthin, astaxanthin and zeaxanthin are the main microbial pigments among carotenoids. Violacein, indigoidine, prodigiosin and other pigments are among them. Therefore, these pigments can be readily used to colour a variety of culinary items, such as cakes, pastries, milk etc., if they are extracted correctly (Ganguly et al. 2019). Certain bio-colours are produced by microbes such as *Arthrospira platensis* (*spirulina* bacteria), *Blakeslea trispora* (β-carotene), *Dunaliella salina* and *Monococous* sp. (carotenoids). *Rhodophyta* and *Chlorophyta* algae produce red and blue hues through phycobiliproteins (Chittaragi and Menon 2023).

Microbial pigments appear to be quite important for pigment creation since they have some benefits over those of plant and animal species. One excellent supplier of pigments is thought to be bacteria. They are also known as 'biocolours' because they are derived from living things. Pigments derived from plants and animals are also accessible in nature, but because of several disadvantages, such as their year-round availability, water solubility, organic–inorganic solvent solubility, and susceptibility to light and heat, the majority of bacterial pigments are favored. Pigments are typically produced by aerobic microorganisms. Their secondary metabolic products are known as biocolours. They have good potential for numerous industrial applications. The names 'dyes' and 'pigments' are both distinct according to their solubility. In general, pigments dissolve in solvents while dyes dissolve in water. From natural environmental sources, *Leifsonia aquatica*, *Pseudomonas aeruginosa* and *Serratia marcescens* were isolated for their respective yellow, green and red colour pigments. The greatest colour output was attained at 25 °C, 150 rpm, and during 72 h of incubation. The use of glycerol was done to improve colour visibility (Mehta and Shah 2016).

2.2.2.2 Fungal Pigments

The best sources of naturally occurring microbial pigments are thought to be a number of yeast species which can also be utilized to synthesize other carotenoids. Melanin is a pigment that is present in all parts of nature and is crucial for UV protection. Many aspects of the chemical *structure* are still unknown due to its insoluble nature. Due to their resistance to radiation, the majority of melanized fungi may flourish in harsh settings like Antarctica and contaminated nuclear reactors (Rosa et al. 2010; Zhdanova et al. 2000). Melanin is synthesized by fungi using 1-3, 4-dihydroxyphenylalanine (L-dopa), a mechanism that is comparable to that of mammalian melanin formation. However, *Cryptococcus neoformans* is an essential part of fungal melanization since it only manufactures melanin via the L-dopa pathway and is entirely reliant on external substrate (Helene and Arturo 2012). The European Union recently approved the use of fungal carotenoids as food colouring agents in the future to create polyketide azaphilone colours. One major advantage of using colourants derived from fungi is that the manufacturer is no longer reliant on the seasonal availability of raw materials, which reduces batch-to-batch variability. Non-toxic fungal strains like *Penicillium* and *Epicoccum* sp. can be used to make food colouring (Gupta et al. 2011; Ganguly et al. 2019).

2.2.2.3 Algal Pigments

Microalgae act as a repository for valuable elements. The production is based on taking advantage of their extraordinarily efficient photosynthetic process. Red algae (*Rhodophyta*) and blue-green algae (*Cyanophyta*) produce a group of bilin or tetrapyrrole-based pigments that are incredibly light-absorbing. The use of these phycobilin proteins as natural colourants, especially as substitutes for artificial colours, in foods, cosmetics and medications. It is recommended to utilize ice candies, frozen treats, sherbets, candies and chewing gum with preparations of pigments in either alcohol or water (Ganguly et al. 2019; Monishdeep et al. 2022).

2.2.3 Animal-Derived Biocolours

The presence of bromine causes the sea fish (*Hexaplex trunculus*) to emit a purple colour from their shell. The textile and fabric industries are its primary users. Cochineal, a reddish- purple carminic acid, is produced by the bodies of certain scale insects (*Dactylopius coccus*). Their bodies are left to dry and crushed in order to obtain the red colour, and they eat cactus plants. These are mostly utilized in Central America, Africa and Spain (Chittaragi and Menon 2023).

2.3 Biochemical Pathways of Biocolour Synthesis

2.3.1 Plant Pigment Biosynthesis

Enzymatic pathways for the production of chlorophyll, carotenoids and anthocyanins are essential biochemical processes in plants that are involved in pigmentation, light absorption and protection from stress. Each pathway is unique in its enzymes, substrates and intermediates. Below is a detailed explanation of the enzymatic pathways for the production of these important pigments:

2.3.1.1 Chlorophyll Production Pathway (Wang and Grimm 2021)

The following steps are involved in the synthesis of chlorophyll:

2.3.1.1.1 Porphyrin Ring Synthesis

The first step involves the synthesis of *phenylalanine ammonia-lyase* (ALA) from glutamate. This is catalyzed by the enzyme *glutamyl-tRNA reductase* (GluTR). ALA is then converted into porphobilinogen (PBG) by the enzyme *ALA dehydratase*. The next step involves the condensation of four molecules of PBG to form hydroxymethylbilane (HMB), catalyzed by PBG deaminase.

2.3.1.1.2 Chlorophyll Production

Chlorophyll a/b Formation Chlorophyllide is further converted into chlorophyll a and chlorophyll b by the enzymes chlorophyll synthase. Chlorophyll a is the principal colourant in photosynthesis, and chlorophyll b supports in light amalgamation. Figure 2.2 shows the complete process of chlorophyll synthesis step-wise.

2.3.1.2 Carotenoid Production Pathway (Sandmann 2022).

Carotenoids are derived from isoprenoids, and the pathway begins with the formation of geranylgeranyl diphosphate (GGPP) through the mevalonate pathway or the non-mevalonate pathway. The complete synthesis process is represented in Fig. 2.3.

2.3.1.3 Anthocyanin Production Pathway (Sunil and Shetty 2022)

The biosynthesis of anthocyanins starts with the amino acid phenylalanine which is originated from the shikimate pathway and converted to cinnamate by *phenylalanine ammonia-lyase* (PAL) followed by its conversion into chalcone (the basic

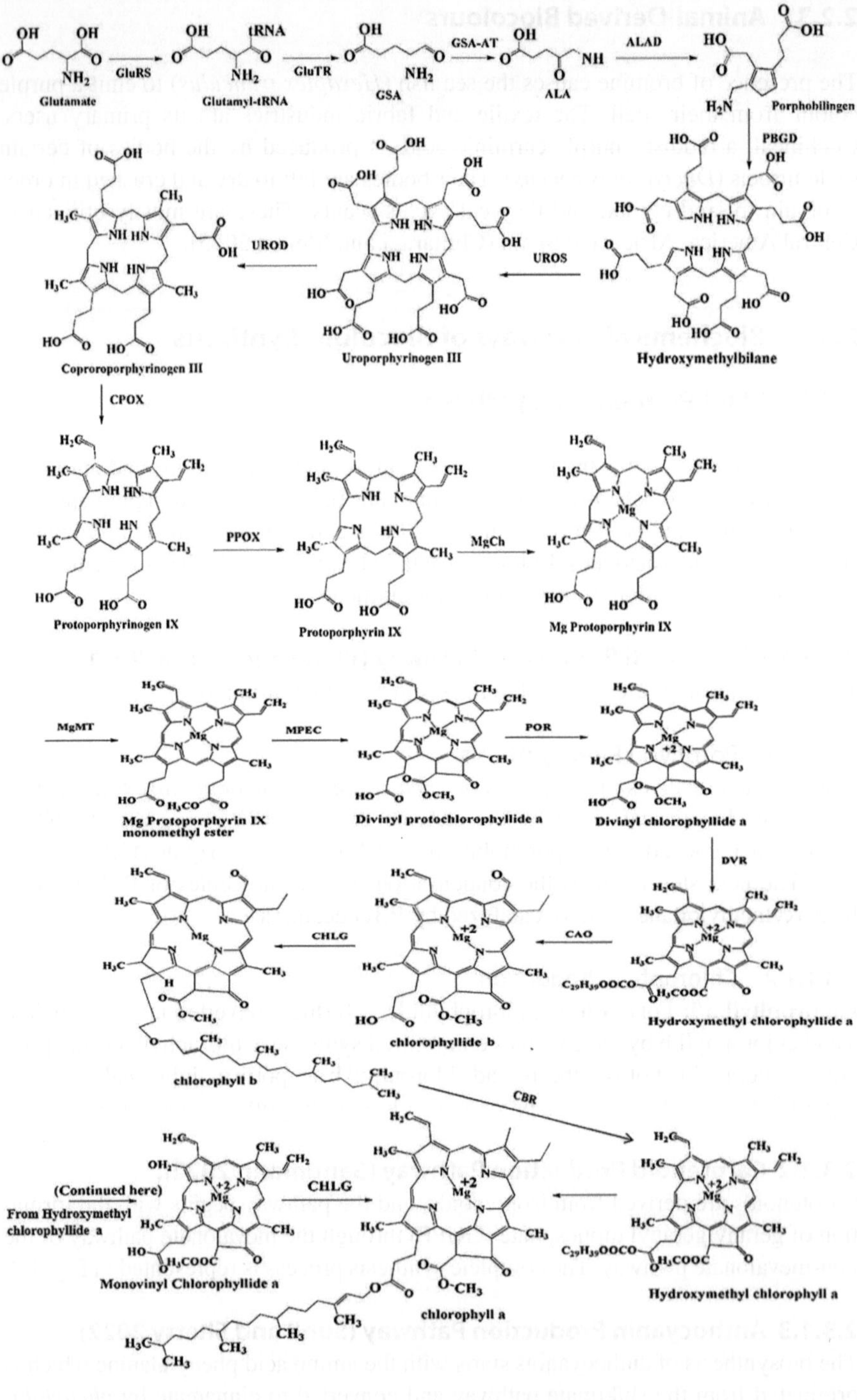

Fig. 2.2 Chlorophyll biosynthesis pathway in higher plants (Eaton-Rye et al. 2012)

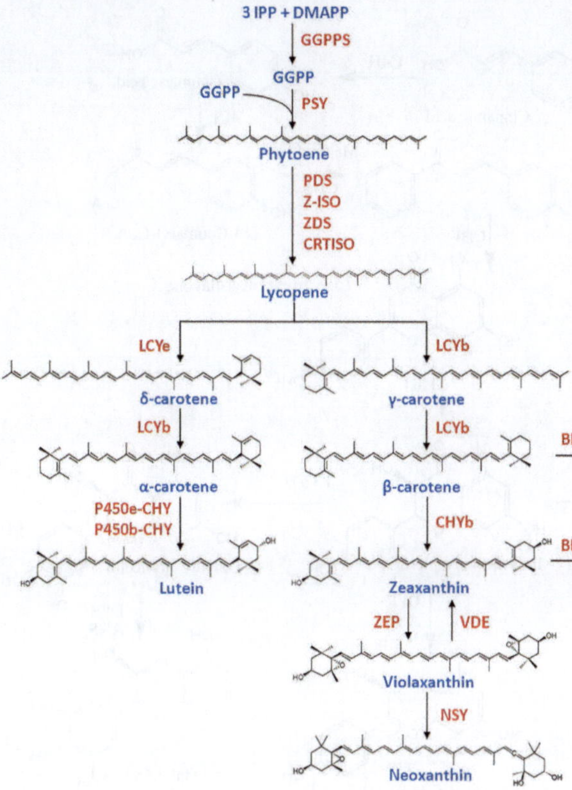

Fig. 2.3 Schematic diagram of the carotenoid biosynthetic pathway in plants and green algae. IPP: isopentenyl pyrophosphate, DMAPP: dimethylallyl pyrophosphate, GGPP: geranylgeranyl pyrophosphate, GGPPS: geranylgeranyl pyrophosphate synthase, PSY: phytoene synthase, PDS: phytoene desaturase, Z-ISO: 15-cis-ζ-carotene isomerase, ZDS: ζ-carotene isomerase, CRTISO: carotene isomerase, LCYe: lycopene ε-cyclase, LCYb: lycopene β-cyclase, P450e-CHY: cytochrome P450ε-hydroxylase, P450b-CHY: cytochrome P450 β-hydroxylase, CHYb: carotene β-hydroxylase, BKT: β-carotene oxygenase, ZEP: zeaxanthin epoxidase, VDE: violaxanthin de-epoxidase and NSY: neoxanthin synthase (Cordero et al. 2012)

skeleton of flavonoids) by the help of enzyme *chalcone synthase* (CHS). The enzyme *dihydroflavonol 4-reductase* (DFR) reduces dihydroflavonol into leucoanthocyanidins which is converted into anthocyanidins (such as delphinidin, cyanidin or pelargonidin) by the enzyme *leucoanthocyanidin dioxygenase*. Anthocyanidins are finally glycosylated to form anthocyanins, facilitated by the enzyme *anthocyanin synthase* (ANS) which adds glucose or rhamnose (sugar molecules) to the anthocyanidins to form anthocyanins. Figure 2.4 depicts all the steps involved in the synthesis of anthocyanin.

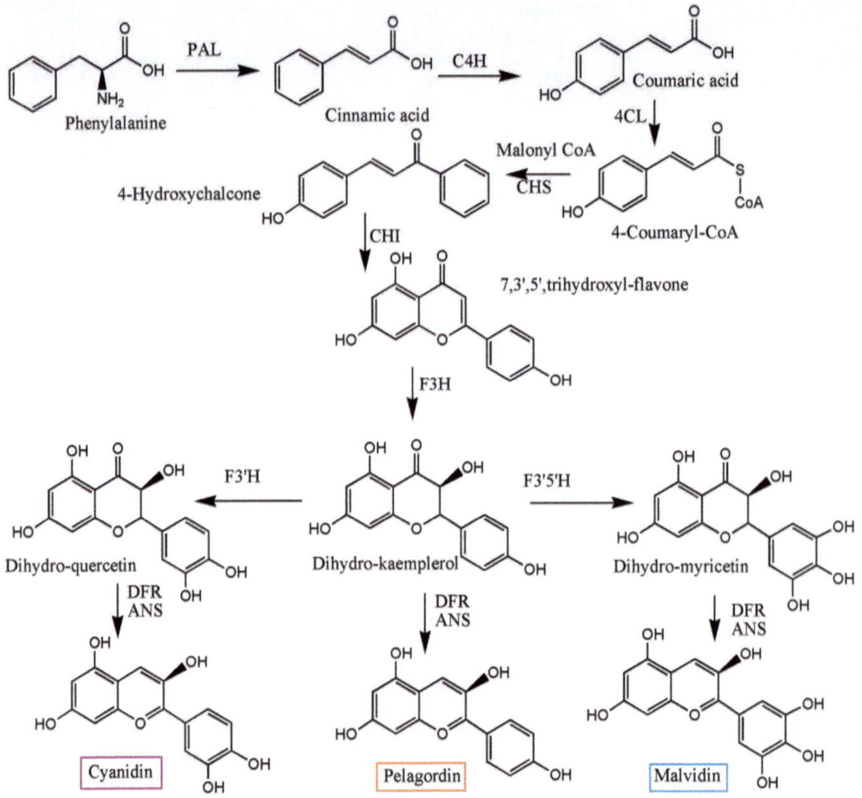

Fig. 2.4 Biochemical pathway for the synthesis of anthocyanidins. PAL: phenylalanine ammonia-lyase, C4H: cinnamic acid 4-hydroxylase, 4CL: 4-coumarate-CoA ligase, CHS: chalcone synthase, CHI: chalcone isomerase, F3H: flavanone 3-hydroxylase, F3′H: flavonoid 3′-hydroxylase, F3′5′H: flavonoid 3′,5′-hydroxylase, DFR: dihydroflavonol reductase and ANS: anthocyanidin synthase (Mannino et al. 2021)

2.3.2 Microbial Pigment Biosynthesis

2.3.2.1 Pathways in Bacteria

The biosynthesis of pigments in bacteria is a fascinating topic because it involves intricate biochemical pathways that enable bacteria to produce colourful compounds, often with unique biological activities. One well-known example of bacterial pigment biosynthesis is the production of prodigiosin, a red pigment produced by various species of *Serratia*, *Streptomyces*, *Vibrio* and other bacteria (Rather et al. 2023). The biosynthesis of prodigiosin in bacteria, such as *Serratia marcescens*, exemplifies the complexity of microbial pigment pathways. The prodigiosin pigment has not only an aesthetic value but also significant bioactivity, making it of interest for biotechnological and pharmaceutical applications. Understanding the biosynthesis of microbial pigments like prodigiosin opens up new avenues for utilizing bacteria in producing valuable compounds with diverse applications.

Prodigiosin is a red pigment that belongs to the class of tripyrrolic compounds, meaning it contains three pyrrole rings. It has gained attention not only for its vibrant colour but also for its bioactivity, including antimicrobial, anticancer and immunosuppressive properties. The biosynthesis of prodigiosin in bacteria involves several key enzymes and intermediate molecules.

2.3.2.1.1 Biosynthesis Pathway of Prodigiosin

The biosynthesis of prodigiosin in bacteria like *Serratia marcescens* is a multi-step process involving the steps described below in Fig. 2.5.

2.3.2.2 Fungal Pigment Biosynthesis

Fungal pigment biosynthesis is an important area of study because fungi produce a variety of bioactive pigments, many of which have ecological, protective and industrial significance. Among the most well-known fungal pigments are melanins, which are produced by many fungi and play a crucial role in their survival and interactions with the environment (Lin and Xu 2022). Melanin is a dark, water-insoluble polymer that is found in the cell walls and extracellular matrices of fungi, offering protection against UV radiation, oxidative stress, and other environmental challenges.

2.3.2.2.1 Melanin Biosynthesis Pathways in Fungi

Melanin biosynthesis in fungi is a well-characterized pathway, and it primarily involves the enzymatic conversion of tyrosine (an amino acid) into complex polymeric structures. There are several types of melanins, including DOPA (dihydroxyphenylalanine)-type, eumelanin and pheomelanin, which are produced through different biochemical routes. Fungal melanins are most commonly of the DOPA type, synthesized from tyrosine or its derivatives. The biosynthesis of fungal melanin involves several key intermediates and enzymatic steps (Gao et al. 2022), primarily starting with tyrosine. Tyrosine (an amino acid) is the primary precursor for melanin biosynthesis and may either be absorbed from the environment or synthesized by the fungi. *Tyrosinase* is the first key enzyme in melanin biosynthesis, which catalyzes the hydroxylation of tyrosine to form DOPA

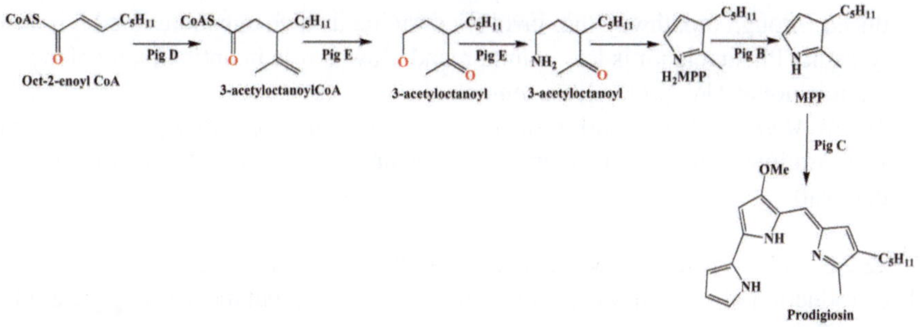

Fig. 2.5 Biosynthetic pathway of prodigiosin (Paul et al. 2022)

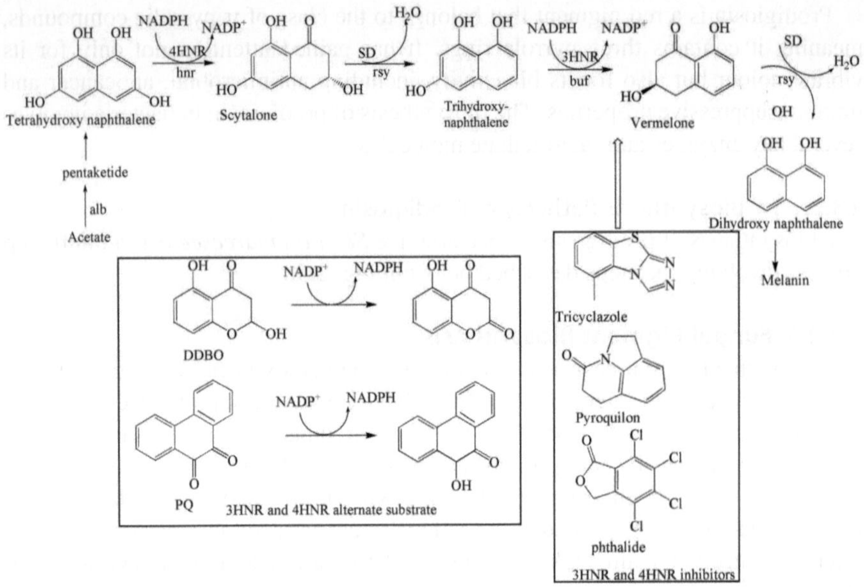

Fig. 2.6 Figure showing the fungal melanin biosynthetic pathway (Thompson et al. 2000)

(dihydroxyphenylalanine). It also catalyzes the oxidation of DOPA into DOPAquinone (the immediate precursor) in the formation of melanins.

In some fungi, the enzyme *laccase* is also involved in melanin polymerization, catalyzing the oxidation of DOPAquinone to form the melanin polymer. The biosynthesis of fungal melanin involves several key intermediates and enzymatic steps (Gao et al. 2022) shown in Fig. 2.6.

2.3.2.2.2 Types of Melanin in Fungi

- *Eumelanin*: It is a black to brown colour pigment and the most common type of melanin consists of polymerized indolic units derived primarily from DOPAquinone. Eumelanin provides protection against UV radiation, desiccation and other environmental stresses.
- *Pheomelanin*: It is a yellow to red pigment, contains sulphur groups and is produced through a pathway that involves cysteine as a co-substrate in addition to tyrosine. Pheomelanin is less common and plays a role in antioxidant defences, but it is not as UV- protective as eumelanin.
- *DOPA Melanin*: It is a dark brown to black pigment, primarily produced from DOPAquinone and involved in structural and protective roles in the fungal cell wall.

2.3.2.2.3 Biosynthesis of Melanin in Specific Fungal Species

The production of melanin varies between fungal species, but the pathway generally follows the same basic steps.

- *Cryptococcus neoformans* (a basidiomycete yeast): *Cryptococcus neoformans*, a pathogenic yeast, produces melanin through the oxidation of L-DOPA (derived from tyrosine) (Firacative et al. 2021). The melanin produced by *C. neoformans* helps the fungus survive in hostile environments and enhances its virulence by protecting it from oxidative damage and the host's immune response. *Tyrosinase* and *laccase* play significant roles in the melanin production pathway in this yeast.
- *Aspergillus fumigatus* (a filamentous fungus): *Aspergillus fumigatus* is another example of a fungus that produces melanin (Liu et al. 2022). It uses L-DOPA as a precursor, and its melanin production is believed to provide resistance against oxidative stress and UV radiation. *Tyrosinase* and other oxidative enzymes are involved in this process.
- *Exophiala* spp. (a black yeast): Some species of the genus *Exophiala* produce melanin in response to environmental stress, especially when exposed to UV light or high temperatures (Haranto et al. 2024). These fungi are often used as model organisms to study melanin production in dark-pigmented fungi. Melanin production in *Exophiala* contributes to its resistance to environmental stresses.

2.4 Regulation of Pigment Production

The biosynthesis of pigments in microorganisms, fungi and plants is governed by intricate genetic and environmental influences (Lyu et al. 2022). The production of pigments like chlorophylls, carotenoids, anthocyanins and melanins is regulated by different gene expression processes that can be affected by environmental cues. These ecological factors consist of light, temperature, pH, oxygen concentration and nutrient accessibility, which significantly influence the regulation of essential pigment production genes.

2.4.1 Gene Expression in Pigment Biosynthesis

The process by which information from a gene is used to produce useful gene products, usually proteins, which often operate as enzymes that aid in metabolic processes like pigment creation, is known as gene expression. The expression of genes involved in pigment formation is influenced by a number of essential enzymes that transform precursor molecules into pigments. For example, *tyrosinase* contributes to the synthesis of melanin (Zhao et al. 2022). Genes are involved in the synthesis of carotenoids, including those linked to *phytoene synthase* and carotenoid biosynthesis. *Chlorophyll synthase* contributes to the synthesis of chlorophyll, while *anthocyanin synthase* is involved in the development of anthocyanins. Specific genes that are either transcriptionally activated or repressed in response to metabolic and environmental cues encode each of these enzymes.

2.4.2 Environmental Factors Influencing Pigment Biosynthesis

Pigment production in organisms can be significantly affected by environmental conditions, and these factors regulate the expression of genes involved in the biosynthetic pathways (Zhan et al. 2022). Here is a breakdown of the key environmental factors.

2.4.2.1 Light

Light is one of the most important environmental factors influencing pigment biosynthesis, especially in plants, algae and photosynthetic microorganisms.

2.4.2.1.1 Chlorophyll Biosynthesis in Plants

Depending on the type of light, such as red or blue light, genes involved in the synthesis of chlorophyll, like *Chl synthase*, express themselves differently. Without light, photosynthesis slows down and plants may produce less chlorophyll. Photosynthetic bacteria (like *Rhodobacter*) can absorb light and use it as energy because light causes the production of genes that produce pigments like bacteriochlorophyll. A photoreceptor called phytochrome is essential for controlling the synthesis of chlorophyll in accordance with cycles of light and dark.

2.4.2.1.2 Carotenoid Biosynthesis in Bacteria and Fungi

In bacteria like *Bacillus* or *Streptomyces*, light can increase the expression of genes like psy (phytoene synthase), which is involved in the manufacture of carotenoid pigments. Fungal organisms like *Aspergillus* can accumulate carotenoids due to gene activation associated with carotenoid production under specific illumination conditions.

2.4.2.1.3 Anthocyanin Biosynthesis in Plants

Particularly, blue light increases the expression of genes involved in the anthocyanin production pathway, such as ANS (*anthocyanidin synthase*) and CHS (*chalcone synthase*). UV radiation also promotes the synthesis of anthocyanins in many plants as a defence against UV-induced damage.

2.4.2.2 Temperature

The generation of pigments is significantly impacted by temperature, especially in bacteria and fungus. Genes involved in the synthesis of melanin can be activated by high or fluctuating temperatures. Genes involved in the formation of melanin may be activated by high or low temperatures in fungal pigments (carotenoids and melanin). When exposed to heat stress, certain *Exophiala* species may create more melanin which would boost their resistance to the negative effects of heat. At increasing temperatures, the carotenoid biosynthesis process produces more pigment. At higher temperatures, *Aspergillus* produces more pigment because temperature changes can trigger the carotenoid biosynthesis pathway. The activity of enzymes like *phytoene synthase* is directly impacted by temperature. In certain bacteria, the production of pigments depends on temperature, with elevated temperatures potentially resulting in increased expression of carotenoid biosynthesis genes. By altering the expression of genes involved in chlorophyll biosynthesis, high temperatures can lower the amount of chlorophyll in plants and prevent photosynthesis. On the other hand, ideal temperatures promote the expression of genes that produce chlorophyll.

2.4.2.3 pH

The growth medium's pH level may have an impact on the expression of genes involved in colour biosynthesis pathways. The acidic pH of their environment may lead some fungus to produce more melanin. For instance, *Neurospora crassa* needs lower pH values in order to generate more melanin which is necessary for its survival in hostile environments. The pH of the vacuole during anthocyanin synthesis influences its colour, with red pigments being preferred in acidic environments. The genes involved in carotenoid production are similarly impacted by pH. A more acidic pH environment may cause carotenoid production in some bacteria, possibly as a stress-reaction mechanism.

2.4.2.4 Nutrient Availability

For pigment formation to be regulated, nutrients—especially carbon and nitrogen sources must be present. Carbon sources such as glucose, glycerol or acetate can control the expression of genes involved in carotenoid production. In certain bacteria and fungi, too much carbon or nitrogen can activate the carotenoid biosynthesis pathway, whereas too little of these elements can reduce the formation of pigments. For instance, nitrate and ammonium levels can alter the expression of genes such as psy (*phytoene synthase*) in bacterial carotenoid pathways. Nitrogen-rich environments may prevent fungus from producing melanin, but a lack of nutrients may cause melanin biosynthesis genes to be expressed as a defence mechanism.

2.4.2.5 Oxygen Levels

Particularly for bacteria that produce colours and depend on oxidative reactions for this, oxygen availability is crucial. Under aerobic conditions, *Streptomyces* species may produce higher quantities of carotenoid because oxygen is necessary for oxidizing precursors like phytoene to make carotenoids. Low oxygen levels or anaerobic environments may hinder this process. Melanin production usually takes place in aerobic circumstances because the generation of melanin requires the oxidative polymerization of intermediates such as DOPAquinone. Inadequate oxygen levels can slow down or impede this process.

2.4.2.6 Osmotic Stress

Osmotic stress, such as high salt or sugar concentrations, can also affect gene expression involved in pigment production. Osmotic stress can induce the carotenoid biosynthetic pathway, particularly in halophilic bacteria that produce carotenoids for protection against salt stress.

2.4.3 Regulation of Gene Expression in Pigment Biosynthesis

Numerous transcription factors and signaling pathways control the expression of genes involved in pigment biosynthesis. Plant production of anthocyanins is regulated by the transcription factors Myb and bHLH. RpoS and other sigma factors in bacteria help regulate pigment production under harsh conditions, such as oxidative stress or dietary constraint (Zhao et al. 2022). The production of pigments in fungi under different dietary conditions is regulated by the cAMP/PKA system. Some fungi's mitogen-activated protein kinase (MAPK signaling system can alter melanin synthesis in response to environmental stressors. Changes like histone acetylation or DNA methylation can affect how colour biosynthesis genes are expressed.

2.5 Molecular Structure and Characteristics of Biocolours

2.5.1 Chemical Structures of Major Biocolours

Biocolours, or natural pigments, come from various biological sources and include a range of different chemical structures. Colour, source, chemical formula, chemical structure description and biological functions of major biocolours are summarized in Table 2.1.

These pigments are derived from plants, algae, and animals, and they serve various biological functions, including attracting pollinators, protecting from UV radiation, and playing roles in metabolism. Fig. 2.7 shows the chemical structures of some important biocolours.

2.5.2 Interaction with Light (Absorption and Reflection Spectra)

Biocolours interact with light in specific ways based on their molecular structures. These interactions depend primarily on the absorption and reflection spectra of the pigments. Resulting colour after absorption and reflection of light with reason is depicted in Table 2.2.

2.5.3 Stability and Degradation of Biocolours

The stability and degradation of biocolours depend on several factors, including the chemical structure of the pigment, environmental conditions (light, temperature, pH etc.) and the presence of chemical or biological agents (Singh et al. 2023).

2.5.3.1 Light

UV and visible light can cause photodegradation, particularly for pigments like anthocyanins, carotenoids and betalains, leading to colour loss.

Table 2.1 Table showing the chemical structures of major biocolours

Biocolour	Colour	Source	Chemical formula	Chemical structure description	Role/ function	References
Chlorophyll	Green	Cyanobacteria, algae and plants	$C_{55}H_{72}MgN_4$	Magnesium ion at the Centre of a porphyrin ring	Photosyn thesis	Bashir et al. (2024)
Carotenoids	Yellow, orange and red	Various plants and algae	$C_{100}H_{160}$	Hydrocarbons with conjugated double bonds, can be xanthophylls (with oxygen) or carotenes (hydrocarbons)	Photosyn thesis, antioxida nts	Devi et al. (2022)
Anthocyanin s	Red, purple and blue	Various plants, notably flowers and fruits	$C_{16}H_{16}O_7$	Flavonoid-based pigments, joined to sugar groups	Pollinator attraction, plant defence	Bashir et al. (2024)
Betalains	Red, yellow	Beetroot, certain other plants	$C_{16}H_{19}N_3O_7$	Nitrogen- containing heterocyclic structure	Pigmenta tion, antioxidant activity	Verma and Kaur (2024)
Carminic acid	Red	Cochineal insects	$C_{19}H_{18}O_7$	Polyphenolic compound	Red food and cosmetic Colourant	Singh et al. (2023)
Indigo	Blue	*Indigofera tinctoria* (and other plants)	$C_{16}H_{10}N_2O_2$	Carbonyl group in the middle of an indole ring	Textile dye	De Mol and Vandamme (2024)
Lutein	Yellow	Xanthophyl l (a type of carotenoid, found in plants)	$C_{18}H_{24}O_2$	Similar to beta-carotene, but with oxygen atoms, particularly hydroxyl groups	Eye health, present in green leafy vegetables	Verma and Kaur (2024)

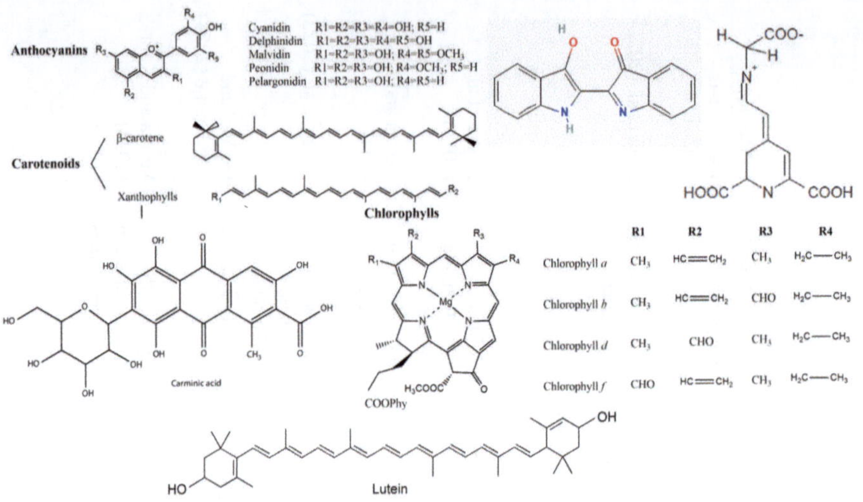

Fig. 2.7 Chemical structures of some important pigments

Table 2.2 Breakdown of how biocolours absorb and reflect light

Biocolour	Absorbed light (wavelengths)	Reflected light (wavelengths)	Resulting colour	Reason for absorption and reflection
Chlorophyll (Wang and Grimm 2021)	Red (660–680 nm), blue (430–450 nm)	Green (500–570 nm)	Green	Conjugated system absorbs red and blue light for photosynthesis, while structure reflects green
Carotenoids (Mascoli et al. 2021)	Blue (400–500 nm), green (500–570 nm)	Yellow, orange and red (480–600 nm)	Yellow, orange and red	Long conjugated system of double bonds absorbs blue and green light and reflects yellow, orange and red
Anthocyanins (Ma et al. 2021)	Blue (400–530 nm), green (500–530 nm)	Red, purple and blue (550–700 nm)	Red, purple and blue	Large conjugated structure absorbs blue and green light, with colour variation based on pH
Betacyanins (Tossi et al. 2021)	Blue (540–600 nm)	Red to violet	Red	Conjugated system absorbs blue light and reflects red to violet due to specific chemical structure
Betaxanthins (Liu et al. 2021)	Blue and green (450–550 nm)	Yellow	Yellow	Conjugated structure absorbs blue and green light, reflects yellow based on specific absorption properties

(continued)

Table 2.2 (continued)

Biocolour	Absorbed light (wavelengths)	Reflected light (wavelengths)	Resulting colour	Reason for absorption and reflection
Carminic Acid (de Sérgio 2021)	Blue and green (400–550 nm)	Red (600–700 nm)	Red	Conjugated molecular system absorbs blue/green light, reflects red light due to specific molecular configuration
Indigo (Xiao and Ahn 2022)	Orange and red (580–700 nm), Yellow and green (weaker)	Blue (420–490 nm)	Blue	Conjugated system absorbs longer wavelengths (orange-red), reflects shorter wavelengths (blue) due to molecular structure
Lutein (Bashir et al. 2024)	Blue and green (400–500 nm)	Yellow (490–550 nm)	Yellow	Conjugated system absorbs blue/green light, reflects yellow light due to molecular structure

2.5.3.2 Temperature

Heat accelerates the degradation of many natural pigments, especially carotenoids and betalains, through processes like oxidation or thermal breakdown.

2.5.3.3 Oxygen

Oxidation is a major cause of degradation for many biocolours, particularly carotenoids, betalains and chlorophyll. This process breaks down the pigment molecules, leading to colour fading.

2.5.3.4 pH

Many natural pigments, such as anthocyanins, betalains and carminic acid, are highly sensitive to pH. Acidic or alkaline conditions can cause the pigments to degrade or change colour.

2.5.3.5 Enzymatic Action

In biological systems, enzymes can catalyze the breakdown of pigments, especially during post-harvest storage of fruits and vegetables. This can accelerate degradation. Major pigments, their stability and causes of degradation are listed in Table 2.3.

2.6 Role of Biocolours in Biological Systems

2.6.1 Biocolours as Antioxidants and Protective Agents

Biocolours, or natural pigments, not only serve to provide colour but also offer a variety of health benefits, particularly due to their antioxidant properties. Many biocolours, such as carotenoids, anthocyanins and betalains, possess significant

Table 2.3 Various pigments, their stability factors and the causes of degradation for each pigment (Vaidya et al. 2021)

Pigment	Colour	Stability factors	Degradation factors
Chlorophyll	Green	Stable at low temperatures, protected by antioxidants in plant cells, more stable under acidic conditions	*Photo-oxidation*: loss of colour, breakdown into chlorophyllides and pheophytin *Heat*: prolonged heat leads to yellowish-brown colour *pH sensitivity*: degrades in acidic environments, turning brown/olive.
Carotenoids	Yellow, orange, and red	Stable at low temperatures, with antioxidants. Less prone to oxidation compared to chlorophyll	*Oxidation*: loss of colour, cleavage products like apocarotenals *Light*: breakdown, reduced colour intensity *Heat*: degradation at high temperatures, colour reduction *pH sensitivity*: Degrades in high acidity
Anthocyanins	Red, purple and blue	Stable with antioxidants, somewhat stable to moderate heat	*Oxidation*: colour loss, formation of quinones, brownish hue *Light*: UV light accelerates degradation *Heat*: prolonged heat causes colour fading
Betalains	Red, yellow	Stable under acidic conditions, less stable at high temperatures	*Oxidation*: colourless or brown products form, especially betanin breakdown *Light*: UV light speeds up degradation *Heat*: thermal degradation leads to colour loss
Carminic acid	Red	Stable under cool, dark conditions, more stable in acidic environments	*Light and heat*: degradation causes fading and brownish products *Oxidation*: loss of colour intensity *pH sensitivity*: degrades under alkaline conditions
Indigo	Blue	Stable under normal storage, neutral pH stability	*Oxidation*: leads to colour change, formation of colourless/brown compounds *Light*: UV light causes degradation *Heat*: high temperatures degrade the pigment
Lutein	Yellow	Stable with antioxidants, most stable in neutral/ slightly acidic environments	*Oxidation*: Loss of colour intensity, formation of luteoxanthin *Light*: Degradation, colour loss *Heat*: Degradation, colour fading in food processing

antioxidant activity, which helps protect cells from oxidative stress and associated damage. Here is an overview of how biocolours function as antioxidants and protective agents (Table 2.4).

2.6.2 Pigments in Plant–Animal Interactions

Pigments in plants play crucial roles in interactions with animals, particularly in processes like pollination and defence mechanisms. These interactions are often mediated through the visual cues provided by pigments, which attract or deter specific animal species. Here is an exploration of how pigments function in plant–animal interactions.

2.6.2.1 Role of Pigments in Pollination

Pigments are vital in pollination, as they help plants attract pollinators like insects, birds and bats. The colours of flowers, often resulting from anthocyanins, carotenoids and flavonoids, play an essential role in these interactions.

2.6.2.1.1 Attracting Pollinators

Because some animal groups, including bees, butterflies and birds, can frequently see the colours of anthocyanins (red, purple and blue pigments) and carotenoids (yellow, orange pigments), these substances are especially crucial for drawing pollinators. Different pollinators are drawn to different colours. For example, bees are more attracted to blue and purple flowers than to red ones, indicating a visual preference for these hues. Hummingbirds in particular are drawn to red and orange blooms which frequently indicate a high nectar content. Yellow, orange and purple are among the many colours that attract butterflies. White or pale flowers that are visible in low light are attractive to bats that pollinate flowers at night.

2.6.2.1.2 Guiding Pollinators

- Patterns and UV reflectance: pollinators, particularly bees, can clearly see patterns or UV-reflective regions in many flowers that are imperceptible to the human eye. Pollinators are directed towards the nectar source by these patterns which are frequently produced by pigments such as flavonoids or anthocyanins. (Kenyon and Martin 2023).
- Nectar guides: to ensure more effective pollination, certain flowers feature particular pigmentation patterns known as nectar guides that help guide pollinators to the nectar supply.

2.6.2.2 Role of Pigments in Defence

Plants also use pigments as part of their defence mechanisms against herbivores, pathogens and environmental stress. Pigments, especially those from the flavonoid, phenolic, carotenoid and betalain groups can have both deterrent and protective roles.

Table 2.4 The pigments, their antioxidant properties, and protective effect for each pigment (Sinha et al. 2024)

Pigment	Colour	Antioxidant properties	Protective effects
Caroten oids	Yellow, orange and red	Neutralize free radicals, protect cells from oxidative damage. Beta- carotene, lutein, zeaxanthin and lycopene are key antioxidants	*Eye health*: lutein and zeaxanthin protect retinal cells from oxidative stress. *Skin protection*: beta-carotene and lycopene protect skin from UV damage, preventing photoageing *Cardiovascular health*: lycopene supports heart health by reducing inflammation and oxidative stress *Cancer and chronic disease protection*: lower the risk by protecting cells from damage and boosting immune function
Anthocyanins	Red, purple and blue	Scavenge free radicals, prevent oxidative damage. Cyanidin is a potent antioxidant	*Anti-inflammatory*: lower pro-inflammatory cytokines and enzymes *Cardiovascular protection*: shield blood vessels, reduce the risk of atherosclerosis. *Neuroprotection*: lower the risk of neurodegenerative diseases like Alzheimer's and Parkinson's *Cancer protection*: prevent oxidative DNA damage, lowering the risk of mutations
Betalains	Red, yellow	Scavenge free radicals, prevent oxidative stress. Betacyanin is a strong antioxidant	*Liver health*: aid detoxification, enhance liver function and reduce inflammation *Heart health*: lower blood pressure, reduce oxidative stress and lower atherosclerosis risk *Cellular protection*: shield proteins, DNA and lipids from oxidative damage, slowing cellular ageing *Anti-inflammatory*: reduce the risk of long-term inflammation
Chlorophyll	Green	Neutralizes free radicals, chelates metal ions to prevent oxidative species formation	*Detoxification*: binds with toxins (heavy metals, carcinogens) and promotes their elimination *Wound healing*: has antimicrobial properties, supporting tissue regeneration *Liver digestive health*: helps promote a healthy liver and digestive system by lowering oxidative damage
Carminic acid	Red	Scavenges free radicals and reduces oxidative stress, though less studied than carotenoids or anthocyanins	*Skin protection*: shields skin from oxidative stress, potentially reducing UV- induced ageing *Chronic disease protection*: helps prevent cellular damage by neutralizing free radicals

(continued)

Table 2.4 (continued)

Pigment	Colour	Antioxidant properties	Protective effects
Indigo	Blue	Some research indicates antioxidant and anti-inflammatory properties, though less studied	*Skin and tissue protection*: may help minimize oxidative stress in tissues exposed to external stressors
Lutein	Yellow	Strong antioxidant properties that protect retinal cells from UV and blue light-induced oxidative damage	*Eye health*: reduces the risk of cataracts and age-related macular degeneration (AMD) *Skin protection*: reduces oxidative stress and inflammation from sun exposure, shielding skin from UV damage

2.6.2.2.1 Camouflage and Concealment

Many plants use pigments to blend into their environment, helping them avoid herbivory. For example, some leaves have a green pigment (chlorophyll) that helps them blend in with their surroundings and remain inconspicuous to herbivores.

2.6.2.2.2 Repellent Effects

Some pigments, such as tannins (types of phenolic compounds), are produced by plants as a defence mechanism to deter herbivores. Tannins can make leaves taste bitter, reducing the likelihood of herbivory. Betalains, produced in some plants like beets, have been shown to have insecticidal effects, deterring herbivores.

2.6.2.2.3 Attracting Herbivore Predators

Some plants have brightly coloured fruits that attract herbivores but also signal predators of these herbivores (e.g., birds) to help reduce herbivore populations. Carotenoids in fruits may signal ripeness and attract frugivores, who help disperse seeds, but the bright colours can also warn potential herbivores about the presence of toxins in the fruit (Miller et al. 2011).

2.6.2.2.4 Pigments as Indicators of Chemical Defences

Secondary metabolites, including as phenolics, terpenoids and alkaloids, are produced by a variety of plants and serve as poisons or herbivore deterrents. Herbivores may be alerted to the existence of these defences by the colours these chemicals frequently possess.

2.6.2.2.5 Antioxidant Properties

In addition to adding colours, certain pigments, including anthocyanins and flavonoids, also contain antioxidant qualities that shield the plant from harm from UV rays, herbivore diet and environmental stressors. Interspecific competition and colour signaling. Pigments can also play a role in *interspecific competition* between plants. Some plants may produce pigments that deter herbivores from feeding on them, while others may use pigments to attract specific pollinators, effectively gaining a competitive edge in the ecosystem (Miller et al. 2011).

2.6.2.3 Biocolours in Visual Signaling and Communication

Biocolours play a crucial role in visual signaling and communication among organisms, including plants, animals and even microorganisms. The ability to perceive and respond to colour allows species to communicate effectively, whether it is for attracting mates, warning predators, signaling the presence of food or facilitating species recognition. Here is an overview of how biocolours are used in visual signaling and communication across different organisms. In plants, biocolours primarily serve to attract pollinators, signal ripeness to fruit-eating animals and sometimes warn herbivores of toxicity (Bashir et al. 2024).

2.6.2.3.1 Fruit Ripening and Seed Dispersal

As fruits ripen, they often change colour due to the conversion of chlorophyll to other pigments like carotenoids (which produce yellow, orange and red colours). This colour change signals to frugivores (fruit-eating animals) that the fruit is ripe and ready to be consumed and dispersed. For example, a tomato turns red due to the accumulation of lycopene (a carotenoid), signaling to birds and mammals that it is ripe and ready to be eaten.

2.6.2.3.2 Defence Mechanisms in Plants

Certain plants have vivid hues, frequently red or yellow, which indicate toxicity or the presence of protective compounds. This is called aposematic colouration, and it alerts herbivores to the possibility that the plant is toxic or unpleasant. For instance, betalains in some species (like beets) serve as both colourants and deterrents, whereas some toxic blooms may be red or yellow in colour, warning herbivores of danger.

2.6.2.3.3 Pigment Signaling in Pathogenic Microbes

A number of bacteria and fungi generate pigments that act as indicators of pathogenicity, defensive mechanisms or biofilm development. For example, the blue-green pigment pyocyanin is produced by the common *bacterium Pseudomonas aeruginosa* to indicate the existence of a biofilm and to prevent the growth of competing bacteria. The pigment hue of certain fungus can indicate the existence of harmful substances that repel diseases or other organisms.

2.6.2.3.4 Humans and Biocolours in Communication

Humans have long used colour as a visual communication tool, both in cultural contexts and in signaling emotions or intentions.

Cultural Signaling Colours are often used in clothing, flags and other societal markers to convey messages related to identity, status or affiliation. For example, red is often used to symbolize love, passion or danger, while blue can represent calmness, trust or sadness.

Emotional and Psychological Signaling Colours in marketing and branding: In marketing, colour psychology is used to evoke emotional responses from consumers. For instance, green can be associated with environmental sustainability, yellow with optimism and black with luxury.

Social and Emotional Cues In human interactions, the colour of an individual's clothing or surroundings can communicate emotional states, such as the use of red to signal excitement or aggression or blue for calm and trust.

2.7 Challenges and Future Directions

2.7.1 Issues Related to Stability, Solubility and Cost of Biocolours

Biocolours, also known as natural or plant-based dyes, have become increasingly popular due to their environmental benefits and safety compared to synthetic dyes. However, they still face several challenges related to stability, solubility and cost (Zhao et al. 2022). Here is an overview of the key issues in these areas and potential future directions.

2.7.1.1 Stability Issues

2.7.1.1.1 Light Stability

Biocolours can degrade under exposure to light (especially UV radiation), leading to fading and loss of colour intensity. For instance, anthocyanins (from fruits like berries) are known to be sensitive to light. One promising solution is the encapsulation of biocolours in protective materials such as nanoparticles, which can protect them from environmental factors like light and oxygen.

2.7.1.1.2 Thermal Stability

Natural dyes can be sensitive to high temperatures which limit their use in products requiring heat treatment or exposure. Research into stabilizing agents that can crosslink biocolours and prevent degradation is ongoing. For example, incorporating natural antioxidants could help extend the shelf life.

2.7.1.1.3 Oxidation

Many natural colourants, like flavonoids and carotenoids, are prone to oxidation which affects both their colour and their functional properties. Combining natural colourants with synthetic stabilizers or new polymer-based matrices could improve stability while maintaining the sustainability aspect.

2.7.1.2 Solubility Issues

2.7.1.2.1 Solvent Dependency

Many biocolours exhibit poor solubility in water, which limits their application in aqueous formulations (e.g., in food or cosmetics). For example, anthocyanins tend to be more soluble in ethanol or methanol than in water. Optimizing extraction techniques to increase the yield and solubility of bioactive compounds in water or other solvents is an area of active research.

2.7.1.2.2 Inconsistent Performance

Variability in solubility across different plant-based colourants makes it difficult to use them uniformly in industrial applications. Researchers are investigating the use of surfactants or co-solvents that can improve the solubility of biocolours, making them more compatible with diverse formulations. Using biopolymers like starch, chitosan or cellulose derivatives may aid in solubilizing and stabilizing biocolours, especially for use in food and textile industries.

2.7.1.3 Cost Issues

2.7.1.3.1 Extraction Costs

Obtaining biocolours from natural sources (plants, insects etc.) can be expensive due to labour-intensive extraction processes and low yields per plant. The use of genetically engineered microorganisms (e.g., bacteria, yeast and algae) to produce biocolours on a large scale is a promising direction. This could significantly reduce the cost and environmental impact of producing natural dyes.

2.7.1.3.2 Limited Availability

Some biocolours, such as saffron or cochineal, are produced in limited quantities, and this scarcity drives up costs. Exploring new and more abundant sources of natural dyes, including food waste, agricultural by-products or algae, could lower costs while increasing sustainability.

2.7.1.3.3 Scalability

Large-scale production of biocolours often faces difficulties in terms of scaling up production while maintaining product quality and reducing costs. Investing in automated and more efficient extraction methods can drive down the costs of biocolour production. Techniques like supercritical fluid extraction (SFE) or microwave-assisted extraction (MAE) are being explored.

2.7.1.4 Future Directions in Research and Development

2.7.1.4.1 Personalization and Tailoring

Tailoring the colour properties of biocolours for specific applications (food, textiles and cosmetics) will require deeper understanding of and innovation in biosynthetic pathways, genetic engineering and polymer chemistry.

2.7.1.4.2 Sustainability Assessment

Research into the life cycle assessment (LCA) of biocolours could help industries understand the environmental benefits or trade-offs in switching from synthetic to natural dyes.

2.7.1.4.3 Integration with Other Sustainable Practices

Biocolours could be part of a larger movement toward sustainable manufacturing practices, including the use of renewable energy, reducing water usage and minimizing waste.

2.8 Conclusions

The commercialization of biocolours (natural dyes from plants, animals and microorganisms) has grown significantly in recent years, driven by increased consumer demand for sustainable and non-toxic alternatives to synthetic dyes. As awareness around the environmental and health impacts of synthetic dyes continues to grow, the future of biocolours holds promising potential across various industries, including food, cosmetics, textiles and pharmaceuticals. Partnerships with universities and research institutions can lead to the development of new biocolour sources, improved extraction methods and more effective use of biocolours in different applications. Corporate social responsibility and sustainability goals will drive partnerships between businesses and organizations working on sustainable biocolour solutions. Companies in food, cosmetics, fashion, and textiles will collaborate with biocolour producers to create new, innovative products that meet the growing consumer demand for environmentally friendly alternatives.

References

Anonymous (2023) E140; Chlorophylls and Chlorophyllins. https://1source.com/ingredients/e140#:~:text=Chlorophyll%20is%20green%20pigment%20found,%2C%20jams%2C%20jellies%20and%20marmalade. Accessed 1 June 2025

Arunkumar S, Kamalakannan S, Haripriya K, Kamalakannan P (2024) Bio-colours from vegetable crops – a review. Indian J Appl Pure Bio 39(2):802–815

Bashir I, Pandey VK, Dar AH, Dash KK, Shams R, Mir SA, Fayaz U, Khan SA, Singh R, Zahoor I (2024) Exploring sources, extraction techniques and food applications: a review on biocolors as next-generation colorants. Phytochem Rev 23(4):1–26

Bora P, Das P, Bhattacharyya R, Barooah MS (2019) Biocolour: the natural way of colouring food. J Pharmacogn Phytochem 8(3):3663–3668

Chaitanya LG (2014) Food coloring: the natural way. Res J Chem Sci 4(2):87–96

Chittaragi D, Menon JS (2023) Bio-Colours from spices. In: Ivanišová E (ed) Herbs and spices – new advances. IntechOpen. https://doi.org/10.5772/intechopen.104270

Cordero F, Baldo C, Inmaculada L, Rosa R, Herminia V, Maria (2012) Isolation and characterization of a lycopene ε-cyclase gene of chlorella (Chromochloris) zofingiensis. Regulation of the carotenogenic pathway by nitrogen and light. Mar Drugs. https://doi.org/10.3390/md10092069

De Mol ML, Vandamme EJ (2024) Arts, cultural heritage, sciences and micro-/bio-/techno–/logy: impact of biomaterials and biocolorants from antiquity till today! J Ind Microbiol Biotechnol 51:kuae049

de Sérgio SMJ (2021) The molecules of colour. In: The real blue indigo: photostability mechanisms, new functional derivatives, and hybrids

Delgado-Vargas F, Jiménez AR, Paredes-Lopez O (2000) Natural pigments: carotenoids, anthocyanins and betalins-characteristics, biosynthesis, processing and stability. Crit Rev Food Sci Nutr 40(3):173–289

Devi MP, Joshi VK, Hemanta L, Aresh J (2022) Biocolor: source, role, and its therapeutic significance. In: Bioreactor technology in food processing. CRC Press, pp 533–566

Eaton-Rye JJ, Tripathy BC, Sharkey TD (2012) Photosynthesis: plastid biology, energy conversion and carbon assimilation. Adv Photosynth Respir 34:63–94

EFSA (2011) Statement summarising the conclusions and recommendations from the opinions on the safety of irradiation of food adopted by the BIOHAZ and CEF panels. Eur Food Saf Authority 2(4):2107

Firacative C, Trilles L, Meyer W (2021) Recent advances in cryptococcus and cryptococcosis. Microorganisms 10(1):13

Ganguly S, Bandyopadhyay K, Dutta P, Pramanik A (2019) A review on microbial-pigment: a good source of biocolour. Pharma Innov J 8(12):416–420

Gao J, Wenderoth M, Doppler M, Schuhmacher R, Marko D, Fischer R (2022) Fungal melanin biosynthesis pathway as source for fungal toxins. MBio 13(3):e00219–e00222

Gupta C, Prakash D, Garg A, Gupta S (2011) Microbes as potential source of Biocolours. Pharmocologyonline 2:1309–1318

Haranto T, Khunnamwong P, Wongkamhaeng K, Lueangjaroenkit P (2024) Exophiala amphawaensis sp. nov., a novel black yeast isolated from the gut of amphipods in Thailand. Int J Syst Evol Microbiol 74(5):006363

Helene EC, Arturo C (2012) Synthesis and assembly of fungal melanin. Appl Microbiol Biotechnol 93(3):931–940

Joshi SS, Shinde NH (2017) Study of different Biocolours and its application in cosmetic products. Int J Sci Eng Res (IJSER) 5(10):171925

Juailova P, Martinkova L, Machet F (1997) The existing genetic diversity in microbes and sophistication of technology has made their choice more feasible. Enzym Microb Technol 16:231–235

Kenyon HL, Martin PR (2023) Color as an interspecific badge of status: a comparative test. Am Nat 202(4):433–447

Kole PL, Jadhav HR, Thakurdesai P, Nagappa AN (2005) Cosmetic products of herbal extracts. Nat Prod Radiance 4:4

Lin L, Xu J (2022) Production of fungal pigments: molecular processes and their applications. J Fungi 9(1):44

Liu Q, He Z, Zeng M, Qin F, Wang Z, Liu G, Chen J (2021) Effects of different food ingredients on the color and absorption spectrum of carminic acid and carminic aluminum lake. Food Sci Nutr 9(1):36–43

Liu S, Le Mauff F, Sheppard DC, Zhang S (2022) Filamentous fungal biofilms: conserved and unique aspects of extracellular matrix composition, mechanisms of drug resistance and regulatory networks in aspergillus fumigatus. npj Biofilms Microbiomes 8(1):83

Lyu X, Lyu Y, Yu H, Chen W, Ye L, Yang R (2022) Biotechnological advances for improving natural pigment production: a state-of-the-art review. Bioresour Bioprocess 9(1):8

Ma Y, Ma X, Gao X, Wu W, Zhou B (2021) Light induced regulation pathway of anthocyanin biosynthesis in plants. Int J Mol Sci 22(20):11116

Mannino G, Gentile C, Ertani A, Serio G, Bertea CM (2021) Anthocyanins: biosynthesis, distribution, ecological role, and use of biostimulants to increase their content in plant foods—a review. Agriculture 11(3):212

Marmion DM (1991) Handbook of US colorants: foods, drugs, cosmetics and medical devices, Publishers 558, 3rd edn. Wiley

Mascoli V, Liguori N, Cupellini L, Elias E, Mennucci B, Croce R (2021) Uncovering the interactions driving carotenoid binding in light-harvesting complexes. Chem Sci 12(14):5113–5122

Mehta M, Shah G (2016) Incorporation of Biocolours in textiles: an ecofriendly approach. Int J Appl Res 2(2):214–218

Miller R, Owens S, Rørslett B (2011) Plants and colour: flowers and pollination. Opt Laser Technol 43:282–294. https://doi.org/10.1016/j.optlastec.2008.12.018

Monishdeep VHW, Despande A, Poshadri, Kumar PBV (2022) A review on potential microbial sourced bio-colours from fungi. Pharma Innov J SP-11(9):3200–3205

Pankaj VR, Kumar R (2016) Microbial pigment as a potential natural colorant for contributing to mankind. Res Trends Mol Biol:85–98

Paul T, Bandyopadhyay T, Mondal A, Tiwari O, Muthuraj M, Bhunia B (2022) A comprehensive review on recent trends in production, purification, and applications of prodigiosin, vol 12. Biomass Convers Biorefinery

Rather LJ, Mir SS, Ganie SA, Li Q (2023) Research progress, challenges, and perspectives in microbial pigment production for industrial applications-a review. Dyes Pigments 210:110989

Rosa LH, Vieira LMA, Santiago IF, Rosa CA (2010) Endophytic fungi community associated with the dicotyledonous plant *Colobanthus quitensis* (Kunth) Bartl. (Caryophyllaceae) in Antarctica. FEMS Microbiol Ecol. https://doi.org/10.1111/j.1574-6941.2010.00872.x

Sandmann G (2022) Carotenoids and their biosynthesis in fungi. Molecules 27(4):1431

Sen T, Barrow CJ, Deshmukh SK (2019) Microbial pigments in the food industry- challenges and the way forward. Front Nutr 6:7

Sharma D (2014) Understanding biocolour-a review. Int J Sci Technol Res 3(1). www.ijstr.org

Singh T, Pandey VK, Dash KK, Zanwar S, Singh R (2023) Natural bio-colorant and pigments: sources and applications in food processing. J Agric Food Res 12:100628

Sinha D, Abid R, Chakraborty W, Rashid M, Gupta LK, Khan B, Datta PN, Noor S, Pomila, Ghazanfar S, Saha U (2024) Effect of abiotic stress on terpene biosynthesis in plants. In: Plant secondary metabolites and abiotic stress, pp 481–524

Sunil L, Shetty NP (2022) Biosynthesis and regulation of anthocyanin pathway genes. Appl Microbiol Biotechnol 106(5):1783–1798

Thompson J, Fahnestock S, Farrall L, Liao, Der I, Valent B, Jordan D (2000) The second naphthol reductase of fungal melanin biosynthesis in Magnaporthe grisea: tetrahydroxynaphthalene reductase. J Biol Chem 275:34867

Tossi VE, Tosar LM, Pitta-Álvarez SI, Causin HF (2021) Casting light on the pathway to betalain biosynthesis: a review. Environ Exp Bot 186:104464

Vaidya H, Upasani N, Wagh PS (2021) Microbial pigments: natural colorants and their industrial applications. Int J Curr Microbiol Appl Sci 5:2319–7706

Venil CK, Zakaria ZA, Ahmad WA (2013) Bacterial pigments and their applications. Process Biochem 48(7):1065–1079

Verma S, Kaur K (2024) Biocolorants from plant sources: a novel approach to enhance the aesthetics and health profile of foods. Bioscene 21(2). www.explorebioscene.com

Wang P, Grimm B (2021) Connecting chlorophyll metabolism with accumulation of the photosynthetic apparatus. Trends Plant Sci 26(5):484–495

Xiao S, Ahn DU (2022) Enhanced lutein stability under UV-light and high temperature by loading it into alginate-chitosan complex. LWT 164:113663

Zhan X, Chen Z, Chen R, Shen C (2022) Environmental and genetic factors involved in plant protection-associated secondary metabolite biosynthesis pathways. Front Plant Sci 13:877304

Zhao X, Zhang Y, Long T, Wang S, Yang J (2022) Regulation mechanism of plant pigments biosynthesis: anthocyanins, carotenoids, and betalains. Metabolites 12(9):871

Zhdanova NN, Zakharchenko VA, Vember VV, Nakonechnaya LT (2000) Fungi from Chernobyl: Mycobiota of the inner regions of the containment structures of the damaged nuclear reactor. Mycol Res 104(12):1421–1426

Molecular Insights into Bacterial Pigments

3

Sayasta Ahmed, Suruchi Rai, Barsha Datta, Pritam Chhetri, and Rakshak Kumar

Abstract

Biocolours derived from natural biological sources have become a promising substitute for synthetic dyes because of their multifunctional applications and environment-friendly qualities. Bacterial pigments have gained a lot of attention due to their diverse range of qualities which include antimicrobial, antioxidant, and anticancer activities. As a result, biocolours are well suited for use in food, pharmaceutical, textile, and cosmetic industries. Although notable advancements have been made in identifying and characterizing various bacterial pigments, a thorough understanding of the biosynthetic pathways of different bacterial pigments including the underlying molecular mechanisms, enzymatic players, genetic frameworks, and regulatory networks is still an evolving study area. This chapter seeks to bridge this knowledge gap by systematically elucidating the molecular basis of biosynthesis in a few selected bacterial pigments. It explores current biotechnological interventions, such as genetic engineering and synthetic biology techniques, to improve bacterial pigment production with a focus on the structure of pigment-associated gene clusters, the characterization of the important enzymes. Furthermore, the chapter covers studies in functional genes and enzymes, comparative pathway analyses, and strategies for host system optimization. It concludes by sharing insights into the future potential of using cutting-edge technologies like genome mining and integrative omics to find novel pigment-producing microbes and develop industrially viable applications for bacterial pigments.

S. Ahmed · S. Rai · B. Datta · P. Chhetri · R. Kumar (✉)
Environmental Molecular Microbiology and Applications Laboratory, Department of Molecular Biology & Bioinformatics, Tripura University (A Central University), Suryamaninagar, Tripura, India
e-mail: rakshakkumar@tripurauniv.ac.in

R. K. Bhatia, A. Walia (eds.), *Biocolours*,
https://doi.org/10.1007/978-981-95-1738-1_3

Keywords

Biocolours · Bacterial pigments · Biosynthetic pathway · Genomics · Genetic engineering

3.1 Introduction to Biocolours

A vital component of food quality is its colour as it influences consumers' view and acceptance. Colouring agents have been used by humans since ancient times to improve food's overall quality, make up for colour loss during processing and increase the visual appeal of food. Concerns about the use of synthetic dyes have led to a rise in the demand for natural food colouring agents. Plants and microbes are biological sources of biocolours, which give off colour and have important health advantages like anticancer, antioxidant, and food-stabilizing qualities (Ganguly et al. 2019). Anthocyanins, chlorophylls, carotenoids, curcuminoids, betalains, anthraquinones, azaphilones, iridoid glycosides, and phycobiliproteins are examples of biocolourants that can be grouped both by their natural sources and by their chemical makeup (Thakur and Modi 2024). Additionally, they can be grouped based on the range of colours they produce, including red, blue, green, yellow, violet, and purple.

Synthetic colourants have been used in the food industry for many years because of their stability, affordability, and ease of manufacturing. However, the majority of artificial dyes are made chemically from petroleum and cold tar, which raises questions about their possible toxicity and negative health effects (Amchova et al. 2015). A move towards natural alternatives has been fuelled by growing consumer awareness, by regulatory constraints and the popularity of nutraceutical food products. Nowadays, research is being done on natural pigments as safer alternatives that have the potential to be bioactive and provide additional health benefits (Rodriguez-Amaya 2016). Natural pigments have historically mostly come from plant and vegetable sources, but their large-scale extraction necessitates substantial agricultural resources, which makes them less sustainable.

In contrast, microorganisms offer a renewable, cost-effective, and scalable source of biocolours (Parmar and Phutela 2015). Microbial pigments have many advantages over synthetic and plant-derived pigments, making them a superior alternative. Microbial pigments are in demand and are mostly preferred because of their stability, consistent availability irrespective of seasonal variations, cost-effectiveness, and high production potential with strain optimization. Moreover, the process of extraction is more efficient, ensuring smoother downstream processing and greater production than traditional methods (Malik et al. 2012). Many bacteria, fungi, and algae naturally produce a variety of pigments that can be utilized in food, pharmaceutical, and industrial applications (Pankaj and Kumar 2016). Their potential as a substitute source of natural colours is further increased by their fast growth, easy cultivation, and ability for genetic modification.

Microorganisms can synthesize a diverse range of biocolours, increasing their value and use as natural food colourants. Biocolours exhibit significant structural diversity, with variations in hydrocarbon chains, oxygen and nitrogen content, and aromatic ring structures, influencing their light absorption and reflection properties (Hendry 1996). The colouration of biological pigments is determined by their ability to specifically absorb particular light wavelengths, which also makes them useful for a variety of physiological and biochemical processes (Ahmad et al. 2012). As research advances, microbial pigments are proving to be a sustainable and versatile substitute for synthetic dyes, reinforcing the global transition towards safer and eco-friendly food colourants.

Microbial pigments have attracted high attention because of their sustainability, stability, and vast area of applications. Bacteria, specifically, serve as a reliable source of biocolourants because of their fast growth rates and capacity to generate a broad spectrum of pigments with functional advantages. Agro-industrial byproducts can be used as substrates to optimize bacterial biocolour production due to lower biomass waste generation, and this further lowers production costs (Venil et al. 2020; Padhan et al. 2021). Bacterial pigments span multiple chemical classes, with key examples including carotenoids (zeaxanthin, astaxanthin, canthaxanthin, beta-carotene, lycopene), violacein, prodigiosin, indigoidine, melanin, riboflavin, and phycocyanin (Bashir et al. 2024). Several bacterial species are known for their pigment production, including *Rheinheimera* sp., which produces a blue pigment; *Chromobacterium* sp., known for synthesizing the violet pigment violacein; *Serratia marcescens*, which generates the red pigment prodigiosin; *Micrococcus* sp., capable of producing carotenoids in various shades; *Agrobacterium aurantiacum*, a producer of the pink-red pigment astaxanthin; and *Flavobacterium* sp., which yields the yellow pigment zeaxanthin (Venil et al. 2013). Moreover, bacteria are capable of forming various additional pigments like carotenoids, bacteriochlorophylls, monascins, pyocyanin, and melanin (Ponjavic et al. 2023). These pigments are valued in the food, cosmetic, and medicine industries because of their beneficial properties, including antioxidant, antimicrobial, and anticancer activities.

Regardless of the detailed study on bacterial pigments and their uses, gaps or questions still remain in understanding their molecular biosynthesis, gene regulation, and biotechnological potential. The biosynthetic pathways of a few pigments have been partially elucidated, yet deeper research is required to understand their genomic mechanisms involved in synthesis of these pigments. In order to improve pigment yield, stability, and functionality, more research and investigation are required to uncover the operon structures, regulatory genes, and metabolic engineering strategies. Approaches like synthetic biology can also be used for optimization of pigment production and expansion of their industrial applications. This chapter explores the types of bacterial pigments and their applications, biosynthesis and genomics of key bacterial pigments such as carotenoids, bacteriochlorophylls, violacein, prodigiosin, melanin, and indigoidine. Focusing on the genetic and molecular aspects of pigment synthesis, this chapter also highlights the regulatory controls of pigment production and their relevance to industrial and biotechnological applications.

3.2 Bacterial Pigments: Types and Applications

A wide variety of naturally occurring pigments are present across plants and microbes, including bacteria and fungi (Sajjad et al. 2020). Natural biocolours play a key role in combating diseases by suppressing microbial proliferation, showing potent anti-inflammatory and anticancer properties. Over the years, various biocolours have been isolated and identified, each having specific advantages. Among these, bacterial pigments stand out as promising alternatives to synthetic dyes due to their unique properties and broad applications (Fried et al. 2022). Bacteria are known to generate a diverse range of secondary metabolites, including various pigments ranging from yellow, red, orange, green, purple, and blue. Bacteria can produce a wide range of pigments that make them valuable across multiple industries. The rapid growth rate and the ease of pigment extraction make bacterial pigments more suitable for large-scale production compared to synthetic pigments. Various bacterial pigments exhibit diverse industrial and biomedical applications (Table 3.1). Beyond their role in food and cosmetics, bacterial pigments also offer significant health benefits, including antimicrobial, antiviral, antioxidant, and anticancer properties (Aman Mohammadi et al. 2022). Pigment-producing bacteria thrive in a variety of environments, from terrestrial to marine ecosystems, including extreme conditions such as deserts and saline regions (Takagi et al. 2023). Additionally, they have been found in organic waste, decomposing fruits and vegetables, industrial wastewater, and landfill sites, showing their adaptability and potential for biotechnological exploitation (Sharma et al. 2020).

While pigment production varies among bacterial species, it is most commonly observed in the phylum Actinobacteria (Li and Lou 2018). Some bacteria continuously produce pigments regardless of their environment or physiological needs (Venil et al. 2013), while others adjust pigment production based on external factors like nutrient shortages, UV exposure, harmful substances, or stress conditions (Mummaleti et al. 2024). For instance, in *Escherichia coli*, pigment production is initiated through a molecular mechanism in which specific genes are activated in response to external stimuli (Ahn et al. 2021). New opportunities in biotechnology can be unlocked by utilizing bacterial pigments, as advancements in strain selection, genetic engineering, and mutation-based improvements can greatly enhance pigment yields compared to naturally occurring strains. The ease of genetic manipulation in bacteria can help in the optimization of pigment biosynthesis and make them a sustainable source of natural colourants (Singh et al. 2021).

3.3 Biosynthesis of Biocolours: Key Biochemical Pathways and Enzymatic Roles in Bacteria

Researchers are using various methods, such as genetic engineering and advanced fermentation techniques, to increase the production of microbial pigments. A key focus of these efforts is improving the biosynthetic pathways in microorganisms to enhance pigment yield. In the following sections, we will explore the molecular characteristics and biosynthetic pathways of some important bacterial pigments.

Table 3.1 Table represents important types of bacterial pigments, various pigment-producing bacterial strains and their key applications

Key bacterial pigments		Reported bacterial source	Applications	Reference
Carotenoids*	β-carotene	*Rhodococcus maris*	Functions as a precursor for vitamin A, possesses anticancer activities, suitable for food industry as food colour	Venil et al. (2013)
	Astaxanthin	*Agrobacterium aurantiacum*; *Paracoccus carotinifaciens*		
	Zeaxanthin	*Flavobacterium* sp.; *Paracoccus zeaxanthinifaciens*; *Staphylococcus aureus*		
	Canthaxanthin	*Bradyrhizobium* sp.		
Bacteriocholorophyll		*Anabaena variabilis*; *Nostoc commune*; *Arthrospira platensis*; *Microcoleus vaginatus*	Used as an antimicrobial agent and as biosensors for indicating water quality	Ohkubo and Miyashita (2017), Suvorov et al. (2021), and Bindu et al. (2024)
Violacein		*Chromobacterium violaceum*; *Duganella violaceinigra*; *Janthinobacterium lividum*; *Microbulbifer* sp.; *Pseudoalteromonas* Sp.; *Collimonas* sp.; *Massilia violaceinigra*	Plays a role in biofilm formation; has pharmaceutical applications (antioxidant, immunomodulatory, antitumor, antiparasitic activities); used as a dye in textiles and cosmetics.	Atalah et al. (2020), and Wang et al. (2018)
Prodigiosin		*Serratia marcescens*; *Streptomyces pilosus*; *Rugamonas rubra*; *Streptoverticillium rubrireticuli*; *Vibrio gaogene*; *Alteromonas rubra*; *Pseudomonas* sp.; *Vitrio* sp.; *Pseudoalteromonas* sp.	Exhibits anticancer, immunosuppressive, and antifungal properties; utilized as a dye in textiles, candles, paper, and ink.	Malik et al. (2012) and Muhammad et al. (2024)

(continued)

Table 3.1 (continued)

Key bacterial pigments	Reported bacterial source	Applications	Reference
Melanin	*Bacillus cereus*; *Bacillus thuringiensis*; *Burkholderia cepacia*; *Pseudomonas stutzeri*; *Streptomyces cyaneus*; *Shewanella colwelliana*; *Vibrio cholera*; *Alteromonas nigrifaciens*; *Tyrosinoxydans* sp.; *Cellulophaga* sp.	Used as a food colourant and exhibits anticancer properties.	El-Naggar and Saber (2022), El-Naggar and El-Ewasy (2017), Tran-ly et al. (2020), Roy and Rhim (2022), and Wang et al. (2020)
Indigoidine	*Corynebacterium insidious*;, *Vogesella indigofera*; *Phaeobacter* sp.; *Streptomyces chromofuscus*; *Pseudomonas fluorescens*	Provides protection against oxidative stress	Venil et al. (2013), Day et al. (2017), Cude et al. (2012), and Yu et al. (2013)

Note: (*) denotes the types of bacterial carotenoids listed in the table

3.3.1 Carotenoid

Carotenoids are naturally occurring and important class of biocolourants found in both plants and microorganisms (Nabi et al. 2020). Carotenoids are naturally occurring biomolecules known for their colour range (yellow to red). Carotenoids belong to the isoprenoid family and are synthesized by various plants, fungi, algae, yeast, and certain bacteria (Ram et al. 2020). While the production of carotenoids has been extensively studied in plants and algae, other microbial sources are also gaining attention (López et al. 2021). Bacterial carotenoids have recently entered the mainstream since bacteria can be grown at short intervals of time, are readily extracted, nontoxic, available in a variety of colours, and yield biodegradable products (Ram et al. 2020). Furthermore, bacterial pigment production results in low biomass waste and is less expensive than plant-derived pigments (Thakur and Modi 2024). Carotenoids are C40 tetraterpenoid hydrocarbon compounds made up of eight isoprene units and a conjugated polyene chain having alternating double bonds (Maoka 2019; Ram et al. 2020). Carotenoids absorb light in the range of 400–500 nm, yielding yellow, orange, and red colours (Ram et al. 2020). In bacteria, carotenoids stabilize the membrane, especially under low-temperature conditions, by controlling the membrane permeability. Photosynthetic and non-photosynthetic bacteria produce various types of carotenoids, like astaxanthin, canthaxanthin, zeaxanthin, and β-carotene being one of the most important. Carotenoids play a vital role in safeguarding the photosynthetic bacteria cells against the damaging action of singlet oxygen generated during the photodynamic reaction (Cogdell et al. 2000). Non-photosynthetic bacteria use these pigments to protect the cell against oxidative stress and UV light and assist in the regulation of membrane fluidity to shift in response to environmental stress. Carotene occurs in three isomers-α, β, and γ-carotene, among which β-carotene is the most active and stable compound (Xu et al. 2018). The carotenoids are usually divided into two groups: carotenes and xanthophylls, depending on their chemical structure. Carotenes, e.g., α-carotene, β-carotene, γ-carotene, and lycopene, are soluble in organic solvents but not in polar solvents, whereas xanthophylls, e.g., lutein, zeaxanthin, and violaxanthin, are soluble in organic and polar solvents (González-Peña et al. 2023).

β-carotene also acts as a strong anticancer antioxidant. It has an important role in improving immunity and preventing the formation of tumours and ageing (Mata-Gómez et al. 2014). β-carotene is structurally composed of several olefinic bonds and is also a precursor for vitamin A and is effective in disease treatment concerning vitamin A deficiency (Qiang et al. 2019). Carotenoids have the ability to produce a broad spectrum of colours from yellow to red, making them appropriate for their application in the food industry as a general natural biocolourant (Wu et al. 2017). *Rhodococcus maris* is a bacterium known to produce β-carotene (Venil et al. 2013). Over the past years, there has been a rising demand for β-carotene because of its numerous physiological advantages, due to its extensive use in health supplements, medicine, cosmetics, and animal nutrition (Zhang et al. 2020). It is a critical component because of its antioxidant activity and provitamin-A activity. Astaxanthin, a red-orange keto-carotenoid, also possesses good antioxidant properties and is widely used in human dietary supplements and feed additives for animal husbandry and aquaculture. Sources of bacteria are *Rhodococcus maris*, *Agrobacterium aurantiacum*, *Paracoccus marcusii*, and *Paracoccus carotinifaciens* (Venil et al. 2013; Mummaleti et al. 2024). Astaxanthin occurs naturally in marine life and is responsible for the red-pink colour of salmon, red trout, shrimp, crabs, and flamingos (Schneider et al. 2012). Canthaxanthin is an orange to deep pink trans-carotenoid with high antioxidant activity that is synthesized by *Bradyrhizobium* species (Chuyen and Eun 2017). It finds application as a food colourant and in animal feed and as protection against photooxidative damage in photosynthetic organisms.

Zeaxanthin, a yellow-orange xanthophyll, imparts colouration to foods like paprika, corn, and wolfberries, and contributes to the aroma and taste of saffron through its breakdown into picrocrocin and safranal. Bacterial producers include *Staphylococcus aureus*, *Bacillus* sp., *Flavobacterium* sp., and *Corynebacterium michiganense* (Schweiggert and Carle 2016; Ram et al. 2020). Additional xanthophyll-producing bacterial genera include *Monascus roseus*, *Brevibacterium*, and *Paracoccus*, which synthesize pigments like zeaxanthin, astaxanthin, and canthaxanthin with applications in cancer prevention, pharmaceutical products, and nutraceuticals because they act as antioxidants and free radical scavengers (Vaidya et al. 2021). Other notable bacterial genera known for carotenoid production include *Serratia marcescens*, *Pseudomonas aeruginosa*, *Micrococcus*, *Chromobacterium*, *Rheinheimera*, and *Arthrobacter* (Ram et al. 2020). These bacterial carotenoids are also being explored for industrial applications in textiles, paper manufacturing, and beyond due to their natural origin, stability, and functional properties (Vaidya et al. 2021; Gul et al. 2017).

Carotenoid biosynthesis occurs via two distinct pathways, namely, the mevalonate (MVA) pathway and the methylerythritol 4-phosphate (MEP) pathway. The MVA pathway was discovered earlier and well-studied in plants and some bacteria, and the MEP pathway is newly discovered and occurs in most bacteria and algae (Ram et al. 2020; Paniagua-Michel et al. 2012; Schweiggert and Carle 2016). The universal precursors for the carotenoids are isopentenyl diphosphate (IPP) and its isomer dimethylallyl diphosphate (DMAPP) (Ram et al. 2020). IPP and DMAPP are formed from acetyl-CoA or pyruvic acid via the MVA or MEP pathway (Maoka 2019). Enzyme Geranylgeranyl pyrophosphate synthase (GGPS), actively encoded by the CrtE gene, catalyses the conversion of IPP and DMAPP into two molecules of geranylgeranyl diphosphate (GGPP) which are then condensed by phytoene synthase (PSY) encoded by the CrtB gene to form a colourless molecule named phytoene (Fig. 3.1). Phytoene step-wisely desaturates to form lycopene, a coloured molecule via phytofluene, zeta (ζ)-carotene, and neurosporene by phytoene desaturase (PDS) and ζ-carotene desaturase (ZDS). Then, from lycopene, lycopene-ε-cyclase (LCY-ε) and lycopene-β-cyclase (LCY-β) produce α-carotene and β-carotene with cyclic terminal end groups, respectively. Further, conversions of carotenes by hydroxylases and epoxidases result in various xanthophylls. The conversion of α-carotene and β-carotene into lutein and zeaxanthin, respectively, is catalysed by α-ring hydroxylase and β-ring hydroxylase. Zeaxanthin is subsequently epoxidized to violaxanthin through the action of zeaxanthin epoxidase (ZEP), which is further reverted to zeaxanthin by violaxanthin de-epoxidase (VDE). Ultimately, violaxanthin is transformed into neoxanthin by the enzyme neoxanthin synthase (NSY). These carotenoid biosynthetic pathways have been described by enzymatic and genetic studies in bacteria (Ram et al. 2020; Maoka 2019).

3.3.2 Bacteriochlorophyll

Chlorophototrophic bacteria consist of bacteriochlorophylls that have a crucial role in photosynthesis excluding oxygen production. These bacteria use chlorophylls or bacteriochlorophylls in the presence of sunlight to undergo chemical synthesis that acts as an energy source. This enables them to capture light with electromagnetic radiation ranging from 350 to 1100 nm and create complex light-harvesting structures. The most prominent colour of the photosynthetic pigment in bacteria is purple. Protoporphyrin IX (PPIX) is the precursor biosynthesis of chlorophyll, and this process is initiated by the addition of Mg^{2+} into the PPIX with the help of different enzymes like magnesium chelatase complex (ChlIDH & Gun4) for the formation of Mg-protoporphyrin IX (MgP). With the help of different enzymes, the first green-coloured intermediate known as divinyl protochlorophyllide is formed (Chen and Hunter 2023). There are seven known types of bacteriochlorophyllides, divinyl chlorophyllide a synthesizes bacteriochlorophyllides a, b and g and chlorophyllides a that get converted into chlorophyllides c, d, e and f (Thweatt et al. 2019). Therefore,

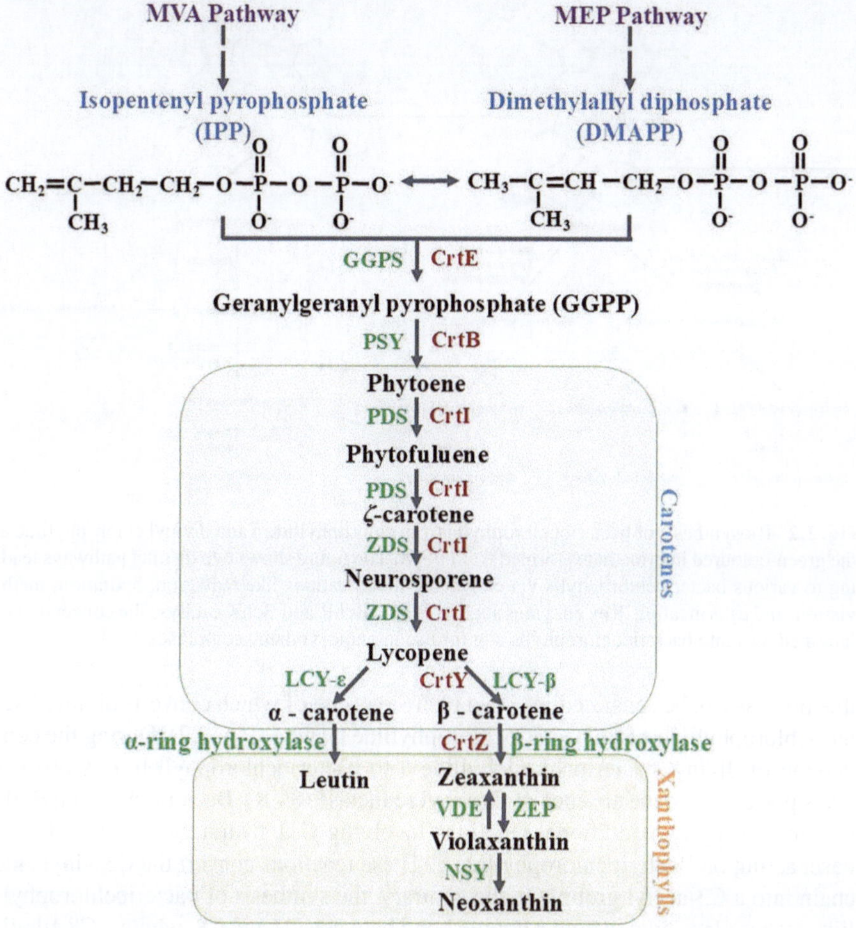

Fig. 3.1 Schematic representation of the carotenoid biosynthesis pathway in bacteria. The pathway begins with the formation of precursors, isopentenyl pyrophosphate (IPP) and dimethylallyl pyrophosphate (DMAPP), synthesized via either the mevalonate (MVA) or methylerythritol 4-phosphate (MEP) pathway. These precursors are condensed by geranylgeranyl pyrophosphate synthase (GGPS), encoded by the CrtE gene, to produce geranylgeranyl pyrophosphate (GGPP). GGPP converted into phytoene by phytoene synthase (PSY), encoded by gene CrtB. Desaturation of phytoene proceeds through a series of intermediates- phytofluene, ζ-carotene, and neurosporene, ultimately forming lycopene. These steps are catalysed by phytoene desaturase (PDS) and ζ-carotene desaturase (ZDS), often encoded by CrtI in bacteria. Lycopene undergoes cyclization via lycopene ε-cyclase (LCY-ε) and lycopene β-cyclase (LCY-β) encoded by CrtY, forming α-carotene and β-carotene, respectively. α-carotene and β-carotene are hydroxylated by α-ring hydroxylase and β-ring hydroxylase (CrtZ) to produce lutein and zeaxanthin, respectively. Zeaxanthin is converted to violaxanthin by zeaxanthin epoxidase (ZEP) and can revert back via violaxanthin de-epoxidase (VDE). Finally, neoxanthin synthase (NSY) converts violaxanthin into neoxanthin. In the figure, genes are labelled in red and the enzyme names are shown in green for visual clarity

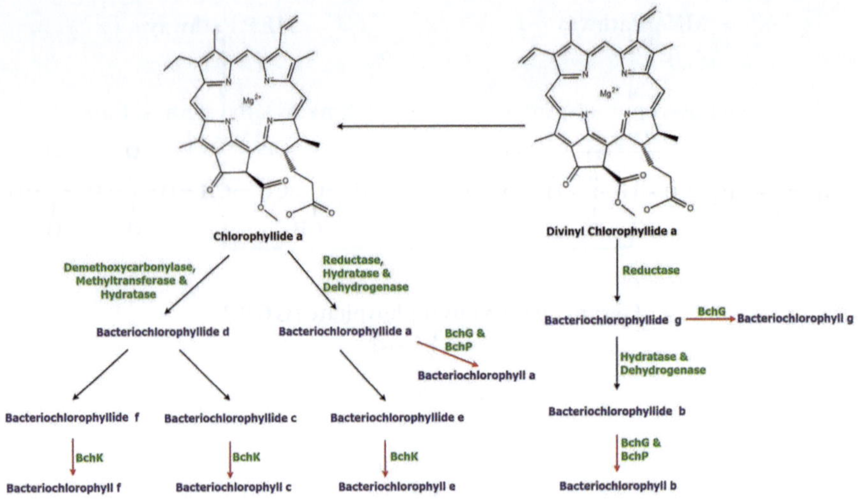

Fig. 3.2 Biosynthesis of bacteriochlorophylls from chlorophyllide a and divinyl chlorophyllide a, the green-coloured intermediates formed from PPIX. The figure shows two distinct pathways leading to various bacteriochlorophylls via enzymatic modifications like reduction, hydration, methylation, and esterification. Key enzymes such as BchG, BchP, and BchK catalyse the conversion of intermediates into bacteriochlorophylls a–g for use in photosynthetic complexes

this process can be separated into two pathways, one of which converts divinyl bacteriochlorophyllide a into bacteriochlorophyllide b and g (Fig. 3.2). During the conversion of divinyl bacteriochlorophyllide a to bacteriochlorophyll b, this process takes place only in the absence of C_8-vinyl reductase (8VR). Bacteriochlorophyllide b is formed by two additional reactions involving C31 hydratase and dehydrogenase, acting on Bacteriochlorophyllide g. These reactions convert the C3-vinyl side chain into a C3-acetyl group. On the contrary, the synthesis of bacteriochlorophyllide g takes place by reducing a double bond between C7 and C8, where a C8 ethylidene group is formed. In the second pathway, bacteriochlorophyllides a, c, d, e, and f are synthesized from chlorophyllide a. Bacteriochlorophyllide a and d are derived from Chlorophyllide a, and after subsequent reduction, hydration, and dehydrogenation, the bacteriochlorophyllide d gets converted into other variants like c, e, and f. Here, the double bond between C7 and C8 gets reduced then the vinyl group undergoes hydration to produce a hydroxyethyl group at C3. The hydroxyethyl group undergoes dehydrogenation to form an acetyl group, which ultimately leads to the synthesis of bacteriochlorophyllide a. Moreover, the formation of bacteriochlorophyllide d starts with the demethoxycarbonylation and methylation of chlorophyllide a. The resulting bacteriochlorophyllide d generated is a combination of species that were methylated after hydration (Yang et al. 2021). The last and final stage in bacteriochlorophyll biosynthesis involves esterification of bacteriochlorophyllides with a long-chain alcohol, typically phytol or geranylgeranyl. The process is catalysed by the enzymes bacteriochlorophyll synthase (BchG) and geranylgeranyl reductase (BchP) for bacteriochlorophyll a and are homologous to their chlorophyll counterparts but have specificity for bacteriochlorophyllide a. Bacteriochlorophyllides e, c, and f get converted into their respective bacteriochlorophylls by BchK enzyme, this enzyme attaches a long-chain alcohol tail to it, and this process is called esterification (Proctor et al. 2022). The resulting bacteriochlorophylls are then integrated into photosynthetic complexes to facilitate light-harvesting and energy conversion (Chen and Hunter 2023). In summary, bacteriochlorophyllides are key intermediates in the biosynthesis of bacteriochlorophylls, and their conversion involves specific enzymatic reactions which result in the formation of diverse types of bacteriochlorophylls found in phototrophic bacteria (Proctor et al. 2022).

Photosynthetic bacteria have been used as a model for molecular genetic analyses of bacteriochlorophyll biosynthesis, and non-sulphur purple bacterial species like *Rhodobacter sphaeroides* and *Rhodobacter capsulatus* are widely utilized due to their flexible growth, genetic manipulation ease and also uncomplicated photosystems (Heyes and Hunter 2009). The derivatives of bacteriochlorophylls like bacteriochlorophyllides have vast applications and can also be used in photodynamic therapy because of their ability to absorb long wavelengths (Kustov et al. 2017). Bacteriochlorophyllides can also be used to explore the lamellar spacing in self-assembling clusters, enhancing our understanding of pigment aggregation production (Pšenčík et al. 2010). Beyond biomedical applications of bacteriochlorophylls, they are also being studied for their application in light energy harvesting, particularly in dye-sensitized solar cells to accelerate the conversion of energy (Tsui et al. 2014). Moreover, bacteriochlorophyll has been identified as a potential diagnostic tool in near-infrared fluorescence imaging, including its role as a mediator in sonodynamic therapy for achieving antitumor effects (Abdulrahman et al. 2020).

3.3.3 Violacein

Violacein is a bis-indole, purple pigment predominantly synthesized by *Chromobacterium violaceum* and certain other Gram-negative bacteria. Structurally, it comprises three main components: a 2-pyrrolidone, an oxindole and a 5-hydroxyindole. Since it incorporates two indole rings, it is classified as a bis-indole compound (Hoshino 2011). It is derived from the condensation of two tryptophan molecules and plays a crucial role in biofilm formation, which is regulated by quorum sensing (Park et al. 2021). The pigment exhibits a wide spectrum of biological activities, including antibacterial effects against *Staphylococcus aureus* and other pathogenic organisms, as well as antiviral, antifungal, and cytotoxic properties against cancerous cells (Moreira et al. 2018). Violacein was first identified in the bacteria *Chromobacterium violaceum*, and since then, several microorganisms have been identified as violacein producers, including *Janthinobacterium* (Xu et al. 2019), *Pseudoalteromonas* (Paulsen et al. 2019), *Massilia* (Wang et al. 2018), and *Duganella* (Choi et al. 2015). However, not all species within these genera produce violacein; for instance, *Janthinobacterium lividum* ERGS5:01 was found to lack the violacein biosynthesis operon, leading to a pink pigmentation instead (Kumar et al. 2018).

Violacein is a secondary metabolite derived from tryptophan through a biosynthetic pathway involving five sequential enzymes coded by the *vioABCDE* operon (Pemberton et al. 1991). The pathway initiates with *vioA* (Fig. 3.3), an L-tryptophan oxidase along with a cofactor-flavin adenine dinucleotide (FAD), catalyses the conversion of L-tryptophan to indole-3-pyruvic acid (IPA) by oxidative deamination (Durán et al. 2016; Park et al. 2021). The polyketide synthase *vioB* then converts IPA into IPA imine dimer (CAP), which is subsequently cyclized by *vioE* to form protodeoxyviolaceinic acid. This intermediate undergoes oxidation by the monooxygenase *vioC* to produce deoxyviolacein. The last step is completed by hydroxylation by tryptophan hydroxylase (*vioD*) that leads to the formation of protoviolaceinic acid, which subsequently leads to the production of violacein with the help of *vioC* (Park et al. 2021). The regulation of violacein biosynthesis is intricately controlled by the coordinated expression of transcripts and proteins, along with the activity of promoters and ribosome binding sites (Lee et al. 2012).

The *vioABCDE* operon was also identified in *Janthinobacterium* sp. B9-8, an extremophile that was cloned and characterized in *Escherichia coli*, leading to

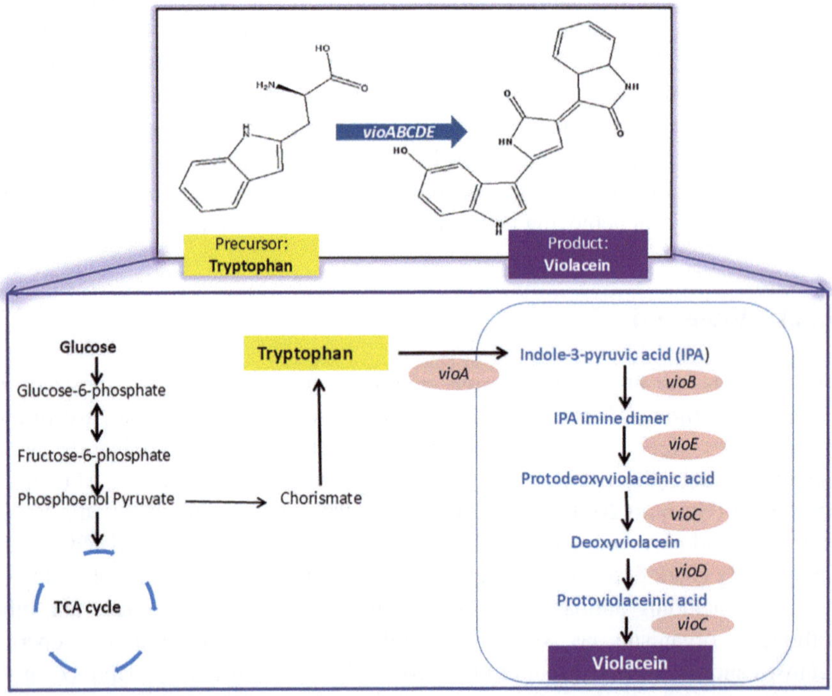

Fig. 3.3 Biosynthetic pathway of violacein in bacteria. The figure illustrates the biosynthesis of violacein, beginning with tryptophan, which is synthesized from a TCA cycle intermediate via activation of the trp operon. Once formed, tryptophan is converted to indole-3-pyruvic acid (IPA) by the enzyme *vioA*. Subsequent enzymatic steps are catalysed by a series of violacein biosynthetic genes—*vioA*, *vioB*, *vioC*, *vioD*, and *vioE*—that lead to the formation of the purple pigment (violacein) within the bacterial cell

improved pigment yield through optimized fermentation for industrial production (Xu et al. 2022). Notably, the researchers have cloned and analysed the violacein synthesis operon from *Pseudoalteromonas* sp. 520P1 in *Escherichia coli*, to enhance violacein production for industrial applications. Different microbial sources exhibit variations in their biosynthetic pathways, regulatory sequences, and gene combinations, leading to differences in production yield and pigment derivatives (Xu et al. 2022; Immanuel et al. 2018). Moreover, modifications of the *vioABCDE* operon can result in the production of various violacein derivatives including deoxyviolacein, oxychromoviridans, deoxychromoviridans and chromoviridans (Immanuel et al. 2018; Rodrigues et al. 2013; Fang et al. 2015; Tong et al. 2021). These derivatives offer potential for industrial and pharmaceutical applications, highlighting the importance of expanding strain resources and exploring novel genes with enhanced biosynthetic capabilities (Xu et al. 2022). Understanding the molecular mechanisms underlying violacein biosynthesis provides insights into its potential applications in biotechnology, medicine, and industrial microbiology. Further research can be done on improving the pigment production by using metabolic engineering and synthetic biology approaches based on the genetic diversity and operon structures of violacein-producing bacteria.

3.3.4 Prodigiosin

A red pigment, alkaloid in nature called Prodigiosin, is produced as a secondary metabolite by various gram-negative bacteria, most notably example being *Serratia marcescens* (Srimathi et al. 2017). Additionally, various c-proteobacteria such as *Pseudoalteromonas rubra*, *Vibrio* sp. *Pseudomonas putida*, and *Janthinobacterium* are known to produce the pigment prodigiosin (Malik et al. 2012; Muhammad et al. 2024). Prodigiosin is comprised of a distinctive structure that consists of three pyrrole rings and a C-4 methoxy pyrrolyl group attached to a pyrromethene skeleton (Williamson et al. 2006). There are several applications of these prodigiosin-producing bacteria; and antifungal, antibacterial, and anticancer properties have been recognized (Paul et al. 2020). The complete biosynthesis of prodigiosin entails two distinct series of reactions that run in parallel, ultimately leading to the formation of tripyrrole pigment through the combination of the products. Prodigiosin biosynthesis proceeds through a bifurcated pathway, which leads to the enzyme-mediated condensation of 4-methoxy-2,2′-bipyrrole-5-carbaldehyde (MBC) with a monopyrrole. The enzymatic union of MBC with 2-methyl-3-pentylpyrrole (MPP) result in the production of prodigiosin (Darshan and Manonmani 2015; Mukhia et al. 2023b).

Prodigiosin synthesis can be seen in bacteria such as *Serratia* spp. *and Streptomyces coelicolour* (Horinouchi et al. 1983), and it can also be witnessed in marine bacteria like *Hahella chejuensis* (Kim et al. 2008) and *Pseudoalteromonas denitrificans* (Kawauchi et al. 1997). Research has documented its antimicrobial potential against various pathogens like *E. coli*, *Streptococcus pyogenes*, *Janthinobacterium sp.*, *Pseudomonas aeruginosa*, *Acinetobacter sp.*, *Enterobacter faecalis*, and *Staphylococcus aureus* (Lapenda et al. 2015; Mukhia et al. 2023a). The red prodigiosin pigment present in *Serratia marcescens* can be utilized as an alternative to synthetic colours and has potential applications in studying anticancer activity in human cervix carcinoma cells (Srimathi et al. 2017). Additionally, undecylprodigiosin is a type of prodigiosin pigment produced by *Streptomyces pilosus which* acts as a potent antimicrobial agent. It can prevent the growth of various harmful pathogens, including *Enterococcus* spp., *Proteus mirabilis*, and *Salmonella* spp. Undecylprodigiosin also demonstrates potential in inhibiting the growth of HeLa cells, indicating its possible application in cancer treatment (Muhammad et al. 2024).

3.3.5 Melanin

Melanin synthesis in bacteria is initially related to the transformation of aromatic amino acids like tyrosine. Whereas in some bacteria melanin can be produced from malonyl-CoA, this process is facilitated by polyketide synthases. Although it was once believed that melanin via the type III polyketide synthase was exclusive to plants, its presence was identified in *Streptomyces griseus* (Funa et al. 1999). Melanins fall into three main types: allomelanins, pheomelanins, and eumelanins; they are typically present as black or brown pigments, but can also exist in other colours (Hill 1992). Some of the species that have been known to produce melanin are *Streptomyces*, *Marinomonas*, *Pseudomonas*, *Serratia*, *Rhizobium*, and *Bacillus* (Martínez et al. 2019). Tyrosinases are copper-containing polyphenol oxidases that have an essential role in the biosynthesis of melanin. This enzyme facilitates the conversion of tyrosine through orthohydroxylation, leading to the production of L-3,4-dihydroxyphenylalanine (L-DOPA). L-DOPA is subsequently oxidized to form dopaquinone (Fig. 3.4), which undergoes further oxidation to yield indole-5,6-quinone (Pavan et al. 2020) . Furthermore, Indole-5,6-quinone, along with its 2-carboxylated form (5,6-dihydroxyindole-2-carboxylic acid, or DHICA), undergoes natural oxidation and polymerization, ultimately resulting in the synthesis of eumelanin or DOPA melanin (Sanchez-Amat et al. 2010).

Tyrosinase activity is identified in numerous bacteria, and the tyrosinase-producing genes from these bacteria have been cloned to understand the genetic factors involved in tyrosinase production (Tseng et al. 1990). Typically, these genes are arranged in a polycistronic operon that consists of the tyrosinase gene and another gene encoding a molecular chaperone vital for the enzyme's copper incorporation and for secreting extracellular tyrosinase (López-Serrano et al. 2004). There are widely known applications of melanin, especially in the field of agriculture, where they are known to enhance plant growth. Additionally, melanins are also helpful in the bioremediation of polluted environments and can be utilized for the extraction of synthetic dyes and harmful metals from contaminated soil and water (Muñoz-Torres et al. 2024). Another interesting application is the production of insecticidal crystal proteins synthesized by *Bacillus thuringiensis*, where the mutant of these bacteria produces melanin. This melanin exhibited increased resistance

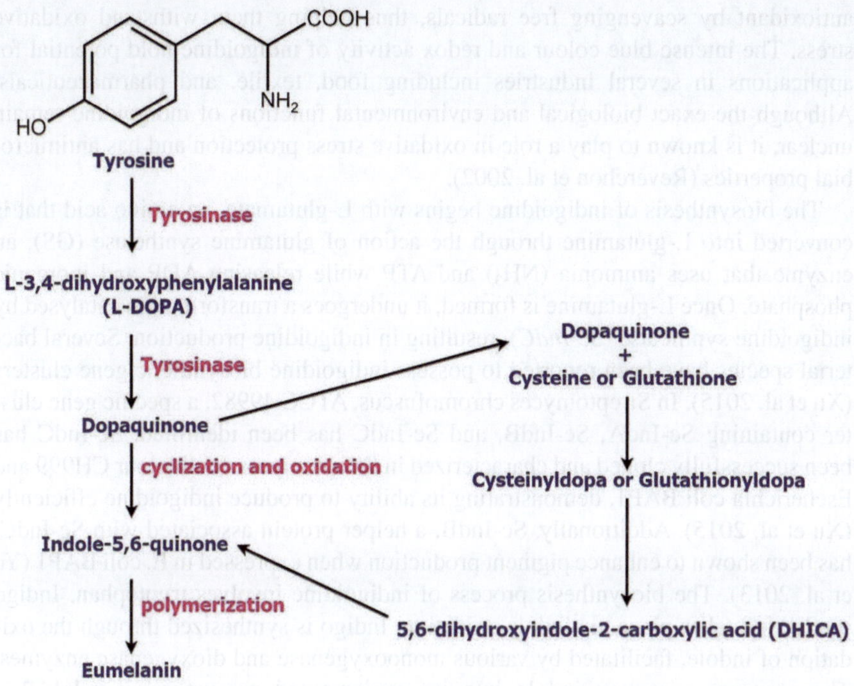

Fig. 3.4 Biosynthesis of Eumelanin from Tyrosine. The figure illustrates the tyrosinase-mediated conversion of tyrosine to L-DOPA and dopaquinone. Dopaquinone undergoes oxidation and cyclization to form indole-5,6- quinone or reacts with cysteine/glutathione to form DHICA. These intermediates polymerize to produce eumelanin, a black-brown pigment involved in microbial pigmentation and UV protection

towards UV radiation, emphasizing its eco-friendly insecticidal properties (Liu et al. 2013). Moreover, dark-pigmented sponges-linked bacteria such as *Sigmadocia pumila*, *Spongia officinalis*, *Callyspongia diffusa*, *Fasciospongia cavernosa*, and *Haliclona pigmentifera* protect animal cells from photo-toxicity due to their secretion of non-cytotoxic melanin that has antioxidant effects (Vijayan et al. 2017).

3.3.6 Indigoidine

Indigoidine or 5,5′-diamino-4,4′-dihydroxy-3,3′-diazadiphenoquinone-(2,2′) is a blue pigment that belongs to the group of azaquinones and is soluble in water. It was first identified in *Erwinia chrysanthemi* (Reverchon et al. 2002) and has since been found in several other bacterial species, including *Streptomyces chromofuscus*, *Pseudomonas fluorescens*, *Phaeobacter sp.*, *Vogesella indigofera*, and *Corynebacterium insidiosum* (Cude et al. 2012; Yu et al. 2013; Day et al. 2017; Ganguly et al. 2019). The unique structure of indigoidine, which contains conjugated carbon–carbon double bonds and carbonyl groups, enables it to act as a potent antioxidant by scavenging free radicals, thus helping them withstand oxidative stress. The intense blue colour and redox activity of indigoidine hold potential for applications in several industries including food, textile, and pharmaceuticals. Although the exact biological and environmental functions of indigoidine remain unclear, it is known to play a role in oxidative stress protection and has antimicrobial properties (Reverchon et al. 2002).

The biosynthesis of indigoidine begins with L-glutamate, an amino acid that is converted into L-glutamine through the action of glutamine synthetase (GS), an enzyme that uses ammonia (NH_3) and ATP while releasing ADP and inorganic phosphate. Once L-glutamine is formed, it undergoes a transformation catalysed by indigoidine synthetase (*Sc-IndC*), resulting in indigoidine production. Several bacterial species have been reported to possess indigoidine biosynthetic gene clusters (Xu et al. 2015). In Streptomyces chromofuscus, ATCC 49982, a specific gene cluster containing Sc-IndA, Sc-IndB, and Sc-IndC has been identified. Sc-IndC has been successfully cloned and characterized in Streptomyces coelicolour CH999 and Escherichia coli BAP1, demonstrating its ability to produce indigoidine efficiently (Xu et al. 2015). Additionally, Sc-IndB, a helper protein associated with Sc-IndC, has been shown to enhance pigment production when expressed in E. coli BAP1 (Yu et al. 2013). The biosynthesis process of indigoidine involves tryptophan. Indigo production takes place via indole oxidation. Indigo is synthesized through the oxidation of indole, facilitated by various monooxygenase and dioxygenase enzymes. These enzymes convert indole into its unpigmented precursor, cis-indole-2,3-dihydrodiol (O'Connor et al. 1997). Several Pseudomonas species have been identified as producers of these enzymes (Ma et al. 2012), although none have been confirmed to synthesize indole directly.

3.4 Genomic Insights into Bacterial Pigment Formation

Bacterial pigments are diverse bioactive compounds with major ecological, industrial, and biomedical applications. Understanding the genetic basis of pigment biosynthesis, gene clusters, and their functional significance can provide valuable insights into the regulatory mechanisms, evolutionary adaptations, and potential biotechnological exploitation of these biocolours. The first genes that are associated with the biosynthesis of carotenoid in bacteria were identified and functionally characterized in *Rhodopseudomonas capsulata* (Yen and Marrs 1976). In bacterial cells, the synthesis of carotene is primarily governed by *crt* genes, which are organized into two different clusters, namely *CrtFECD* and *CrtBIA*, separated by a gene encoding a signal transduction protein, *tspO* (Scolnik et al. 1980). The biosynthetic pathway begins with the *CrtE* gene, which encodes geranylgeranyl pyrophosphate (GGPP) synthase, an enzyme responsible for synthesizing GGPP as reported in *Rhodobacter capsulatus* SB1003 (Fig. 3.5-1A). This intermediate is then converted into phytoene by phytoene synthase, encoded by *CrtB* (Gerjets et al. 2009; Sandmann 2023). The next step involves phytoene desaturation, catalysed by the *CrtI* gene, which determines the pathway's progression towards either neurosporene or lycopene synthesis (Giuliano et al. 1986; Raisig et al. 1996). The desaturation process varies across bacterial species, for example, in *Rhodobacter* species, three desaturation steps lead to the formation of neurosporene, while in *Rubrivivax gelatinosus* an additional fourth step results in lycopene production. In *Rubrivivax gelatinosus* S1, carotenoid synthesis is governed by the same set of *crt* genes but arranged in a different pattern (Fig. 3.5-1B) (Sandmann 2023). In bacteria that produce spirilloxanthin, such as *Rhodospirillum rubrum*, the *CrtA* gene encodes a ketolase that helps in the conversion of spheroidene into 2-ketospirilloxanthin (Takaichi 2009). Additionally, in certain cyanobacteria and purple bacteria, the *CrtY* gene is responsible for the cyclization of lycopene into β-carotene (Vogl and Bryant 2012). Furthermore, carotenoid biosynthesis in purple bacteria has been closely linked to photosynthetic gene clusters, which have been subject to horizontal gene transfer among species (Sandmann 2023). These bacteria primarily synthesize spheroidene and spirilloxanthin (Fig. 3.5) through distinct yet interconnected biosynthetic pathways. The genes involved in carotenoid biosynthesis are typically organized in

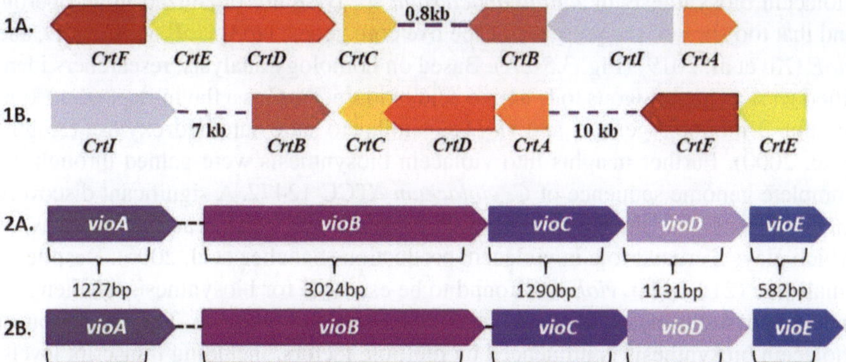

Fig. 3.5 Physical map representing biosynthetic gene clusters in some bacterial strains responsible for producing pigments. 1. Carotenoid biosynthesis gene cluster reported in A. *Rhodobacter capsulatus* SB1003 and B. *Rubrivivax gelatinosus* S1 (adapted from Sandmann 2023; Gerjets et al. 2009); 2. Violacein biosynthesis gene cluster reported in: A. *Janthinobacterium* sp. B9–8 (adapted from Xu et al. 2019), and B. *Pseudoalteromonas* sp. 520P1 (adapted from Zhang et al. 2011) 3. Prodigiosin biosynthesis gene cluster reported in A. *Serratia marcescens* ATCC: 274 and B. *Janthinobacterium* sp. ERMR3:09 (adapted from Mukhia et al. 2023a). The arrows denote the direction of transcription, and the dotted lines represent the spacers

clusters, which has facilitated their molecular characterization and enabled their heterologous expression in other organisms. The cloning of photosynthesis gene clusters from purple bacteria has also provided key insights into their carotenoid biosynthetic pathways (Swingley et al. 2009). Carotenoid biosynthesis is highly regulated by environmental factors such as oxygen availability and light intensity. The *tspO* gene plays a crucial regulatory role by modulating *crt* gene expression, thereby adjusting carotenoid levels in response to oxidative stress (Steiger et al. 2003). Under aerobic conditions, the expression of *CrtA* is upregulated, leading to the ketolation of spheroidene into spheroidenone (Takaichi and Shimada 1999).

The biosynthesis of violacein is governed by a complex five-gene operon structure, which requires the coordinated expression of multiple genes (Hoshino 2011). The violacein biosynthetic gene cluster, originally cloned from the *Chromobacterium violaceum* UQM51 genome, was successfully expressed in *E. coli* and other Gram-negative bacteria (Pemberton et al. 1991). Further investigations showed that the gene cluster contains four ORFs (*vioABCD*) and spans 8 kb in length with three of these enzymes (*vioA*, *vioC*, *and vioD*) containing the GXGXXG sequence motif required for flavin cofactor binding (August et al. 2000). The genes responsible for violacein biosynthesis in *Janthinobacterium* sp. B9-8 are organized in an operon, and that too showed the presence of the five core genes: *vioA*, *vioB*, *vioC*, *vioD*, and *vioE* (Xu et al. 2019) (Fig. 3.5-2A). Based on homology analysis, researchers identified *vioA* as homologous to L-amino acid oxidase, *vioC* as a flavin-dependent kynurenine 3-monooxygenase, and *vioD* as similar to salicylate hydroxylase (August et al. 2000). Further insights into violacein biosynthesis were gained through the complete genome sequence of *C. violaceum* ATCC 12472. A significant discovery in this regard was the identification of an additional open reading frame, *vioE*, which plays a crucial role in violacein production (Sánchez et al. 2006). Despite its small size (21.8 kDa), *vioE* was found to be essential for biosynthesis, challenging previous assumptions that it was not involved in the pathway. The regulation of violacein biosynthesis is influenced by multiple factors, including transcript levels, protein interactions, promoter activity, and ribosome binding site dynamics (Lee et al. 2012). Sequence analysis of *Pseudoalteromonas* sp. 520P1 revealed the presence of the same violacein biosynthetic gene cluster organized as five open reading frames (Fig. 3.5-2B) with a total length of 7383 base pairs (Zhang et al. 2011). Understanding these regulatory mechanisms is essential for optimizing violacein production and exploring its potential applications.

The melanin biosynthetic pathway in microorganisms primarily involves the enzyme tyrosinase, a copper-containing oxidase encoded by the *tyr* gene. Tyrosinase catalyses the initial processes in the formation of melanin pigment including the formation of the two crucial intermediates that lead to melanin polymerization—the conversion of L-tyrosine to L-DOPA through hydroxylation and of L-DOPA to dopaquinone by oxidation (Saini et al. 2020). Furthermore, melanin production was successfully achieved in *Vibrio natriegens*, which is a fast-growing marine bacterium by heterologous expression of melanin synthesizing genes—*tyr1* and *hpaBC* genes derived from other bacterial sources (Wang et al. 2020). The *tyr1* gene from *Bacillus aryabhattai* encodes a tyrosinase enzyme that catalyses the conversion of

L-tyrosine to *L-DOPA* and then to dopaquinone. In addition, the *hpaBC* gene cluster from *Escherichia coli*, which encodes a two-component monooxygenase system also capable of converting *L-tyrosine* to *L-DOPA*, was also examined. However, the *tyr1* gene exhibited significantly higher efficiency in melanin production compared to *hpaBC*, making it the preferred choice for engineering melanin biosynthesis in *V. natriegens*. In addition to the classical melanin biosynthesis pathway involving *tyr*, several other genes contribute to the pyomelanin biosynthetic pathway, which begins with L-tyrosine degradation. The *phhA* gene encodes phenylalanine hydroxylase, which converts L-phenylalanine to L-tyrosine, providing the primary precursor for melanin formation. Once tyrosine is available, the genes—*tyrB* and *aspC* which encode aromatic amino acid aminotransferases, convert L-tyrosine into 4-hydroxyphenylpyruvate (4-HPP). Subsequently, *hppD*, encoding 4-hydroxyphenylpyruvate dioxygenase, catalyses the conversion of 4-HPP to homogentisic acid (HGA), a central intermediate in pyomelanin biosynthesis (Wang et al. 2015).

Prodigiosin was originally identified in bacterial genera such as *Serratia*, *Streptomyces* and *Hahella*. However, more recent studies have revealed its production in additional Gram-negative bacteria, including members of the genus *Janthinobacterium* (Lin et al. 2020). The biosynthesis of prodigiosin is governed by a set of key genes typically organized in a conserved cluster, including *pigA*, *pigB*, *pigC*, *pigD*, *pigE*, *pigF*, *pigG*, *pigH*, *pigI*, *pigJ*, *pigK*, *pigL*, *pigM*, *and pigN* (Harris et al. 2004). These genes function sequentially to drive the synthesis of the pigment. Notably, in the *Janthinobacterium* strains, the *pigN* gene that is present in other prodigiosin-producing bacteria is absent (Fig. 3.5-3B), indicating a divergence in the genetic architecture. Additionally, in *Janthinobacterium sp.* ERMR3:09, the *pigH* and *pigM* genes appear fragmented into two distinct segments, differing from their continuous forms in other bacterial genomes (Mukhia et al. 2023b). Comparative sequence analysis of the *pig* clusters from *Serratia marcescens* ATCC 274 (Fig. 3.5-3A) and *Serratia* sp. ATCC 39006 revealed 14 and 15 open reading frames, respectively. Both strains share a conserved gene arrangement from *pigA* to *pigN*, and the encoded proteins exhibit strong similarity in size and amino acid identity (Harris et al. 2004).

The species such as *Pseudomonas tolaasii*, *P. libanensis*, and *Pseudomonas fluorescens* are reported to also be involved in indigoidine production. Notably, *P. libanensis* was found to produce blue pigment cheese products (Ganguly et al. 2019. Additionally, twelve specific strains of *P. fluorescens* (ps_1, ps_2, ps_5, ps_6, ps_13, ps_22, ps_25, ps_75, ps_76, ps_77, ps_78, and ps_79) were reported as blue pigment producers, with the colour intensifying from light blue to deep grey, dark blue, and eventually deep brown or black (Andreani et al. 2015). All blue-pigmented strains were found to have a phylogenetic cluster that is different from each strain and known as the 'blue branch,' indicating a strong correlation between pigment production and the core genome, despite phenotypic variations often being associated with accessory genes (Andreani et al. 2015). A pangenome study of *P. fluorescens* revealed that blue pigment-producing strains possess two copies of the tryptophan biosynthesis genes (*trpABCDF*) and exhibit upregulation of iron uptake genes, which correlates with pigment production. Chromogenic reactions suggested that the precursor for the pigment is an indole-derivative, while mass spectrometry indicated that the pigment is a hydrophobic indigo analogue (Andreani et al. 2015). Further pangenome analysis identified the core genome of *P. fluorescens* as consisting of 4065 coding DNA sequences (CDSs), shared among all studied strains. Analysis of two blue-pigmenting strains (Ps_22 and Ps_77) and two non-pigmenting strains (Ps_20 and Ps_40) revealed that a *trp* gene cluster is present in the strains that are involved in tryptophan biosynthesis (Xu et al. 2015). In *Streptomyces chromofuscus* ATCC 49982, the gene cluster consists of Sc-IndA, Sc-IndB, and Sc-IndC, along with two additional open reading frames (*orf1* and *orf2*). Notably, Sc-IndC was recognized as the enzyme indigoidine synthetase and was cloned and expressed in *Escherichia coli* BAP1 and *Streptomyces coelicolour* CH999, leading to the successful reconstitution of indigoidine biosynthesis (Xu et al. 2015).

3.5 Improvement in Genetic Engineering Technology of Microbial Strain for Large-Scale Production of Biocolours

Conventional methods of pigment extraction from sources like microalgae and plants face multiple challenges, such as low yields, seasonal inconsistency, and resource-demanding procedures for large-scale production. The demand for eco-friendly and sustainable dyes grows, and developing biocolours gives an alternative to overcome these limitations. These natural pigments, sourced from microorganisms, present a greener and eco-friendly substitute for artificial dyes. With the use of innovative techniques in genetic engineering and modern biological methods, microbial genes can be modified to produce pigments of interest. Advancements in fermentation and the reduction in the cost of production have helped the selection of important pigment-producing strains via mutagenesis (Rao et al. 2017). The advancements in metabolic engineering have helped protein engineering through rational and random approaches, helping in the biological improvement of pigment production. Genetic engineering has been efficient in producing good-quality bacterial pigments at a low cost (Numan et al. 2018). Genetic engineering has improved the production of biocolours by facilitating specific manipulation of bacterial genomes. Directed modifications in biosynthetic pathways have improved yields, shortened production times, and increased the palette of existing biocolours. Understanding biosynthesis pathways of pigments is very important for modifying genes to improve pigment yield (Gharibzahedi et al. 2013). For selecting the microbial strain for enhanced pigment production, the microbial strain should be non-pathogenic, nontoxic, should be able to utilize various nutrient sources and should be capable of withstanding a range of environmental conditions.

With an improvement in the understanding of microbial pigment biosynthetic pathways, identification and genetic engineering of key genes to enhance the production of biocolours have become easier. Cheaper and efficient methods for large-scale production of bicolour involve injecting the gene that is responsible for the production of pigment into the bacterial hosts such as *Pichia pastoris*,

Corynebacterium glutamicum, *Pseudomonas putida*, *Bacillus subtilis*, and *E. coli* (Sen et al. 2019). Various examples highlighted that genetic engineering has led to an increase in the production of pigment by microbial strains. The carotene biosynthesis genes from *Agrobacterium aurantiacum* and *Xanthophyllomyces dendrorhous* have been modified to improve the synthesis of carotenoids like β-carotene, astaxanthin, and lycopene (Heider et al. 2014). In a study by Nam et al. (2017), the three key genes (*mvaK1*, *mvaK2*, and *mvaD*) were cloned from *Enterococcus faecalis* VTCC-B-935, a strain isolated from Vietnam, which were co-transformed with the *CrtEIBY* gene cluster from *Pantoea ananatis* into *E. coli* BL21 (DE3) for further study. This genetic modification increased the production of β-carotene by three times. In another study by Zhao et al. (2013), they further highlighted the significance of regulatory control in metabolic pathways by adjusting the expression of *dxr*, *ispG*, and *ispH* genes using regulatory elements of varying potential. Pigment-producing bacterial strain was modified to overproduce pigments, alter colour and structure. Actinorhodin, a blue pigment, originally produced by *Streptomyces coelicolour*, was genetically modified to yield the bright yellow pigment. This engineered strain is used for the production of anthraquinone-type pigments (Venil et al. 2020). Similarly, cyanidin 3-O-glucoside (C3G), a type of anthocyanin, is produced by metabolic engineering of *Corynebacterium glutamicum*.

Furthermore, *Petunia hybrid* converts catechin to an intermediate compound by an enzyme called anthocyanidin synthase (ANS), to an intermediate compound which is then attached to a glucose molecule by 3-O-glucosyltransferase (3GT) from *Arabidopsis thaliana* to form C3G (Zha et al. 2018). The production of C3G was significantly increased, and when the supply of UDP-glucose was increased, it significantly yielded higher C3G (~40 mg/L) which is a more than 100-fold increase compared to the unmodified bacterial strain (Zha et al. 2018). Studies have explored bacterial strains other than *E. coli* and *B. subtilis* that produced riboflavin. *Leuconostoc falkenbergense* and *Weissella cibaria* are examples of utilized lactic acid bacteria that are used to achieve amplified production of riboflavin (Llamas-Arriba et al. 2021).

Similarly, an acetic acid bacterium named *Gluconobacter oxydans* FBFS97 was reported to produce significantly higher riboflavin compared to other genetically modified strains (Noman et al. 2020; Averianova et al. 2020). In another study, MEP and MVA pathways were optimized in *E. coli* for the production of β-carotene. The modified strain, YJM49, produced an impressive β-carotene yield in fed-batch fermentation-113 times higher than YJM39 and 1.7 times more than YJM45 (Yang and Guo 2014). Under optimized culture conditions, further improvements could be observed in metabolic engineering in microbial systems offering a safer alternative for the production of sustainable β-carotene, an alternative to synthetic sources. The research highlights the advantage of combining number of biosynthetic pathways in engineered microbes. Researchers identified and studied a gene cluster in *Streptomyces chromofuscus* ATCC 49982, to reconstruct the biosynthesis of indigoidine, which comprises genes like *Sc-IndA*, *Sc-IndB*, and *Sc-IndC*, along two other open reading frames. The *Sc-IndC* gene, functioning as indigoidine synthetase, was expressed heterologously in *Streptomyces coelicolour* CH999 and *E. coli*

BAP1, which successfully restored pigment production in these hosts (Xu et al. 2015). Studies demonstrated that *Sc-IndB* along with *Sc-IndC*, when expressed in *E. coli* BAP1 enhanced blue pigment production (Yu et al. 2013). L-glutamine, which is a direct precursor in the biosynthetic pathway for the production of indigoidine, was added to the fermentation medium of genetically modified *E. coli* strains that expressed *Sc-IndC* and *Sc-IndB*. This supplementation significantly enhanced the production of pigment, showing the impact of precursor availability on biosynthetic output (Xu et al. 2015). The exact role of *Sc-IndB* is still unclear, but its co-expression with *Sc-IndC* notably enhances the indigoidine synthesis, highlighting that it functions as a helper protein in the biosynthetic process (Yu et al. 2013).

Metabolic engineering is considered one of the advanced fields of biotechnology which is coming up as a powerful means used for genetically modifying microorganisms for increasing the production of important compounds, including pigments (Mao et al. 2023; Pramastya et al. 2021). The combination of genetic engineering and microbial biotechnology has opened new possibilities for sustainable pigment production. With the modification of regulatory mechanisms, enhancement of biosynthetic pathways and using advanced fermentation technologies, a significantly large amount of pigments can be produced (Gharibzahedi et al. 2013). These developments offer an eco-friendly alternative to synthetic dyes along with cheaper solutions for large-scale industrial applications. As research progresses, bicolours produced through genetically modified microorganisms could become the basis for sustainable dye industries in future, supporting conservation and reducing the dependence on synthetic chemical dyes.

3.6 Conclusion and Future Potential

The environmentally friendly, natural pigments are in demand, compared to those of synthetic pigments. The researchers are focusing on biologically derived alternative biocolours produced by microorganisms. Biocolours offer vibrant pigments suitable for a wide range of industrial applications along with significant bioactive properties including antioxidant, anticancer, and antimicrobial properties. Among multiple microbial sources, bacteria are highly advantageous for pigment production as they grow very fast, they are easy to cultivate, and have the capacity for genetic modification. With the advancement of molecular and synthetic biology tools and techniques, various biosynthetic pathways for pigment in bacteria have been studied and engineered for high yield. Pigments like violacein, indigoidine, melanin, carotenoids, and prodigiosin have been successfully studied at the molecular level. It will enable the development of recombinant strains with higher production and tailored pigment profiles. Despite having these advancements, the commercialization of microbial pigments faces various challenges, such as product stability, scalability, and economic feasibility of downstream processing. Addressing these shortcomings is important for their industrial adoption. Further research should emphasize genome mining and functional characterization of pigment-producing genes, particularly in unexplored bacterial species. characterization of pigment-producing genes, particularly in unexplored bacterial species.

The incorporation of synthetic biology tools such as CRISPR-Cas systems, metabolic flux optimization, and microbiome engineering can help in the creation of high-performance microbial production systems. Furthermore, combining omics technologies with bioprocess optimization will enhance yield, consistency, and the discovery of novel pigment derivatives. A deeper molecular understanding of pigment biosynthesis, supported by interdisciplinary research, will be the way to understand the optimum potential of bacterial pigments in eco-friendly industrial applications.

References

Abdulrahman J, AbdElsamie G, Al-Rawi RA, Kaream SAE (2020) Anti-tumor synergistic activity of nano-chlorophyll with sonophotodynamic on Ehrlich ascites carcinoma in mice. Zanco J Med Sci 24(1):132–142

Ahmad WA, Wan Ahmad WY, Zakaria ZA, Yusof NZ, Ahmad WA, Ahmad WYW et al (2012) Application of bacterial pigments as colourant. Springer, Berlin/Heidelberg, pp 57–74

Ahn SY, Jang S, Sudheer PDVN, Choi KY (2021) Microbial production of melanin pigments from caffeic acid and L-tyrosine using streptomyces glaucescens and FCS-ECH-expressing Escherichia coli. Int J Mol Sci 22(5):2413

Aman Mohammadi M, Ahangari H, Mousazadeh S, Hosseini SM, Dufossé L (2022) Microbial pigments as an alternative to synthetic dyes and food additives: a brief review of recent studies. Bioprocess Biosyst Eng 45(1):1–12

Amchova P, Kotolova H, Ruda-Kucerova J (2015) Health safety issues of synthetic food colourants. Regul Toxicol Pharmacol 73(3):914–922

Andreani NA, Carraro L, Martino ME, Fondi M, Fasolato L, Miotto G, Cardazzo B (2015) A genomic and transcriptomic approach to investigate the blue pigment phenotype in Pseudomonas fluorescens. Int J Food Microbiol 213:88–98

Atalah J, Blamey L, Muñoz-Ibacache S, Gutierrez F, Urzua M, Encinas MV, Blamey JM (2020) Isolation and characterization of violacein from an Antarctic Iodobacter: a non-pathogenic psychrotolerant microorganism. Extremophiles 24:43–52

August PR, Grossman TH, Minor C, Draper MP, MacNeil IA, Pemberton JM, Call KM, Holt D, Osburne MS (2000) Sequence analysis and functional characterization of the violacein biosynthetic pathway from Chromobacterium violaceum. J Mol Microbiol Biotechnol 2:513–519

Averianova LA, Balabanova LA, Son OM, Podvolotskaya AB, Tekutyeva LA (2020) Production of vitamin B2 (riboflavin) by microorganisms: an overview. Front Bioeng Biotechnol 8:570828

Bashir I, Pandey VK, Dar AH, Dash KK, Shams R, Mir SA et al (2024) Exploring sources, extraction techniques and food applications: a review on biocolours as next-generation colourants. Phytochem Rev:1–26

Bindu A, Bhadra S, Nayak S, Khan R, Prabhu AA, Sevda S (2024) Bioelectrochemical biosensors for water quality assessment and wastewater monitoring. Open Life Sci 19(1):20220933

Chen GE, Hunter CN (2023) Engineering chlorophyll, bacteriochlorophyll, and carotenoid biosynthetic pathways in Escherichia coli. ACS Synth Biol 12(8):2236–2244

Choi SY, Kim S, Lyuck S, Kim SB, Mitchell RJ (2015) High-level production of violacein by the newly isolated Duganella viola ceinigra str. NI28 and its impact on Staphylococcus aureus. Sci Rep 5(1):1–12

Chuyen HV, Eun JB (2017) Marine carotenoids: bioactivities and potential benefits to human health. Crit Rev Food Sci Nutr 57(12):2600–2610

Cogdell RJ, Howard TD, Bittl R, Schlodder E, Geisenheimer I, Lubitz W (2000) How carotenoids protect bacterial photosynthesis. Philosophical Transactions of the Royal Society of London. Series B: Biological Sciences 355(1402):1345–1349

Cude WN, Mooney J, Tavanaei AA et al (2012) Production of the antimicrobial secondary metabolite indigoidine contributes to competitive surface colonization by the marine roseobacter Phaeobacter sp. strain Y4I. Appl Environ Microbiol 78(14):4771–4780

Darshan N, Manonmani HK (2015) Prodigiosin and its potential applications. J Food Sci Technol 52(9):5393–5407

Day PA, Villalba MS, Herrero OM et al (2017) Formation of indigoidine derived-pigments contributes to the adaptation of Vogesella sp. strain EB to cold aquatic iron-oxidizing environments. Antonie Van Leeuwenhoek 110(3):415–428

Durán N, Justo GZ, Durán M, Brocchi M, Cordi L, Tasic L, Castro GR, Nakazato G (2016) Advances in Chromobacterium violaceum and properties of violacein-its main secondary metabolite: a review. Biotechnol Adv 34(5):1030–1045

El-Naggar NEA, El-Ewasy SM (2017) Bioproduction, characterization, anticancer and antioxidant activities of extracellular melanin pigment produced by newly isolated microbial cell factories Streptomyces glaucescens NEAE-H. Sci Rep 7(1):42129. https://doi.org/10.1038/srep42129

El-Naggar NEA, Saber WIA (2022) Natural melanin: current trends, and future approaches, with especial reference to Microbial source. Polymers 14(7):1339

Fang MY, Zhang C, Yang S, Cui JY, Jiang PX, Lou K, Wachi M, Xing XH (2015) High crude violacein production from glucose by Escherichia coli engineered with interactive control of tryptophan pathway and violacein biosynthetic pathway. Microb Cell Factories 14(1):1–13

Fried R, Oprea I, Fleck K, Rudroff F (2022) Biogenic colourants in the textile industry–a promising and sustainable alternative to synthetic dyes. Green Chem 24(1):13–35

Funa N, Ohnishi Y, Fujii I, Shibuya M, Ebizuka Y, Horinouchi S (1999) A new pathway for polyketide synthesis in microorganisms. Nature 400(6747):897–899

Ganguly S, Bandyopadhyay K, Dutta P, Pramanik A (2019) A review on microbial- pigment: a good source of biocolour. Pharma Innov J 8:416–420

Gerjets T, Steiger S, Sandmann G (2009) Catalytic properties of the expressed acyclic carotenoid 2-ketolases from Rhodobacter capsulatus and Rubrivivax gelatinosus. Biochimica et Biophysica Acta (BBA) Mol Cell Biol Lipids 1791(2):125–131

Gharibzahedi SMT, Razavi SH, Mousavi SM (2013) Microbial canthaxanthin: perspectives on biochemistry and biotechnological production. Eng Life Sci 13(4):408–417

Giuliano G, Pollock D, Scolnik PA (1986) The gene crtI mediates the conversion of phytoene into coloured carotenoids in Rhodopseudomonas capsulata. J Biol Chem 261:12925–12929

González-Peña MA, Ortega-Regules AE, De Parrodi CA, Lozada-Ramírez JD (2023) Chemistry, occurrence, properties, applications, and encapsulation of carotenoids—a review. Plants 12(2):313

Gul R, Kumar R, Sharma AK (2017) The microbial world of biocolour production. In: Biopigmentation and biotechnological implementations. Wiley, Somerset, pp 263–277

Harris AK, Williamson NR, Slater H, Cox A, Abbasi S, Foulds I et al (2004).The Serratia gene cluster encoding biosynthesis of the red antibiotic, prodigiosin, shows species-and strain-dependent genome context variation. Microbiology 150(11):3547–3560

Heider SA, Peters-Wendisch P, Wendisch VF, Beekwilder J, Brautaset T (2014) Metabolic engineering for the microbial production of carotenoids and related products with a focus on the rare C50 carotenoids. Appl Microbiol Biotechnol 98:4355–4368

Hendry GAF (1996) Natural pigments in biology. In: Hendry GAF, Houghton JD (eds) Natural food colourants, 2nd edn. Springer, Boston, pp 1–39

Heyes DJ, Hunter CN (2009) Biosynthesis of chlorophyll and bacteriochlorophyll. Springer eBooks, pp 235–249

Hill HZ (1992) The function of melanin or six blind people examine an elephant. BioEssays 14(1):49–56

Horinouchi S, Hara O, Beppu T (1983) Cloning of a pleiotropic gene that positively controls biosynthesis of A-factor, actinorhodin, and prodigiosin in Streptomyces coelicolour A3(2) and Streptomyces lividans. J Bacteriol 155(3):1238–1248

Hoshino T (2011) Violacein and related tryptophan metabolites produced by Chromobacterium violaceum: biosynthetic mechanism and pathway for construction of violacein core. Appl Microbiol Biotechnol 91(6):1463–1475

Immanuel SRC, Banerjee D, Rajankar MP, Raghunathan A (2018) Integrated constraints based analysis of an engineered violacein pathway in Escherichia coli. BioSyst 171:10–19

Kawauchi K, Shibutani K, Yagisawa H, Kamata H, Nakatsuji S, Anzai H, Yokoyama Y, Ikegami Y, Moriyama Y, Hirata H (1997) A possible immunosuppressant, cycloprodigiosin hydrochloride, obtained from pseudoalteromonas denitrificans. Biochem Biophys Res Commun 237(3):543–547

Kim D, Kim JF, Yim JH, Kwon SK, Lee CH, Lee HK (2008) Red to red – the marine bacterium Hahella chejuensis and its product prodigiosin for mitigation of harmful algal blooms. PubMed 18(10):1621–1629

Kumar R, Acharya V, Singh D, Kumar S (2018) Strategies for high altitude adaptation revealed from high-quality draft genome of non-violacein producing Janthinobacterium lividum ERGS5: 01. Stand Genomic Sci 13(1):1–13

Kustov AV, Belykh DV, Smirnova NL, Venediktov EA, Kudayarova TV, Kruchin SO, Khudyaeva IS, Berezin DB (2017) Synthesis and investigation of water-soluble chlorophyll pigments for antimicrobial photodynamic therapy. Dyes Pigments 149:553–559

Lapenda JC, Silva PA, Vicalvi MC, Sena KXFR, Nascimento SC (2015) Antimicrobial activity of prodigiosin isolated from Serratia marcescens UFPEDA 398. World J Microbiol Biotechnol 31(2):399–406

Lee JW, Na D, Park JM, Lee J, Choi S, Lee SY (2012) Systems metabolic engineering of microorganisms for natural and non-natural chemicals. Nat Chem Biol 8(6):536–546

Li G, Lou HX (2018) Strategies to diversify natural products for drug discovery. Med Res Rev 38(4):1255–1294

Lin SR, Chen YH, Tseng FJ, Weng CF (2020) The production and bioactivity of prodigiosin: quo vadis?. Drug Discovery Today 25(5):828–836

Liu F, Yang W, Ruan L, Sun M (2013) A *Bacillus thuringiensis* host strain with high melanin production for preparation of light-stable biopesticides. Ann Microbiol 63:1131–1135

Llamas-Arriba MG, Hernández-Alcántara AM, Mohedano ML, Chiva R, Celador- Lera L, Velázquez E, Prieto A, Dueñas MT, Tamame M, López P (2021) Lactic acid bacteria isolated from fermented doughs in Spain produce dextrans and riboflavin. Foods 10(9):2004

López G, Álvarez-Rivera G, Carazzone C, Ibáñez E, Leidy C, Cifuentes A (2021) Bacterial carotenoids: extraction, characterization, and applications. Crit Rev Anal Chem 53(6):1239–1262

López-Serrano D, Solano F, Sanchez-Amat A (2004) Identification of an operon involved in tyrosinase activity and melanin synthesis in Marinomonas mediterranea. Gene 342(1):179–187

Ma Q, Qu Y, Tang H, Yu H, Ma F, Shi S, Zhang X, Zhou H, Zhou J, Xu P (2012) Genome sequence of a novel indigo-producing strain, Pseudomonas monteilii QM. J Bacteriol 194:4459–4460

Malik K, Tokkas J, Goyal S (2012) Microbial pigments: a review. Int J Microbial Res Technol 1(4):361–365

Mao S, Yu H, Ye L (2023) Enhanced astaxanthin production in S. cerevisiae by combinatorial engineering of gene targets outside the synthetic pathway. Biochem Eng J 200:109097

Maoka T (2019) Carotenoids as natural functional pigments. J Nat Med 74(1):1–16

Martínez LM, Martinez A, Gosset G (2019) Production of melanins with recombinant microorganisms. Front Bioeng Biotechnol 7

Mata-Gómez LC, Montañez JC, Méndez-Zavala A, Aguilar CN (2014) Biotechnological production of carotenoids by yeasts: an overview. Microb Cell Factories 13:1–11

Moreira MD, Melo MM, Coimbra JM, dos Reis KC, Schwan RF, Silva CF (2018) Solid coffee waste as alternative to produce carotenoids with antioxidant and antimicrobial activities. Waste Manag 82:93–99

Muhammad M, Basit A, Majeed M, Shah AA, Ullah I, Mohamed HI, Khan A, Ghanaim AM (2024) Bacterial pigments and their applications. In: Bacterial secondary metabolites: synthesis and applications in agroecosystem. Elsevier, Amsterdam, pp 277–298

Mukhia S, Kumar A, Kumar R (2023a) Red bioactive pigment from Himalayan Janthinobacterium sp. ERMR3:09: optimization, characterization, and potential applications. Arch Microbiol 206(1)

Mukhia S, Kumar A, Kumar R (2023b) Antioxidant prodigiosin-producing cold- adapted Janthinobacterium sp. ERMR3:09 from a glacier moraine: Genomic elucidation of cold adaptation and pigment biosynthesis. Gene 857:147178

Mummaleti G, Udo T, Mohan A, Kong F (2024) Synthesis, characterization and application of microbial pigments in foods as natural colours. Crit Rev Food Sci Nutr:1–30

Muñoz-Torres P, Cárdenas-Ninasivincha S, Aguilar Y (2024) Exploring the agricultural applications of microbial melanin. Microorganisms 12(7):1352

Nabi F, Arain MA, Rajput N, Alagawany M, Soomro J, Umer M et al (2020) Health benefits of carotenoids and potential application in poultry industry: a review. J Anim Physiol Anim Nutr 104(6):1809–1818

Nam VH, Trang HHT, Huong TLT et al (2017) Enhanced ? -carotene biosynthesis in recombinant Escherichia coli harboring the bottom portion of mevalonate pathway of enterococcus faecium VTCC-B-935 isolated in Vietnam. J Appl Biotechnol Bioeng 3(4):374–379

Noman AE, Al-Barha NS, Sharaf A-AM, Al-Maqtari QA, Mohedein A, Mohammed HHH, Chen F (2020) A novel strain of acetic acid bacteria Gluconobacter oxydans FBFS97 involved in riboflavin production. Sci Rep 10(1):13527

Numan M, Bashir S, Mumtaz R, Tayyab S, Rehman NU, Khan AL et al (2018). Therapeutic applications of bacterial pigments: a review of current status and future opportunities. 3 Biotech 8:1–15

O'Connor KE, Dobson AD, Hartmans S (1997) Indigo formation by microorganisms ex pressing styrene monooxygenase activity. Appl Environ Microbiol 63:4287–4291

Ohkubo S, Miyashita H (2017) A niche for cyanobacteria producing chlorophyll f within a microbial mat. ISME J 11(10):2368–2378

Padhan B, Poddar K, Sarkar D, Sarkar A (2021) Production, purification, and process optimization of intracellular pigment from novel psychrotolerant *Paenibacillus* sp, BPW19, vol 29. Biotechnology Reports, Amsterdam, p e00592

Paniagua-Michel J, Olmos-Soto J, Ruiz MA (2012) Pathways of carotenoid biosynthesis in bacteria and microalgae. Methods Mol Biol:1–12

Pankaj VR, Kumar R (2016) Microbial pigment as a potential natural colourant for contributing to mankind. Res Trends Mol Biol:85–98

Park HA, Park SA, Yang YH, Choi KY (2021) Microbial synthesis of violacein pigment and its potential applications. Crit Rev Biotechnol 41(6):879–901

Parmar M, Phutela UG (2015) Biocolours: the new generation additives. Int J Curr Microbiol App Sci 4(7):688–694

Paul T, Bandyopadhyay TK, Mondal A, Tiwari ON, Muthuraj M, Bhunia B (2020) A comprehensive review on recent trends in production, purification, and applications of prodigiosin. Biomass Convers Biorefinery 12:1409

Paulsen SS, Strube ML, Bech PK, Gram L, Sonnenschein EC (2019) Marine chitinolytic Pseudoalteromonas represents an untapped reservoir of bioactive potential. mSystems 4(4):e00060–e00019

Pavan ME, López NI, Pettinari MJ (2020) Melanin biosynthesis in bacteria, regulation and production perspectives. Appl Microbiol Biotechnol 104(4):1357–1370

Pemberton JM, Vincent KM, Penfold RJ (1991) Cloning and heterologous expression of the violacein biosynthesis gene cluster from Chromobacterium violaceum. Curr Microbiol 22(6):355–358

Ponjavic M, Malagurski I, Lazic J, Jeremic S, Pavlovic V, Prlainovic N et al (2023) Advancing PHBV biomedical potential with the incorporation of bacterial biopigment prodigiosin. Int J Mol Sci 24(3):1906

Pramastya H, Xue D, Abdallah II, Setroikromo R, Quax WJ (2021) High level production of amorphadiene using Bacillus subtilis as an optimized terpenoid cell factory. New Biotechnol 60:159–167
Proctor MS, Sutherland GA, Canniffe DP, Hitchcock A (2022) The terminal enzymes of (bacterio) chlorophyll biosynthesis. R Soc Open Sci 9(5)
Pšenčík J, Torkkeli M, Zupčanová A, Vácha F, Serimaa RE, Tuma R (2010) The lamellar spacing in self-assembling bacteriochlorophyll aggregates is proportional to the length of the esterifying alcohol. Photosynth Res 104(2–3):211–219
Qiang S, Su AP, Li Y, Chen Z, Hu CY, Meng YH (2019) Elevated β-carotene synthesis by the engineered Rhodobacter sphaeroides with enhanced CrtY expression. J Agric Food Chem 67(34):9560–9568
Raisig A, Bartley G, Scolnik P, Sandmann G (1996) Purification in an active state and properties of the 3-step phytoene desaturase from Rhodobacter capsulatus after overexpression in Escherichia coli. J Biochem 119:559–564
Ram S, Mitra M, Shah F, Tirkey SR, Mishra S (2020) Bacteria as an alternate biofactory for carotenoid production: a review of its applications, opportunities and challenges. J Funct Foods 67:103867
Rao MN, Xiao M, Li WJ (2017) Fungal and bacterial pigments: secondary metabolites with wide applications. Front Microbiol 8(1113):10–3389
Reverchon S, Rouanet C, Expert D, Nasser W (2002) Characterization of indigoidine biosynthetic genes in Erwinia chrysanthemi and role of this blue pigment in pathogenicity. J Bacteriol 184(3):654–665
Rodrigues AL, Trachtmann N, Becker J, Lohanatha AF, Blotenberg J, Bolten CJ, Korneli C, de Souza Lima AO, Porto LM, Sprenger GA (2013) Systems metabolic engineering of Escherichia coli for production of the antitumor drugs violacein and deoxyviolacein. Metab Eng 20:29–41
Rodriguez-Amaya DB (2016) Natural food pigments and colourants. Curr Opin Food Sci 7:20–26
Roy S, Rhim JW (2022) New insight into melanin for food packaging and biotechnology applications. Crit Rev Food Sci Nutr 62(17):4629–4655
Saini R, Rani P, Nayan R et al (2020) Molecular characterization of melanin and its biosynthetic pathway in microorganisms for industrial production. J Genet Eng Biotechnol 18:1–14
Sajjad W, Din G, Rafiq M, Iqbal A, Khan S, Zada S et al (2020) Pigment production by cold-adapted bacteria and fungi: colourful tale of cryosphere with wide range applications. Extremophiles 24:447–473
Sánchez C, Braña AF, Méndez C, Salas JA (2006) Reevaluation of the violacein biosynthetic pathway and its relationship to indolocarbazole biosynthesis. Chembiochem 7(8):1231–1240
Sanchez-Amat A, Solano F, Lucas-Elío P (2010) Finding new enzymes from bacterial physiology: a successful approach illustrated by the detection of novel oxidases in Marinomonas mediterranea. Mar Drugs 8(3):519–541
Sandmann G (2023) Genes and pathway reactions related to carotenoid biosynthesis in purple bacteria. Biology 12(10):1346
Schneider T, Graeff-Hönninger S, French WT, Hernandez R, Claupein W, Holmes WE, Merkt N (2012) Screening of industrial wastewaters as feedstock for the microbial production of oils for biodiesel production and high-quality pigments. J Comb 2012:1–9
Schweiggert RM, Carle R (2016) Carotenoid production by bacteria, microalgae, and fungi. In: Carotenoids: nutrition, analysis and technology. Wiley-Blackwell, Chichester, pp 217–240
Scolnik P, Walker AMA, Marrs BL (1980) Biosynthesis of carotenoid derivatives derived from neurosporene in Rhodopseudomonas capsulata. J Biol Chem 255:2427–2432
Sen T, Barrow CJ, Deshmukh SK (2019) Microbial pigments in the food industry—challenges and the way forward. Front Nutr 6:7
Sharma P, Gaur VK, Kim SH, Pandey A (2020) RETRACTED: Microbial strategies for bio-transforming food waste into resources. Bioresour Technol 299:122580
Singh S, Nimse SB, Mathew DE, Dhimmar A, Sahastrabudhe H, Gajjar A, Ghadge VA, Kumar P, Shinde PB (2021) Microbial melanin: recent advances in biosynthesis, extraction, characterization, and applications. Biotechnol Adv 53:107773

Srimathi R, Priya R, Nirmala M, Malarvizhi A (2017) Isolation, identification, optimization of prodigiosin pigment produced by Serratia Marcescens and its applications. Int J Latest Eng Manag Res 02(09):11–21

Steiger S, Mazet A, Sandmann G (2003) Heterologous expression, purification, and enzymatic characterization of the acyclic carotenoid 1,2-hydratase from Rubrivivax gelatinosus. Arch Biochem Biophys 414:51–58

Suvorov N, Pogorilyy V, Diachkova E, Vasil'ev Y, Mironov A, Grin M (2021) Derivatives of natural chlorophylls as agents for antimicrobial photodynamic therapy. Int J Mol Sci 22(12):6392

Swingley WD, Blankenship RE, Raymond J (2009) Evolutionary relationships among purple photosynthetic bacteria and the origin of proteobacterial photosynthetic systems. In: Hunter CN, Daldal F, Thurnauer MC, Beatty JT (eds) The purple phototrophic bacteria. Springer, Dordrecht, pp 17–29

Takagi T, Aoyama K, Motone K, Aburaya S, Yamashiro H, Miura N, Inoue K (2023) Mutualistic interactions between dinoflagellates and pigmented bacteria mitigate environmental stress. Microbiol Spectr 11(1):e02464–e02422

Takaichi S (2009) Distribution and biosynthesis of carotenoids. In: Hunter CN, Daldal F, Thurnauer MC, Beatty JT (eds) The purple phototrophic bacteria. Springer, New York, pp 97–117

Takaichi S, Shimada K (1999) Pigment composition of two pigment-protein complexes derived from anaerobically and semi-aerobically grown *Rubrivivax gelatinosus*, and identification of a new keto-carotenoid. Plant Cell Physiol 40:613–617

Thakur M, Modi VK (2024) Biocolourants in food: sources, extraction, applications and future prospects. Crit Rev Food Sci Nutr 64(14):4674–4713

Thweatt JL, Canniffe DP, Bryant DA (2019) Biosynthesis of chlorophylls and bacteriochlorophylls in green bacteria. In: Advances in botanical research. Elsevier, San Diego, pp 35–89

Tong Y, Zhou J, Zhang L, Xu P (2021) A golden-gate based cloning toolkit to build violacein pathway libraries in Yarrowia lipolytica. ACS Synth Biol 10(1):115–124

Tran-Ly AN, Reyes C, Schwarze FWMR, Ribera J (2020) Microbial production of melanin and its various applications. World J Microbiol Biotechnol 36(11):1–9

Tseng H, Lin C, Hsu B, Leu W, Lee YW, Chiou S, Hu N, Chen C (1990) The melanin operon of Streptomyces antibioticus: expression and use as a marker in Gram- negative bacteria. Gene 86(1):123–128

Tsui L, Huang J, Sabat M, Zangari G (2014) Visible light sensitization of TIO2 nanotubes by Bacteriochlorophyll-C dyes for photoelectrochemical solar cells. ACS Sustain Chem Eng 2(9):2097–2101

Vaidya H, Upasani N, Wagh PS (2021) Microbial pigments: natural colourants and their industrial applications. Int J Curr Microbiol Appl Sci 5:2319–7706

Venil CK, Zakaria ZA, Ahmad WA (2013) Bacterial pigments and their applications. Process Biochem 48(7):1065–1079

Venil CK, Dufossé L, Renuka Devi P (2020) Bacterial pigments: sustainable compounds with market potential for pharma and food industry. Front Sustain Food Syst 4:100

Vijayan V, Jasmin C, Anas A, Kuttan SP, Vinothkumar S, Subrayan PP, Nair S (2017) Sponge-associated bacteria produce non-cytotoxic melanin which protects animal cells from phototoxicity. Appl Biochem Biotechnol 183(1):396–411

Vogl K, Bryant DA (2012) Biosynthesis of the biomarker okenone:-ring formation. Geobiology 10:205–215

Wang H, Qiao Y, Chai B, Qiu C, Chen X (2015) Identification and molecular characterization of the homogentisate pathway responsible for pyomelanin production, the major melanin constituents in Aeromonas media WS. PLoS One 10(3):e0120923

Wang H, Zhang X, Wang S, Zhao B, Lou K, Xing XH (2018) *Massilia violaceinigra* sp. nov., a novel purple-pigmented bacterium isolated from glacier permafrost. Int J Syst Evol Microbiol 68(7):2271–2227

Wang Z, Tschirhart T, Schultzhaus Z, Kelly EE, Chen A, Oh E, Vora GJ (2020) Melanin produced by the fast-growing marine bacterium Vibrio natriegens through heterologous biosynthesis: characterization and application. Appl Environ Microbiol 86(5):e02749–e02719

Williamson NR, Fineran PC, Leeper FJ, Salmond GP (2006) The biosynthesis and regulation of bacterial prodiginines. Nat Rev Microbiol 4(12):887–899

Wu T, Ye L, Zhao D, Li S, Li Q, Zhang B et al (2017) Membrane engineering-a novel strategy to enhance the production and accumulation of β-carotene in Escherichia coli. Metab Eng 43:85–91

Xu F, Gage D, Zhan J (2015) Efficient production of indigoidine in Escherichia coli. J Ind Microbiol Biotechnol 42(8):1149–1155

Xu Y, Ibrahim IM, Wosu CI, Ben-Amotz A, Harvey PJ (2018) Potential of new isolates of *Dunaliella salina* for natural β-carotene production. Biology 7(1):14

Xu X, Tian L, Zhang S, Jiang L, Zhang Z, Huang H (2019) Complete genome sequence of Janthinobacterium sp. B9–8, a violacein producing bacterium isolated from low- temperature sewage. Microb Pathog 128:178–183

Xu X, Chu X, Du B, Huang C, Xie C, Zhang Z, Jiang L (2022) Functional characterization of a novel violacein biosynthesis operon from Janthinobacterium sp. B9-8. Appl Microbiol Biotechnol 106(8):2903–2916

Yang J, Guo L (2014) Biosynthesis of β-carotene in engineered E. coli using the MEP and MVA pathways. Microb Cell Factories 13:1–11

Yang C, Huang K, Wang Y, Shaw J (2021) A review of Bacteriochlorophyllides: chemical structures and applications. Molecules 26(5):1293

Yen H-C, Marrs B (1976) Map of genes for carotenoid and bacteriochlorophyll biosynthesis in Rhodopseudomonas capsulata. J Bacteriol 126:619–629

Yu D, Xu F, Valiente J et al (2013) An indigoidine biosynthetic gene cluster from Streptomyces chromofuscus ATCC 49982 contains an unusual IndB homologue. J Ind Microbiol Biot 40(1):159–168

Zha J, Zang Y, Mattozzi M, Plassmeier J, Gupta M, Wu X et al (2018) Metabolic engineering of Corynebacterium glutamicum for anthocyanin production. Microb Cell Factories 17:1–13

Zhang X, Enomoto K (2011) Characterization of a gene cluster and its putative promoter region for violacein biosynthesis in Pseudoalteromonas sp. 520P1. Applied microbiology and biotechnology 90(6):1963–1971

Zhang XK, Wang DN, Chen J, Liu ZJ, Wei LJ, Hua Q (2020) Metabolic engineering of β-carotene biosynthesis in Yarrowia lipolytica. Biotechnol Lett 42:945–956

Zhao J, Li Q, Sun T, Zhu X, Xu H, Tang J et al (2013) Engineering central metabolic modules of Escherichia coli for improving β-carotene production. Metab Eng 17:42–50

Plant and Algae-Based Biocolors

4

Gurjoban Singh, Priyanjana Bose, Kamakshi, Rekha Rani, and Preeti Mehta Kakkar

Abstract

Commercially produced organic dyes and their derivatives are widely used in a variety of industries, including textile production, paper, cosmetics, medicines, photography, and petroleum. Such industries produce a significant volume of wastewater effluents containing toxic azo dyes (AD), which causes an inevitable state of extreme contamination and a major risk to human health. Due to increasing environmental concerns and awareness of the harmful effects of synthetic dyes, there is a need for sustainable alternatives. Plant- and algal-based biocolors, derived from natural resources, are emerging as eco-friendly and safe options for various industries. These biocolorants, such as carotenoids, flavonoids, anthocyanins, chlorophyll, and betalain, not only provide vibrant colors but also offer functional benefits, including antimicrobial and antioxidant properties. However, achieving sustainability in natural dyeing processes remains a critical challenge, requiring advancements in extraction techniques, dye stability, and industrial applications. This chapter provides a comprehensive overview of plant- and algal-based biocolors, their sources, characteristics, extraction methods, and industrial applications, and their role in promoting sustainability. Furthermore, it explores the technological and economic considerations required to integrate biocolorants into mainstream industrial practices while addressing the environmental and health-related concerns of synthetic dyes.

Keywords

Organic dyes · Wastewater effluents · Environmental contamination · Biocolorants · Plant-based dyes · Algal-based dyes · Carotenoids

---

G. Singh · P. Bose · Kamakshi · R. Rani · P. M. Kakkar (✉)
Amity Institute of Biotechnology, Amity University, Noida, Uttar Pradesh, India
e-mail: pmkakkar@amity.edu

R. K. Bhatia, A. Walia (eds.), *Biocolours*,
https://doi.org/10.1007/978-981-95-1738-1_4

4.1 Introduction

Color has significantly influenced the development of cultures worldwide throughout history. It affects every aspect of our lives, including furnishings in our houses and the clothes we wear, and is used by many industries such as food, pharmaceuticals, cosmetics, etc. (Yusuf et al. 2017). Natural colors are healthy, of high quality, and preferred by consumers over chemical dyes; their availability and use have grown significantly in recent years. Biocolorants are coloring compounds derived from biological sources. Carotenoids, anthocyanidins, and other pigments are the primary sources of biocolorants (Rymbai et al. 2011). Natural dyes, discovered through the creativity and determination of our ancestors, can stay vibrant for hundreds or even thousands of years. These dyes come from unexpected sources like plant roots, tiny insects, and snail secretions. The distinction between dye and pigment is helpful since dyes are often organic compounds that dissolve in a solvent, and pigments are often minerals and inorganic substances that are insoluble in paint mediums like water or oil (Melo 2009). On the other hand, synthetic dyes composed of azo functional groups and aromatic rings are used by various industries and release these harmful dyes into the environment, having the potential to cause allergic reactions, asthma episodes, and hyperactivity in humans. Excessive use of azo dye causes intestinal cancer (Singh et al. 2023).

In general, dyes are divided into three categories: nonionic, cationic, and anionic. In terms of dyes used for fiber coloring, the textile industry ranks first in their use. In 1871, Woulfe created the first synthetic organic dye, picric acid, by reacting the natural dye indigo with nitric acid. The chemical makeup of the specific chromophoric group in the synthetic dyes can be used to determine their names. As an illustration, consider azo dyes, oxazine chemicals, xanthene compounds, triphenylmethane compounds, and diphenylmethane derivatives (Affat 2021). The textile sector uses dyes that have the worst effects on the planet. The azo dyes are strong, inexpensive, and colorfast. The dyeing performance, ease of access, and affordability of anthraquinones make them the second most popular dyes. Lipsticks, blushers, eye shadows, eyeliners, and nail polish are examples of cosmetics that contain one or more colorants (dyes or pigments) in a concentration of 1–25% to produce the required colors. Hair dye manufacture, which makes up almost 80% of cosmetics in Europe, is the industry's most significant operation (Ardila-Leal et al. 2021). In the food business, dyes make food look more appealing or make up for color changes that occur after processing (Tkaczyk et al. 2020). Carbonated beverages, fruit drinks, energy drinks, sweets, cereals, desserts, snacks, and more are all colored with dyes (Stevens et al. 2014). Among more than 60 synthetic dyes currently available for food use, the anionic dyes Sunset Yellow, Tartrazine, and Ponceau 4R (azo class), Solid Green FCF (triphenylmethane class), and Quinoline Yellow (quinethazones) are among the most widely used (Tikhomirova et al. 2018). As synthetic dyes are used in various industries, these industries discharge a lot of dye effluents and aqueous wastes from the dyeing process into water bodies, which results in long-term high biological oxygen demand (BOD) and other unfavorable environmental and aesthetic effects. During the dyeing and finishing operations, around 103 tons of dyes are created annually, and approximately 2 × 105 tons of these synthetic dyes are discharged into water bodies (Srivastava and Sofi 2020).

Untreated dyeing effluents in water bodies utilized directly for agricultural purposes have a major impact on both human health and the environment. Industrial effluents contain synthetic colors that reduce light penetration, raise the biochemical oxygen demand, and affect aquatic flora's photosynthetic activity and, consequently, aquatic creatures' food sources. The discharge of artificial dyes into aquatic environments has resulted in mutagenic, carcinogenic, and genotoxic impacts in addition to aesthetic harm, which has caused environmentalists to become extremely concerned (Sharma et al. 2021).

Therefore, synthetic additives have been heavily criticized in recent decades, and customers are reluctant to use these products; as a result, they prefer to utilize natural colorants. Customers' interest in natural colorants has grown greatly as a result, while the number of approved artificial colors has decreased significantly (Rymbai et al. 2011). A study on dyes and organic pigments projects that the global demand for organic colorants would rise from 4.9% in 2003 to $10.6 billion in 2008. The global market for natural food coloring has reached $1 billion in investment and is still expanding due to consumer preference for natural food coloring over artificial color (Parmar and Phutela 2015). Growing food output is essential as the world's population is expected to reach 10.4 billion people by 2067 (Chojnacka et al. 2019; Van Dijk et al. 2021), particularly in Asia and Africa. Without sacrificing production, sustainable agronomic techniques such as effective crop protection, irrigation, and fertilization can lessen dependency on artificial fertilizers, fresh water, and pesticides (Young et al. 2021). By improving food security and lowering hunger while protecting natural resources and minimizing crop management costs, bio-based approaches not only promote ecological functions and economic sustainability but also help achieve the UN Sustainable Development Goals (SDGs) (Chojnacka et al. 2019).

4.2 Sources of Biocolors

Since natural dyes can be obtained from a wide range of natural sources, their chemical compositions vary, affecting their properties, stability, and solubility. Natural colors have been categorized according to their sources, which include microbes, minerals, and plants (Singh et al. 2023). The various sources of biocolors, along with chemical and biological, are given below:

4.2.1 Plant Source

From plants, pigment can be obtained from various parts of plants, such as pigments in fruits and vegetables (e.g., anthocyanins, betalains), leaf pigments (e.g., chlorophyll), root pigments (e.g., carotenoids from carrots, beetroot betalain), and flower pigments (e.g., flavonoids, carotenoids).

4.2.1.1 Betalains

Betalains are pigments that contain nitrogen and are soluble in water. They include the yellow betaxanthins and the red violet betacyanins. In plants from most Caryophyllales families, betalains are found in flowers, fruits, and sometimes in vegetative cells (Strack et al. 2003). About 80 plant betalains have been thoroughly studied, and they are all known to include betalamic acid [4-(2-oxoethylidene)-1,2,3,4-tetrahydropyridine-2,6-dicarboxylic acid]. Yellow to orange betaxanthins result from the condensation of betalamic acid with α-amino acids or amines, whereas reddish to violet betacyanins are derived from betanidin, which is an iminium adduct of cyclo-dioxyphenylalanine and betalamic acid (Sadowska-Bartosz and Bartosz 2021).

In food industries, the synthetic red dye is frequently replaced with betanin pigment in candy, yogurts, ice cream, salad dressings, icings, meat alternatives, drinks, and gelatin powders. Because betalains have three times the dyeing abilities, are more soluble in water, and are more stable between pH values of 3 and 7, they are good dyes for foods that are neutral or low in acidity (Carreón-Hidalgo et al. 2022). Numerous medical investigations have proven their beneficial effects on human health, with notable anti-inflammatory, anticarcinogenic, and antioxidant (because of their high electron-donating ability) qualities being identified. Furthermore, plant extracts that include betalain have already been patented in cancer treatment formulations (Sokolova et al. 2024).

4.2.1.2 Anthocyanins

Botanical species, especially blossoms, fruits, and root vegetables, have pigments known as anthocyanins that appear blue, red, or purple. Anthocyanin is highly water soluble and polar solvents. In acidic environments, they show red pigment and blue color in alkaline conditions. Despite having a positive charge at the oxygen atom of the C-ring of the fundamental flavonoid structure, anthocyanin is regarded as a flavonoid. The flavylium ion is the second name for it (Khoo et al. 2017). There are around 25 distinct anthocyanidins that vary from one another due to the presence of methoxy ($-OCH_3$) and hydroxyl (−OH) groups. Thus, anthocyanidins are classified as O-methylated anthocyanidins, 3-hydroxyanthocyanidins, and 3-deoxyanthocyanidins. The three nonmethylated anthocyanidins that are most prevalent in nature are cyanidin (Cy), delphinidin (Dp), and pelargonidin (Pg) (Mannino et al. 2021). Nearly every plant species contains an abundance of anthocyanins, which are found in all portions of fruit plants, including cherries, raspberries, blackberries, oranges, and grapes, as well as in vegetables like red cabbage and purple sweet potatoes. Compared to vegetables, fruits have a significantly higher anthocyanin concentration (Oancea and Oprean 2011). Not only can anthocyanins be added to food and drink to provide visually appealing natural coloring, but they can also be employed to make pharmaceutical and cosmetic items. Reduced risk of heart disease and some types of cancer have been linked to anthocyanins. People can also consume anthocyanins by eating fresh, colorful foods that have been processed into drinks and food (Ananga et al. 2013).

4.2.1.3 Carotenoids

The tetraterpene family includes carotenoids, which come in over 600 known natural structural variations (Tapiero et al. 2004). Based on their chemical makeup, these are separated into two classes: xanthophylls and carotenes. α-carotene, β-carotene, and lycopene are examples of carotenes, while lutein, zeaxanthin, and astaxanthin are examples of xanthophylls (Batchu et al. 2012). The primary sources of carotenoids are fruits and vegetables, which contain these pigments as microcomponents and give them their yellow, orange, and red tones. Lycopene is a lipid-soluble antioxidant that belongs to the carotenoid family of phytochemicals. It is an open, straight-chain hydrocarbon that is highly unsaturated and has two unconjugated and 11 conjugated double bonds (Rao and Rao 2007). It can also be obtained from microalgae; carotenoids found in microalgae include astaxanthin, β-carotene, lutein, lycopene, and canthaxanthin, which are the most valued and commercially available. Fucoxanthin primarily occurs in brown seaweeds such as *Laminaria japonica*, *Undaria pinnatifida*, *Hizikia usiformis*, *Sargassum* spp., and *Fucus* spp., based on research findings. *Haematococcus pluvialis*, *Chlorella zofingiensis*, and *Chlorococcum* sp. are the microalgae that are the best sources of astaxanthin. *Scenedesmus* species, *Chlorella* species, Rhodophyta species, and *Spirulina* species are the main sources of lutein and zeaxanthin (Kumar et al. 2023). Consuming carotenoids has been linked to a few health advantages, such as a lower risk of cataracts, coronary heart disease, and some types of cancer (Eggersdorfer and Wyss 2018). Furthermore, the macular pigment carotenoids (MP), which consist of lutein, zeaxanthin, and meso-zeaxanthin, are often recommended as dietary supplements to help prevent vision loss caused by age-related macular degeneration (AMD) and various other eye conditions (Bernstein et al. 2016).

4.2.1.4 Chlorophyll

Chlorophyll is extensively found in nature and is essential for the survival of both plant and animal kingdoms due to its crucial "light harvesting" function in photosynthesis. Chlorophyll's green hue and selective light absorbance of specific wavelengths make it a suitable coloring agent. Due to shifts in market demands and legislation, there is now a requirement for natural coloring agents to be used in food products instead of artificial colorings (Hosikian et al. 2010). Chlorophylls are primarily sourced from spinach and mustard leaves. Plants that have a greater chlorophyll concentration demonstrate significant free radical scavenging activity and anticancer effects. It may be water-soluble through the oleoresin saponification process, resulting in chlorophyllin. This pigment is used across various industries, including pharmaceuticals, cosmetics, and food production, for its antioxidant properties and as a coloring agent. Additionally, it can serve as a biomordant to enhance dyeing processes in the textile industry and to create dyes with antimicrobial characteristics (Nabi et al. 2023).

4.2.2 Algal Sources

4.2.2.1 Chlorella and Spirulina as Microalgae Sources

Spirulina and *Chlorella* are two of the most well-known genera of microalgae. Spirulina is a filamentous cyanobacterium that is multicellular and lives in freshwater, while *Chlorella* is unicellular. The genera *Chlorella* and Spirulina are recognized as some of the most prominent microalgae because of their significant production of long-chain polyunsaturated fatty acids, phenolic compounds, volatile compounds, sterols, proteins, amino acids, peptides, vitamins, polysaccharides, pigments, and food (Andrade et al. 2018).

The primary categories of pigments in microalgae-carotenoids, chlorophylls, and phycobiliproteins play a crucial role in their photosynthesis and cellular development. In contrast to other natural pigment sources such as fruits, plants, and animals, pigments obtained from microalgae exhibit remarkable physiological properties, including antioxidant and antibacterial effects, and offer a diverse array of health-related benefits (Sun et al. 2023).

4.2.2.2 Macroalgae and Seaweed Pigments

The oceans, which cover over 70% of the Earth's surface, preserve a vast array of species, including marine algae. Algae have served as a traditional food source for a long period, owing to their global availability, accessibility, diversity, and nutritional benefits. Additionally, they are valuable sources of structurally diverse bioactive compounds, such as polysaccharides, phenolic compounds, and natural pigments (Osório et al. 2020). Microalgae that produce pigments are known as seaweed. They are categorized into groups such as brown algae (Phaeophyceae, Heterokontophyta), red algae (Rhodophyta), and green algae (Chlorophyta). Seaweed pigments consist of chlorophyll and carotenoids, including carotenes (β-carotene) and xanthophylls (fucoxanthin, violaxanthin, antheraxanthin, zeaxanthin, lutein, neoxanthin). Red algae, brown algae, and green algae are examples of seaweeds that provide a wide variety of natural food colors, enhancing the sensory qualities and health benefits of food (Aryee et al. 2018).

4.2.2.3 High-Value Pigments (e.g., Phycocyanin, Astaxanthin, Fucoxanthin)

Microalgae provide natural colorants like phycocyanin, phycoerythrin, and astaxanthin, which are attracting increasing interest compared to synthetic alternatives. Phycocyanin and phycoerythrin are photosynthetic pigments found in microalgae, cyanobacteria, rhodophytes, and cryptomonads. They have blue and red colors, respectively. These pigments belong to the phycobiliprotein family. Phycoerythrin and phycocyanin are intensely colored, water-soluble molecules. These molecules are responsive to light and various factors, such as low pH levels, high ionic strengths, increased temperatures, and the presence of alcohol (Hsieh-Lo et al. 2019).

C-Phycocyanin (C-PC), a vibrant blue Bili protein sourced from blue-green algae, is hydrophilic in nature. It consists of a protein component and the chromophore phycocyanobilin. The protein component includes α and β subunits with molecular weights of approximately 18,000 and 20,000 Da, respectively. This colorant demonstrates high stability within the pH range of 5–8 and shows a strong red fluorescence when in its native form (Park et al. 2018). Phycocyanin is not only a safe natural colorant but also a bioactive compound that possesses antioxidant, anti-inflammatory, anticancer, antiviral, immunity boosting, and various other therapeutic features. It has numerous applications in biotechnology across food, cosmetic, pharmaceutical, and medical industries, functioning as a functional food, colorant, fluorescent marker, therapeutic agent, and photosensitizer in cancer photodynamic treatment. The US Food and Drug Administration (FDA) classifies phycocyanin as a naturally derived coloring agent for food and has recognized it as GRAS. Phycocyanin stands out as one of the few natural blue food coloring agents that has received approval in the USA, Europe, and Asia (Lauceri et al. 2023).

4.2.2.4 Astaxanthin

Astaxanthin, a potent lipophilic keto-carotenoid [3,3-dihydroxy-β, β′-carotene-4,4′-dione], is not soluble in water but can dissolve in nonpolar solvents like dichloromethane (~ 30 g/L), chloroform (~ 10 g/L), dimethyl sulfoxide (~ 0.5 g/L), and acetone (~ 0.2 g/L). The AST exhibits a close structural similarity to other carotenoids, including β-carotene, zeaxanthin, and lutein. It is an optically active compound, consisting of a long nonpolar conjugated polyene backbone with two polar β-ionone rings on either side. Every ring possesses one hydroxyl (–OH) and one keto (C=O) moiety. Astaxanthin's polar-nonpolar-polar linear structural arrangement allows it to bind to and span the cell membrane (Patil et al. 2022). Astaxanthin has generated considerable interest because of the traits associated with carotenoids, particularly its efficacy as a strong antioxidant in biological contexts; around 10 times more effective than β-carotene and lutein, and about 100 times higher than alpha-tocopherol. It guards against the oxidation of fatty acids and contributes to metabolic enhancement. In aquaculture, it is commonly used to give fish like salmon or trout a characteristic pinkish hue to their flesh and serves as a nutritional component that enhances survival rates during the juvenile stages (Martínez-Delgado et al. 2017).

4.2.2.5 Fucoxanthin

Fucoxanthin, the dominant carotenoid that accounts for nearly 10% of all carotenoids found on the planet, serves as a pigment for capturing light in various species of Heterokont and Haptophyte. Fucoxanthin is closely associated with chlorophyll, antenna protein, diadinoxanthin, and diatoxanthin to form a fucoxanthin-chlorophyll a/c-protein (FCP) complex. The pigment makeup of the FCP complex contrasts with that of the light-harvesting complex found in terrestrial plants and green algae, where accessory pigments include chlorophyll b and carotenoids that consist of lutein (Wang et al. 2021). Fucoxanthin was first isolated in Germany in 1914 from the brown seaweeds (Phaeophyceae) Dictyota, Fucus, and Laminaria. Two primary groups can be used to classify algae that generate fucoxanthin: macroalgae (multicellular organisms) and microalgae (unicellular organisms). Fucoxanthin is produced on a commercial scale from the waste material of brown macroalgae. The macroalgae had a fucoxanthin content that varied from 0.001% to 0.356% DW. It has been shown that the fucoxanthin content of microalgae is as much as 100 times greater than that of macroalgae (Khaw et al. 2022). Numerous scientific studies have emphasized the biological and therapeutic activities associated with this carotenoid, highlighting its antioxidant, anticancer, antihypertensive, anti-inflammatory, antidiabetic, antiobesity, and antiangiogenic properties, as well as its photoprotective effects (Lourenço-Lopes et al. 2021).

4.3 Cultivation of Plant and Algal Sources of Biocolors

Natural plant-derived pigments, which are well-known for their antioxidant qualities and capacity to substitute manufactured pigments, have recently seen a notable increase in demand worldwide, especially in the food, pharmaceutical, and cosmetics sectors. Numerous synthetic or petroleum-based pigments have been shown to have detrimental effects on human health, including the possibility of allergic reactions and hyperactivity, particularly in young children and other delicate people (Rafiq et al. 2022). The typical biologically derived colors are taken from various plant parts (fruits, vegetables, seeds, and roots) as well as microorganisms (Gupta et al. 2015). Plant-based colors can be found in annatto, turmeric, carmine, grape skin extract, caramel, paprika, and beets (Vijetha et al. 2017).

The annatto dye made from the seeds of *Bixa Orellana* is frequently linked to the plant. The invention of synthetic dyes has put an end to their previous use for dyeing textiles (Vilar et al. 2014; Koul et al. 2003). The primary ingredient, bixin, is currently in high demand as a nontoxic food coloring that dissolves fat (Lewinsohn et al. 2005; Shamina et al. 2008). The dairy industry uses annatto dye to color margarine, cheese, and butter. Additionally, ice cream, baked goods, and culinary oils include it. Additionally, the cosmetics sector uses annatto dye in lipsticks and suntan preparations. It is also used to dye leather as a component of floor wax, shoe and furniture polish, glass lacquer, brass lacquer, etc. (Hwang and Bowen 2002; Sathiya Mala et al. 2015).

A naturally occurring pigment with a deep red color, lycopene is especially prevalent in tomatoes and watermelons (Vilar et al. 2014; Lewinsohn et al. 2005). Tomatoes can contain anywhere between 15 and 56 mg/kg of lycopene. A potent antioxidant, lycopene may help prevent stomach, lung, and prostate cancers as well as cancers of the bladder, breast, cervix, pancreas, colon, rectum, and oral cavity (Shamina et al. 2008; Hwang and Bowen 2002). The food industry uses lycopene extensively as a coloring ingredient because of its safe and advantageous properties. It is typically found as oleoresin, which is a suspension in natural lipids. Because of the lipids and natural antioxidants in this form, lycopene is protected from oxidation (crystalline lycopene is extremely unstable) and bacterial destruction (Sathiya Mala et al. 2015; Hackett et al. 2004).

Microalgae are a repository for economically valuable elements. The manufacturing is based on taking advantage of their extraordinarily efficient photosynthetic mechanism. The bilin or tetrapyrol skeleton of red algae (Rhodophyta) and blue-green algae (Cyanophyta) produces a group of highly light-absorbing pigments. These phycobilin proteins are used as natural coloring agents, especially as substitutes for artificial dyes, in foods, cosmetics, and medications (Monishdeep et al. 2022).

4.3.1 Algal Cultivation in Photobioreactors and Open Ponds

There are two commonly used cultivation systems: closed PBRs and open ponds.

4.3.1.1 Open Ponds

Open pond cultivation offers a simple and cost-effective approach (Gupta et al. 2015). The idea of the open pond was derived from the use of artificial lagoons and oxidation ponds in wastewater treatment (Sharma et al. 2013). The most commonly used open systems include large shallow ponds, tanks, circular ponds, and raceway ponds (Gupta et al. 2015). A raceway pond is most often a rectangular canal with algal culture current flowing from a supply end to an exit end (Chisti 2007). The length-to-width ratio is an important parameter in designing a raceway pond. Larger width may result in weak current speed, which is not desirable for mixing and mass transfer. The length and depth are determined by the light penetration and the amount of culture volume a unit can hold. Each pond contains a paddle wheel to make the water flow continuously around the circuit (Gupta et al. 2015). One of the major advantages of open ponds is that they are easier to construct and operate than most closed systems. However, major constraints include poor light utilization by the cells, water loss due to evaporation, diffusion of CO_2 to the atmosphere, and large space requirements (Chisti 2007). Moreover, contamination due to microbes and other fast-growing heterotrophs may result in impure culture growth and limit the production of the algae. To overcome the inherent disadvantages of using open cultivation, numerous closed PBRs of various volumes and shapes have been designed. The principal final goal of any PBR is a reduction in biomass production costs. This can be done by improving the design and shape of the PBR, controlling environmental parameters, and favoring minimal contamination risk. Overcoming these limitations makes monocultures and the production of pharmaceutical and food goods possible (Gupta et al. 2015).

4.3.1.2 Photobioreactors (PBR)

Closed systems are designed to overcome the problems associated with open pond cultivation systems (Chisti 2007). A PBR is defined as a closed (or mostly closed) vessel for phototrophic production, where energy is supplied via electric lights (Andersen 2005). A PBR design should use light efficiently; illumination should be uniform, reduce mutual shading, and should provide a fast mass transfer of CO_2 and O_2 (Gupta et al. 2015). A typical PBR is comprised of a four-phase system: solid phase (microalgal cells), liquid phase (growth medium), gaseous phase (CO_2 and O_2), and superimposed light-radiation field (Posten 2009). Hence, to design an efficient PBR, an understanding of the complex interaction between biomass production and associated environmental parameters (e.g., fluid dynamics and light transfer) within the reactor is required.

PBRs are categorized as flat plate (Slegers et al. 2011), tubular (Molina et al. 2001), and column (Eriksen 2008). Based on their mode of liquid flow, PBRs can be grouped as stirred type, bubble column, and airlift reactor. Ideal PBRs should have a highly transparent surface, minimal nonilluminated part, high mass transfer rates, and should attain high biomass growth (Gupta et al. 2015). Moreover, PBR design should be suitable for the cultivation of various microalgal species universally and prevent fouling of the reactor. Various types of PBRs are discussed in the following section.

4.4 Extraction Methods

Interest in the use of unconventional extraction techniques has grown recently. These unconventional extraction techniques have quick extraction times, little solvent usage, improved recovery, and more selectivity, among other benefits (Jaime et al. 2009; Pasquet et al. 2011; Khoo et al. 2020; Chen et al. 2016; Molino et al. 2018; Kapoore et al. 2018; Fabrowska et al. 2018; Chronopoulou et al. 2019). Methods used for the extraction of natural dyes:

- Solvent-based extraction
- Enzymatic and ultrasonic-assisted extraction
- Subcritical fluid extraction (Rafiq et al. 2022)

4.4.1 Solvent-Based Extraction

Carotenoids are extracted from microalgae using conventional solvent extraction, which uses organic solvents. The polarity, solubility, and chemical stability of carotenoids determine whether organic or aqueous solvents are used. The solvent used to extract carotenoids is determined by the polarity of the target molecule. Nonpolar solvents such as n-hexane, dichloromethane, dimethyl ether, and diethyl ether, as well as polar solvents such as acetone, methanol, ethanol, and numerous biphasic organic combinations, can be utilized appropriately. (Chini Zittelli et al. 2023). Carotenoids can be extracted from microalgae using green solvents that are both safe and nontoxic, such as ethanol, limonene, and water-organic solvent combinations (Fig. 4.1). However, the technique is currently not economically viable due to high manufacturing costs and solvent requirements (Chini Zittelli et al. 2023).

4.4.2 Enzymatic and Ultrasonic-Assisted Extraction

Compared to traditional methods, Enzymatic-Assisted Extraction (EAE) is a quick and promising way to extract FPSs (Fucose-containing Polysaccharides) with the highest yield. The foundation of EAE is the selectivity of enzyme reactions that may

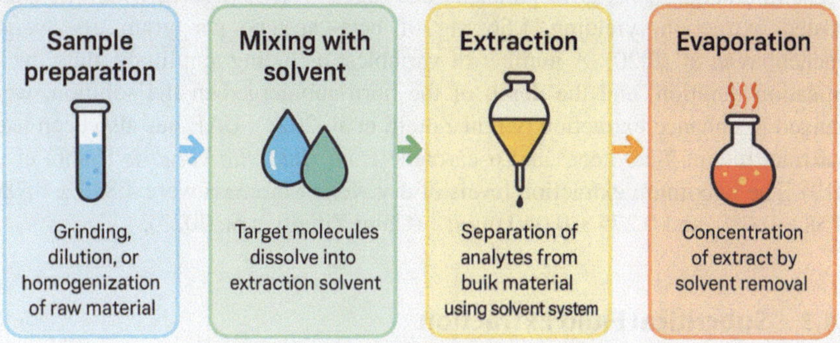

Fig. 4.1 Methods of solvent-based extraction

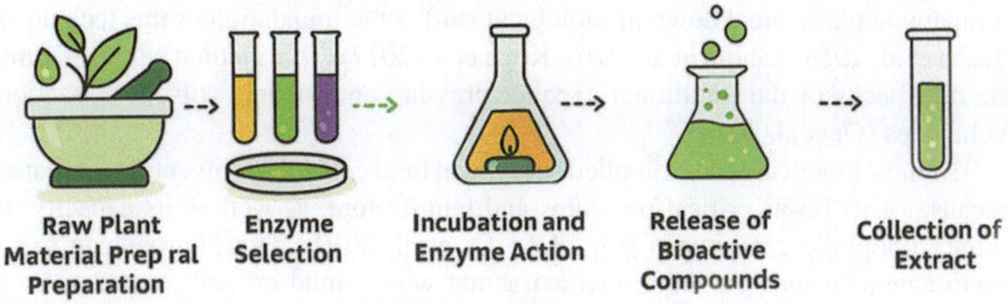

Fig. 4.2 Methods of enzymatic-based extraction

degrade particular cell wall membranes while using a desired amount of solvent (Rafiq et al. 2022). Since microalgal cells are not easy to disrupt, recovering physical or enzymatic pretreatment can aid carotenoids before the extraction. Protease, pectinase, pectin esterase, cellulase, hemicellulase, cellobiose, α-amylase, and fructosyl transferase are among the enzymes that are used in EAE to hydrolyze cell walls and liberate phytochemicals that are attached to lipid and carbohydrate chains within the cell (Puri et al. 2012). Following this pretreatment, volatile chemicals, hydrophilic and hydrophobic pigments, phenolic compounds, and other bioactive components are extracted from plant samples using solvent extraction or pressurized hot water extraction (Fig. 4.2) (Rafiq et al. 2022).

Ultrasound-assisted extraction is based on ultrasound cavitation (Chini Zittelli et al. 2023). High-strength, low-frequency ultrasound waves are used to perform ultrasonic extraction. When utilized as sources of natural compounds, ultrasound-aided extraction is an intriguing method for obtaining highly valuable compounds and may raise the value of some food by-products (Rafiq et al. 2022). Its short operating time and low solvent consumption allow for high yields; nevertheless, because PBPs (phycobiliproteins) are temperature-sensitive molecules, it is important to manage and prevent temperature increases in PBP samples to maintain their functionalities (Kannaujiya et al. 2017; Jaeschke et al. 2021).

The microalga *Arthrospira* platensis was successfully extracted with ultrasonic-assisted extraction, yielding 4.66 mg of beta-carotene per gram dry weight (Bachchhav et al. 2020). A number of variables, including amplitude, duty cycle, sonication duration, and the depth of the horn submerged in the solution, were changed to enhance extraction (Chini Zittelli et al. 2023). UAE has also been used to extract lutein, β-carotene, and α-carotene from *Chlorella vulgaris* (Şahin et al. 2019). The maximum extraction levels of dry weight biomass were 4.844 ± 0.780, 0.258 ± 0.020, and 0.275 ± 0.040 mg g^{-1} (Chini Zittelli et al. 2023).

4.4.3 Subcritical Fluid Extraction

The subcritical fluid extraction method is a revolutionary extraction technology that has a high extraction yield, a low cost, moderate conditions, and easy product separation (Zhao et al. 2019a, b). It also doesn't require high-pressure equipment. The comparable phase breakdown of biological stuff is the foundation of this technique (Duba et al. 2015; Zanqui et al. 2016; Koyu et al. 2017). This method can overcome the drawbacks of the traditional expeller pressing and organic solvent extraction techniques (Qi et al. 2019).

The most practical fluid in applied subcritical fluid extraction solvents is n-butane, because of its lower critical pressures and temperature, as well as its capacity to extract lipophilic compounds (Fig. 4.3) (Qi et al. 2019). Tetrafluoroethene is an environmentally benign subcritical extractant with a mild critical temperature of 101.1 °C and a critical pressure of 4.06 MPa. It has high application prospects in everyday chemical, medical, and health care goods, is nonflammable, and does not affect the ozone layer (Herrero et al. 2015; Rocha et al. 2013). The essential oil of N. alba var. was extracted under ideal conditions utilizing ultrasound-enhanced subcritical fluid extraction. The technique was done four times, each at 35 °C for 30 minutes. The key parameters were the frequency of 20 kHz, the ultrasonic power of 250 W/L, and the material-to-liquid ratio of 1:3. In these ideal circumstances, the essential oil extraction rate was 0.315%. The 11 constituents of the essential oil from N. alba var. had a content of more than 1%, according to the GC-MS result (Zhao et al. 2019a, b).

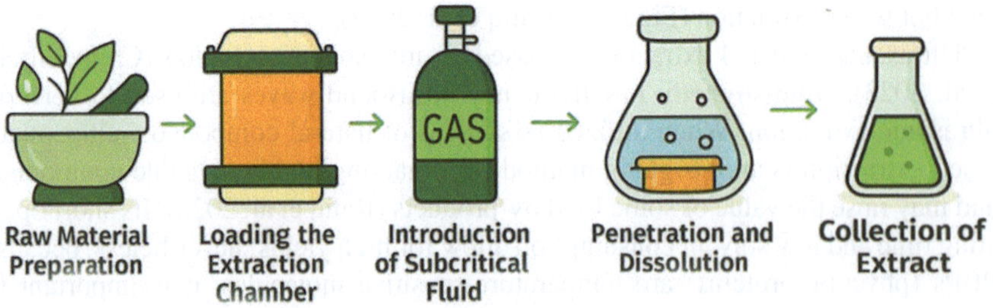

Fig. 4.3 Methods of subcritical fluid extraction

4.5 Stability and Sensitivity to Environmental Factors (Light, pH, Temperature)

Natural colorants are sourced selectively from natural matrices, including plants, algae, and mineral substances, and their stability depends on various parameters, as explained below in Table 4.1.

Table 4.1 Impact of various environmental factors on the stability of different biocoloring pigments

Compound	Effect of temperature	Effect of pH	Effect of light	References
Anthocyanins	Anthocyanin derived from flower petals demonstrates good thermal stability at 60 °C and 70 °C, whereas pigment obtained from fruits degrades at these temperatures.	At pH 1 (red color) At pH 2–4 (red & purple color) At pH 5–6 (colorless compound) At pH above 7 degradations of pigment	In light conditions, anthocyanins begin to break down, whereas their stability improves when stored in the dark.	Oancea (2021), Enaru et al. (2021), Li et al. (2023)
Betalain	Remains stable when kept at 4 °C and is even more stable when frozen (at −30 °C), but is heat-sensitive and deteriorates at temperatures exceeding 50 °C.	It is relatively stable within a pH range of 3–7, with the optimal range for maximum betanin stability being 5–6. Low acidity results in staining from red to purple, while alkaline conditions cause degradation.	The electrons of double bonds in betacyanin's molecules were excited by light, leading to increased degradation of betacyanin and a slower degradation rate of pigments indark (Kumar et al. 2016)	Sadowska-Bartosz and Bartosz (2021), Kumar et al. (2016)
Chlorophyll	Mostly stable at temperature −80 °C and 25 °C and degrade above 25 °C	Acidic Conditions (Low pH): rapid degradation occurs, resulting in a color change from vibrant green to olive-brown. At (pH 7–9) stays constant while maintaining its bright green color.	Chlorophyll remains more stable in subdued light or darkness. Prolonged exposure to intense light can cause chlorophyll degradation due to photooxidation.	Chen et al. (2024), Koca et al. (2007), Ferreira et al. (2023)

(continued)

Table 4.1 (continued)

Compound	Effect of temperature	Effect of pH	Effect of light	References
Carotenoids	The pigment remains stable at temperatures below or equal to 40 °C; temperatures exceeding 40 °C will compromise the pigment's stability	Under neutral and slightly basic conditions (pH levels of 8 and 7), the total carotenoid content decreased by 26%. Conversely, under acidic conditions (pH levels of 6, 5, 4, and 3), the total carotenoid content increased by 18%, 22%, and 27%.	Pigment is mostly stable in dark conditions and slightly degrades in indoor light, and mostly degrades in outdoor light.	Zheng et al. (2016), Bell et al. (2016), Zheng et al. (2016)
Phycocyanin	Maximum synthesis of phycocyanin is often supported by ideal growing conditions, which are 25–35 °C. Damage to phycocyanin is brought on by high temperatures (over 40 °C).	The breakdown of phycocyanin can occur at pH values below 6 or over 11. The best conditions for the formation of phycocyanin are slightly alkaline (pH 8–10).	For the formation of phycocyanin in many species, light intensity is ideal. Photoinhibition can be brought on by intense light. Different organisms have various optimal levels of light.	Fernandes et al. (2023), Klepacz-Smółka et al. (2020)
Astaxanthin	The astaxanthin may degrade at temperatures above 30 °C. Temperature range: 20–30 °C is the ideal condition. Degradation happens at higher temperatures because of heat stress.	Degradation can occur in environments that are excessively basic or too acidic. The ideal pH range is 6–8.	Too much light may degrade astaxanthin. Light exposure should be moderate, ideally red or blue. Degradation might result from overexposure.	Liu et al. (2024), Anarjan and Tan (2013), Liu et al. (2024)
Fucoxanthin	Higher temperatures often increase its production to an optimal level. Fucoxanthin production tends to decrease at very high or very low temperatures.	For certain algae, the ideal pH range is slightly alkaline (pH 7–8). Because pH impacts enzyme performance, fucoxanthin production tends to decline in extremely acidic or basic environments.	Light intensity and spectrum (e.g., blue light) can significantly promote fucoxanthin accumulation in algae. Too little light can decrease synthesis, while too much light can cause photoinhibition.	Pajot et al. (2022), Zhao et al. (2019a, b), Khaw et al. (2022)

4.6 Applications of Biocolors

Biocolorants are becoming more and more popular due to concerns about environmental issues, fashion, pharmaceuticals, nutrition, hygiene, and health (Chittaragi and Menon 2023). Biocolors have a variety of uses, which are explained below.

4.6.1 Food and Beverage Industry

The identification and advancement of new natural colorants have sparked significant interest, with food producers and consumers increasingly inclined to utilize natural pigments for coloring their food (Shahid and Mohammad 2013). Most natural biocolorants have antimicrobial properties that prevent food from being spoiled by bacteria, viruses, and fungi (Rymbai et al. 2011). The most often used natural pigments for food applications include ANC, carotenoids, phycobiliproteins, betanins, and chlorophylls, even though a large variety of natural pigments have been employed in the food sector (Luzardo-Ocampo et al. 2021). Carotenoids are naturally occurring pigments responsible for the yellow, orange, and red colors found in certain fruits and vegetables (Náthia-Neves and Meireles 2018). Chlorophylls are used in soups, dairy products, and fruit juices, and they are also used as a natural food coloring ingredient and have antimutagenic and antioxidant properties (Nabi et al. 2023). Beets were long considered to be the only source of betanin, and the only color additive on the FDA list that contains betanin is dried beets (Rodriguez-Amaya 2016).

4.6.2 Cosmetics and Pharmaceutical Applications

These days, plants and algae are commonly utilized for the production of natural cosmetics and pharmaceuticals (Bujak et al. 2022). In traditional medicine, saffron is used for its properties as an antispasmodic, digestive aid, gum sedative, antimucosal agent, nerve relaxant, gas expeller, perspiration inducer, mucus-clearing agent, energizer, stomach supporter, libido enhancer, and menstrual stimulant, and for use in the cosmetics sector, natural colors were produced from the rhizomes of Curcuma longa L. (Zingiberaceae), *Trigonella foenum graecum* (Fabaceae) seeds, and *Nerium oleander* (Apocyanaceae) flowers (Bujak et al. 2010; Selvam et al. 2015).

4.6.3 Textile and Dyeing Industry

Natural dyes are regarded as environmentally friendly because they are biodegradable and derived from renewable resources. Studies have indicated that a class of active biomolecules called phytochemicals, which vary depending on the plant being studied and their mode of action, may give textiles a number of useful qualities, including antibacterial, antifungal, UV protective, insect repellent, and aromatic qualities (Selvam et al. 2023). Catechins, rosmarinic acid, flavonoids, carotenoids, ascorbic acid, and anthocyanin groups are structural components of natural pigments that have natural antioxidant properties, and to improve their environmentally beneficial qualities, natural dyes are mostly used to color fabrics made of natural fibers (Ahsan et al. 2020).

4.6.4 Emerging Applications in Biotechnology

Numerous health advantages, including anticancer, antioxidant, antibacterial, immunosuppressive, anti-inflammatory, and antiproliferative properties, are provided by natural pigments. Betalain, chlorophyll, and carotenoids: these natural pigments have various approaches in the field of biotechnology (Ahsan et al. 2020).

4.7 Conclusion & Future Aspects

Natural colors from plants and algae are a viable and sustainable alternative to synthetic dyes, diminishing ecological harm and danger to human health. These colors are found in textiles, food, cosmetics, and pharmaceutical industries because of their interesting bioactivity. However, factors such as low yield, stability, and scalability need to be solved in order for improved commercial viability. Future research should look into utilizing biorefineries, genetic engineering, nanotechnology, and improved extraction processes to enhance pigment stability and manufacturing efficiency. If an advantageous extraction process and units of scale can be combined, then cost-effective, larger-scale production of biocolors may be seen due to biotechnological advances. In addition, increasing new sources for biocolors and optimizing their constituent properties will invite more industries to commercialize biocolors.

References

Affat SS (2021) Classifications, advantages, disadvantages, toxicity effects of natural and synthetic dyes: a review. Univ Thi-Qar J Sci 8(1):130–135

Ahsan R, Masood A, Sherwani R, Khushbakhat H (2020) Extraction and application of natural dyes on natural fibers: an eco-friendly perspective. Rev Educ Admin LAW 3(1):63–75

Ananga A, Georgiev V, Ochieng J, Phills B, Tsolova V (2013) Production of anthocyanins in grape cell cultures: a potential source of raw material for pharmaceutical, food, and cosmetic industries. In: The Mediterranean genetic code-grapevine and olive, vol 1. IntechOpen, pp 247–287

Anarjan N, Tan CP (2013) Effects of storage temperature, atmosphere and light on chemical stability of astaxanthin nanodispersions. J Am Oil Chem Soc 90(8):1223–1227

Andersen RA (ed) (2005) Algal culturing techniques. Academic press

Andrade LM, Andrade CJ, Dias M, Nascimento C, Mendes MA (2018) Chlorella and spirulina microalgae as sources of functional foods. Nutra Food Supplements 6(1):45–58

Ardila-Leal LD, Poutou-Piñales RA, Pedroza-Rodríguez AM, Quevedo-Hidalgo BE (2021) A brief history of colour, the environmental impact of synthetic dyes and removal by using laccases. Molecules 26(13):3813

Aryee AN, Agyei D, Akanbi TO (2018) Recovery and utilization of seaweed pigments in food processing. Curr Opin Food Sci 19:113–119

Bachchhav MB, Kulkarni MV, Ingale AG (2020) Process-intensified extraction of phycocyanin followed by β-carotene from Spirulina platensis using ultrasound-assisted extraction. Sep Sci Technol 55(5):932–944

Batchu SN, Chaudhary KR, Wiebe GJ, Seubert JM (2012) Bioactive compounds in heart disease. In: Bioactive food as dietary interventions for cardiovascular disease: bioactive foods in chronic disease states, vol 22. Elsevier, p 431

Bell T, Alamzad R, Graf BA (2016) Effect of pH on the chemical stability of carotenoids in juice. Proc Nutr Soc 75(OCE3):E94. https://doi.org/10.1017/S0029665116001099

Bernstein PS, Li B, Vachali PP, Gorusupudi A, Shyam R, Henriksen BS, Nolan JM (2016) Lutein, zeaxanthin, and meso-zeaxanthin: the basic and clinical science underlying carotenoid-based nutritional interventions against ocular disease. Prog Retin Eye Res 50:34–66

Bujak T, Zagórska-Dziok M, Ziemlewska A, Nizioł-Łukaszewska Z, Lal K, Wasilewski T, Hordyjewicz-Baran Z (2022) Flower extracts as multifunctional dyes in the cosmetics industry. Molecules 27(3):922

Carreón-Hidalgo JP, Franco-Vásquez DC, Gómez-Linton DR, Pérez-Flores LJ (2022) Betalain plant sources, biosynthesis, extraction, stability enhancement methods, bioactivity, and applications. Food Res Int 151:110821

Chen CY, Hsieh C, Lee DJ, Chang CH, Chang JS (2016) Production, extraction and stabilization of lutein from microalga Chlorella sorokiniana MB-1. Bioresour Technol 200:500–505

Chen C, Wang L-X, Li M-Y, Tan G-F, Liu Y-H, Liu P-Z, Li Y-P, Liu H, Zhuang J, Tao J-P, Xiong A-S (2024) Effect of temperature on photosynthetic pigment degradation during freeze–thaw process of postharvest of celery leaves. Horticulturae 10(3):267. https://doi.org/10.3390/horticulturae10030267

Chini Zittelli G, Lauceri R, Faraloni C, Silva Benavides AM, Torzillo G (2023) Valuable pigments from microalgae: phycobiliproteins, primary carotenoids, and fucoxanthin. Photochem Photobiol Sci 22(8):1733–1789

Chisti Y (2007) Biodiesel from microalgae. Biotechnol Adv 25(3):294–306

Chittaragi D, Menon JS (2023) Bio-Colours from spices. In: Herbs and spices-new advances. IntechOpen

Chronopoulou L, Dal Bosco C, Di Caprio F, Prosini L, Gentili A, Pagnanelli F, Palocci C (2019) Extraction of carotenoids and fat-soluble vitamins from Tetradesmus obliquus microalgae: an optimized approach by using supercritical CO2. Molecules 24(14):2581

Chojnacka K, Moustakas K, Witek-Krowiak A (2019) Bio-based fertilizers: a practical approach towards circular economy. Bioresour Technol 295:122223. https://doi.org/10.1016/j.biortech.2019.122223

Duba KS, Casazza AA, Mohamed HB, Perego P, Fiori L (2015) Extraction of polyphenols from grape skins and defatted grape seeds using subcritical water: experiments and modeling. Food Bioprod Process 94:29–38. https://doi.org/10.1016/j.fbp.2015.01.001

Eggersdorfer M, Wyss A (2018) Carotenoids in human nutrition and health. Arch Biochem Biophys 652:18–26

Enaru B, Drețcanu G, Pop TD, Stănilă A, Diaconeasa Z (2021) Anthocyanins: factors affecting their stability and degradation. Antioxidants 10(12):1967

Eriksen NT (2008) The technology of microalgal culturing. Biotechnol Lett 30:1525–1536

Fabrowska J, Messyasz B, Szyling J, Walkowiak J, Łęska B (2018) Isolation of chlorophylls and carotenoids from freshwater algae using different extraction methods. Phycol Res 66(1):52–57

Fernandes R, Campos J, Serra M, Fidalgo J, Almeida H, Casas A, Toubarro D, Barros AIRNA (2023) Exploring the benefits of Phycocyanin: from spirulina cultivation to its widespread applications. Pharmaceuticals (Basel, Switzerland) 16(4):592. https://doi.org/10.3390/ph16040592

Ferreira AS, Pereira L, Canfora F, Silva TH, Coimbra MA, Nunes C (2023) Stabilization of natural pigments in Ethanolic solutions for food applications: the case study of *Chlorella vulgaris*. Molecules 28(1):408. https://doi.org/10.3390/molecules28010408

Gupta PL, Lee SM, Choi HJ (2015) A mini review: photobioreactors for large-scale algal cultivation. World J Microbiol Biotechnol 31:1409–1417

Hackett MM, Lee JH, Francis D, Schwartz SJ (2004) Thermal stability and isomerization of lycopene in tomato oleoresins from different varieties. J Food Sci 69(7):536–541

Herrero M, del Pilar Sánchez-Camargo A, Cifuentes A, Ibáñez E (2015) Plants, seaweeds, microalgae and food by-products as natural sources of functional ingredients obtained using pressurized liquid extraction and supercritical fluid extraction. TrAC Trends Anal Chem 71:26–38

Hosikian A, Lim S, Halim R, Danquah MK (2010) Chlorophyll extraction from microalgae: a review on the process engineering aspects. Int J Chem Eng 1:391632

Hsieh-Lo M, Castillo G, Ochoa-Becerra MA, Mojica L (2019) Phycocyanin and phycoerythrin: strategies to improve production yield and chemical stability. Algal Res 42:101600

Hwang ES, Bowen PE (2002) Can the consumption of tomatoes or lycopene reduce cancer risk? Integr Cancer Ther 1(2):121–132

Jaeschke DP, Teixeira IR, Marczak LDF, Mercali GD (2021) Phycocyanin from spirulina: a review of extraction methods and stability. Food Res Int 143:110314

Jaime L, Rodríguez-Meizoso I, Cifuentes A, Santoyo S, Suarez S, Ibáñez E, Señorans FJ (2009) Pressurized liquids as an alternative process to antioxidant carotenoids' extraction from Haematococcus pluvialis microalgae. LWT Food Sci Technol 43(1):105–112. https://doi.org/10.1016/j.lwt.2009.06.023

Kannaujiya VK, Sundaram S, Sinha RP (2017) Phycobiliproteins: recent developments and future applications. Springer Singapore, Singapore, pp 83–98

Kapoore RV, Butler TO, Pandhal J, Vaidyanathan S (2018) Microwave-assisted extraction for microalgae: from biofuels to biorefinery. Biology 7(1):18

Khaw YS, Yusoff FM, Tan HT, Noor Mazli NAI, Nazarudin MF, Shaharuddin NA et al (2022) Fucoxanthin production of microalgae under different culture factors: a systematic review. Mar Drugs 20(10):592. https://doi.org/10.3390/md20100592

Khoo HE, Azlan A, Tang ST, Lim SM (2017) Anthocyanidins and anthocyanins: colored pigments as food, pharmaceutical ingredients, and the potential health benefits. Food Nutr Res 61(1). https://doi.org/10.1080/16546628.2017.1361779

Khoo KS, Chew KW, Yew GY, Manickam S, Ooi CW, Show PL (2020) Integrated ultrasound-assisted liquid biphasic flotation for efficient extraction of astaxanthin from Haematococcus pluvialis. Ultrason Sonochem 67:105052

Klepacz-Smółka A, Pietrzyk D, Szeląg R, Głuszcz P, Daroch M, Tang J, Ledakowicz S (2020) Effect of light colour and photoperiod on biomass growth and phycocyanin production by Synechococcus PCC 6715. Bioresour Technol 313:123700

Koca N, Karadeniz F, Burdurlu HS (2007) Effect of pH on chlorophyll degradation and colour loss in blanched green peas. Food Chem 100(2):609–615

Koul VK, Koul S, Tikoo CL (2003) Process optimization for extraction and purification of bixin from annatto. Indian J Chem Technol 10(5):545–547

Koyu H, Kazan A, Ozturk TK, Yesil-Celiktas O, Haznedaroglu MZ (2017) Optimizing subcritical water extraction of Morus nigra L. fruits for maximization of tyrosinase inhibitory activity. J Supercrit Fluids 127:15–22

Kumar SS, Manoj P, Nimisha G, Giridhar P (2016) Phytoconstituents and stability of betalains in fruit extracts of Malabar spinach (*Basella rubra* L.). J Food Sci Technol 53(11):4014–4022. https://doi.org/10.1007/s13197-016-2404-8

Lauceri R, Cavone C, Chini Zittelli G, Kamburska L, Musazzi S, Torzillo G (2023) High purity grade phycocyanin recovery by decupling cell lysis from the pigment extraction: an innovative approach. Food Bioprocess Technol 16(1):111–121

Lewinsohn E, Sitrit Y, Bar E, Azulay Y, Meir A, Zamir D, Tadmor Y (2005) Carotenoid pigmentation affects the volatile composition of tomato and watermelon fruits, as revealed by comparative genetic analyses. J Agric Food Chem 53(8):3142–3148

Li Q, Zhang F, Wang Z, Feng Y, Han Y (2023) Advances in the preparation, stability, metabolism, and physiological roles of Anthocyanins: a review. Foods 12(21):3969. https://doi.org/10.3390/foods12213969

Liu X, Li W, Yue Z, Qian J, Zhu W, Dai H, Wang J, Pi F (2024) Evaluation of astaxanthin stability under varying temperatures and ultraviolet irradiation durations based on Raman spectroscopy. Food Chem: X 24:101947. https://doi.org/10.1016/j.fochx.2024.101947

Lourenço-Lopes C, Fraga-Corral M, Jimenez-Lopez C, Carpena M, Pereira AG, García-Oliveira P et al (2021) Biological action mechanisms of fucoxanthin extracted from algae for application in food and cosmetic industries. Trends Food Sci Technol 117:163–181

Luzardo-Ocampo I, Ramírez-Jiménez AK, Yañez J, Mojica L, Luna-Vital DA (2021) Technological applications of natural colorants in food systems: a review. Foods 10(3):634

Mannino G, Gentile C, Ertani A, Serio G, Bertea CM (2021) Anthocyanins: biosynthesis, distribution, ecological role, and use of bio stimulants to increase their content in plant foods—a review. Agriculture 11(3):212

Martínez-Delgado AA, Khandual S, Villanueva–Rodríguez SJ (2017) Chemical stability of astaxanthin integrated into a food matrix: effects of food processing and methods for preservation. Food Chem 225:23–30

Melo MJ (2009) History of natural dyes in the ancient mediterranean world. In: Handbook of natural colorants, vol 8. Wiley, p 1

Molina E, Fernández J, Acién FG, Chisti Y (2001) Tubular photobioreactor design for algal cultures. J Biotechnol 92(2):113–131

Molino A, Mehariya S, Iovine A, Larocca V, Di Sanzo G, Martino M et al (2018) Extraction of astaxanthin and lutein from microalga Haematococcus pluvialis in the red phase using CO2 supercritical fluid extraction technology with ethanol as CO-solvent. Mar Drugs 16(11):432

Monishdeep V, Deshpande H, Poshadri A, Kumar B (2022) A review on potential microbial sourced bio-colours from fungi. Pharm Innov J 11:3200–3205

Nabi BG, Mukhtar K, Ahmed W, Manzoor MF, Ranjha MMAN, Kieliszek M et al (2023) Natural pigments: Anthocyanins, carotenoids, chlorophylls, and betalains as colorants in food products. Food Biosci 52:102403

Náthia-Neves G, Meireles MAA (2018) Genipap: a new perspective on natural colorants for the food industry. Food Public Health 8(1):21–33

Oancea S (2021) A review of the current knowledge of thermal stability of anthocyanins and approaches to their stabilization to heat. Antioxidants 10(9):1337

Oancea S, Oprean L (2011) Anthocyanins, from biosynthesis in plants to human health benefits. Acta Universitatis Cinbinesis, Series E Food Technol 15(1)

Osório C, Machado S, Peixoto J, Bessada S, Pimentel FB, Alves CR, Oliveira MBP (2020) Pigments content (chlorophylls, fucoxanthin and phycobiliproteins) of different commercial dried algae. Separations 7(2):33

Pajot A, Hao Huynh G, Picot L, Marchal L, Nicolau E (2022) Fucoxanthin from algae to human, an extraordinary bioresource: insights and advances in up and downstream processes. Mar Drugs 20(4):222. https://doi.org/10.3390/md20040222

Park WS, Kim HJ, Li M, Lim DH, Kim J, Kwak SS et al (2018) Two classes of pigments, carotenoids and c-phycocyanin, in spirulina powder and their antioxidant activities. Molecules 23(8):2065

Parmar M, Phutela UG (2015) Biocolors: the new generation additives. Int J Curr Microbiol Appl Sci 4(7):688–694

Pasquet V, Chérouvrier JR, Farhat F, Thiéry V, Piot JM, Bérard JB et al (2011) Study on the microalgal pigments extraction process: performance of microwave-assisted extraction. Process Biochem 46(1):59–67

Patil AD, Kasabe PJ, Dandge PB (2022) Pharmaceutical and nutraceutical potential of natural bioactive pigment: astaxanthin. Nat Prod Bioprospect 12(1):25

Posten C (2009) Design principles of photo-bioreactors for cultivation of microalgae. Eng Life Sci 9(3):165–177

Puri M, Sharma D, Barrow CJ (2012) Enzyme-assisted extraction of bioactives from plants. Trends Biotechnol 30(1):37–44
Qi Z, Xiao J, Ye L, Chuyun W, Chang Z, Shugang L, Fenghong H (2019) The effect of the subcritical fluid extraction on the quality of almond oils: compared to conventional mechanical pressing method. Food Sci Nutr 7(7):2231–2241
Rafiq N, Gupta N, Bhat A, Singh J, Bandral J, Sood M et al (2022) Extraction techniques and utilization of natural pigments. Chem Sci Rev Lett 11:239–245
Rao AV, Rao LG (2007) Carotenoids and human health. Pharmacol Res 55(3):207–216
Rocha DL, Batista AD, Rocha FR, Donati GL, Nobrega JA (2013) Greening sample preparation in inorganic analysis. TrAC Trends Anal Chem 45:79–92
Rodriguez-Amaya DB (2016) Natural food pigments and colorants. Curr Opin Food Sci 7:20–26
Rymbai H, Sharma RR, Srivastav M (2011) Bio-colorants and its implications in health and food industry–a review. Int J Pharmacol Res 3(4):2228–2244
Sadowska-Bartosz I, Bartosz G (2021) Biological properties and applications of betalains. Molecules 26(9):2520
Şahin S, Nasir NTBM, Erken İ, Çakmak ZE, Çakmak T (2019) Antioxidant composite films with chitosan and carotenoid extract from Chlorella vulgaris: optimization of ultrasonic-assisted extraction of carotenoids and surface characterization of chitosan films. Mater Res Express 6(9):095404
Sathiya Mala K, Prabhakara Rao P, Prabhavathy MB, Satyanarayana A (2015) Studies on application of annatto (Bixa orellena L.) dye formulations in dairy products. J Food Sci Technol 52(2):912–919
Selvam RM, Athinarayanan G, Nanthini AUR, Singh AR, Kalirajan K, Selvakumar PM (2015) Extraction of natural dyes from Curcuma longa, Trigonella foenum graecum and Nerium oleander, plants and their application in antimicrobial fabric. Ind Crop Prod 70:84–90
Shahid M, Mohammad F (2013) Recent advancements in natural dye applications: a review. J Clean Prod 53:310–331
Shamina A, Shiva KN, Parthasarathy VA (2008) Food colours of plant origin. CABI Rev:25
Sharma NK, Rai AK, Stal LJ (2013) Cyanobacteria: an economic perspective. Wiley
Sharma J, Sharma S, Soni V (2021) Classification and impact of synthetic textile dyes on aquatic Flora: a review. Reg Stud Mar Sci 45:101802
Singh T, Pandey VK, Dash KK, Zanwar S, Singh R (2023) Natural bio-colorant and pigments: sources and applications in food processing. J Agric Food Res 12:100628
Slegers PM, Wijffels RH, van Straten G, Van Boxtel AJB (2011) Design scenarios for flat panel photobioreactors. Appl Energy 88(10):3342–3353
Sokolova DV, Shvachko NA, Mikhailova AS, Popov VS, Solovyeva AE, Khlestkina EK (2024) Characterization of Betalain content and antioxidant activity variation dynamics in table beets (*Beta vulgaris* L.) with differently colored roots. Agronomy 14(5):999. https://doi.org/10.3390/agronomy14050999
Srivastava R, Sofi IR (2020) Impact of synthetic dyes on human health and environment. In: Impact of textile dyes on public health and the environment. IGI Global, pp 146–161
Stevens LJ, Burgess JR, Stochelski MA, Kuczek T (2014) Amounts of artificial food colors in commonly consumed beverages and potential behavioral implications for consumption in children. Clin Pediatr 53:133–140
Strack D, Vogt T, Schliemann W (2003) Recent advances in betalain research. Phytochemistry 62(3):247–269
Sun H, Wang Y, He Y, Liu B, Mou H, Chen F, Yang S (2023) Microalgae-derived pigments for the food industry. Mar Drugs 21(2):82
Tapiero H, Townsend DM, Tew KD (2004) The role of carotenoids in the prevention of human pathologies. Biomed Pharmacother 58(2):100–110
Tikhomirova TI, Ramazanova GR, Apyari VV (2018) Effect of nature and structure of synthetic anionic food dyes on their sorption onto different sorbents: peculiarities and prospects. Microchem J 143:305–311

Tkaczyk A, Mitrowska K, Posyniak A (2020) Synthetic organic dyes as contaminants of the aquatic environment and their implications for ecosystems: a review. Sci Total Environ 717:137222

Van Dijk M, Morley T, Rau ML, Saghai Y (2021) A meta-analysis of projected global food demand and population at risk of hunger for the period 2010–2050. Nat Food 2(7):494–501. https://doi.org/10.1038/s43016-021-00322-9

Vijetha P, Nadh RV, Naidu KR, Sundari TS (2017) A study on extraction of plant based bio colours. J Rural Dev Soc Trends Sci Technol 1(1)

Vilar DDA, Vilar MSDA, Moura TFADLE, Raffin FN, Oliveira MRD, Franco CFDO et al (2014) Traditional uses, chemical constituents, and biological activities of Bixa orellana L.: a review. Sci World J 1:857292

Wang S, Wu S, Yang G, Pan K, Wang L, Hu Z (2021) A review on the progress, challenges and prospects in commercializing microalgal fucoxanthin. Biotechnol Adv 53:107865

Young MD, Ros GH, De Vries W (2021) Impacts of agronomic measures on crop, soil, and environmental indicators: a review and synthesis of meta-analysis. Agric Ecosyst Environ 319:107551. https://doi.org/10.1016/j.agee.2021.107551

Yusuf M, Shabbir M, Mohammad F (2017) Natural colorants: historical, processing and sustainable prospects. Nat Prod Bioprospect 7:123–145

Zanqui AB, da Silva CM, de Morais DR, Santos JM, Ribeiro SAO, Eberlin MN et al (2016) Sacha inchi (Plukenetia volubilis L.) oil composition varies with changes in temperature and pressure in subcritical extraction with n-propane. Ind Crop Prod 87:64–70

Zhao Y, Fan YY, Yu WG, Wang J, Lu W, Song XQ (2019a) Ultrasound-enhanced subcritical fluid extraction of essential oil from Nymphaea alba var and its antioxidant activity. J AOAC Int 102(5):1448–1454

Zhao D, Yu D, Kim M, Gu MY, Kim SM, Pan CH, Kim GH, Chung D (2019b) Effects of temperature, light, and pH on the stability of fucoxanthin in an oil-in-water emulsion. Food Chem 291:87–93. https://doi.org/10.1016/j.foodchem

Zheng L, Xin J, Yao T, Wang Z, Gao S (2016) Study on the stability of carotenoid pigments. In: International conference on electronics, mechanics, culture and medicine. Atlantis Press, pp 548–552

Microbial and Fungal Biocolors

5

Ravi Kant Bhatia, Abhishek Walia, and Laurent Dufossé

Abstract

Pigment-producing microorganisms are widespread in nature, but transforming their potential from lab-scale studies to commercial applications has been a complex journey. Twenty-five years ago, researchers debated whether microbial pigments would remain scientific curiosities or evolve into viable industrial products. Today, the picture is mixed; some microbial processes, using fungi, bacteria, or yeasts, have achieved industrial-scale production of pigments like carotenoids and phycocyanin. However, commercialization involves several challenges. For instance, Monascus-derived red pigments are widely consumed in Asia, with over a billion consumers, yet remain banned in several other countries due to safety concerns. Broader acceptance, especially in Europe and North America, hinges on the development and global availability of strains proven to be toxin-free, such as those genetically engineered to delete citrinin-producing genes. Bringing new microbial pigments to market is resource-intensive. Each new pigment or microbial source demands extensive experimentation, process refinement, safety testing, and regulatory clearance. The future of pigments such as azaphilones or anthraquinones depends on the outcomes of ongoing investments.

R. K. Bhatia
Department of Biotechnology, Himachal Pradesh University, Summer Hill, Shimla, Himachal Pradesh, India

A. Walia
Department of Microbiology, College of Basic Sciences, CSKHPKV, Palampur, Himachal Pradesh, India

L. Dufossé (✉)
Laboratoire de Chimie et Biotechnologie des Produits Naturels (CHEMBIOPRO), Université de La Réunion, ESIROI Agroalimentaire, Saint-Denis Cedex, France
e-mail: laurent.dufosse@univ-reunion.fr

R. K. Bhatia, A. Walia (eds.), *Biocolours*,
https://doi.org/10.1007/978-981-95-1738-1_5

Emerging strategies include combinatorial biosynthesis, targeted gene deletions, and the microbial production of rare pigments, and these advances may shape the next generation of natural colorants.

Keywords

Carotenoid · Azaphilone · Anthraquinone · Phycoerythrin · Zeaxanthin · Isorenieratene · Bacteria · Fungi · Yeast

5.1 Introduction

Microbial fermentation is increasingly pivotal in producing a diverse array of ingredients used across industries such as food, animal feed, cosmetics, and pharmaceuticals. This technology leverages microorganisms to synthesize valuable compounds, offering sustainable and versatile alternatives to traditional chemical or plant-based production methods. Key products include polysaccharide-based thickeners and gelling agents like xanthan gum, curdlan, and gellan gum, which are widely used to improve texture and stability in food products. Flavor enhancers, such as yeast hydrolysate and monosodium glutamate (MSG), enhance sensory profiles, while polyunsaturated fatty acids (PUFAs) contribute to nutritional value in functional foods and supplements. Additionally, volatile flavor compounds like γ-decalactone, diacetyl, and methyl ketones provide distinctive aromas, and essential nutrients, including vitamins, amino acids, and acidulants like lactic and citric acid, serve critical roles in food fortification and preservation (Mannazzu et al. 2015).

A significant challenge in microbial fermentation is the production cost of microbial pigments, which often struggle to compete economically with synthetic dyes or plant-derived colorants (Dufossé et al. 2005; Kim et al. 2025). Despite this, ongoing research and process optimization efforts are focused on improving the cost-effectiveness of these bio-based pigments. The successful commercialization of natural pigments, such as β-carotene, lutein, and astaxanthin, highlights the potential for market growth. These pigments, sourced from nonconventional platforms like algae or conventional sources like flowering plants, have found applications as food colorants and dietary supplements. Their adoption is driven by a growing consumer preference for "natural and health-promoting ingredients," with many willing to pay a premium for products perceived as safer and more sustainable (Jaswir et al. 2011; Zhou et al. 2015). This trend has spurred the development of niche markets catering to health-conscious and environmentally aware consumers.

Among nonconventional microbial sources, filamentous fungi are particularly promising due to their metabolic versatility and ability to produce a wide range of pigments. These include major chemical classes such as carotenoids (e.g., β-carotene), melanins, azaphilones, flavins, phenazines, and quinones, as well as specialized bioactive compounds like monascins, violacein, and indigo (Dufossé 2006). These pigments not only serve as natural colorants but also offer functional benefits, such as antioxidant or antimicrobial properties, making them attractive for applications beyond food, including cosmetics and pharmaceuticals. The diversity of fungal metabolites, combined with advances in genetic engineering and fermentation technology, positions filamentous fungi as a cornerstone of innovation in the bioeconomy.

The rise of microbial fermentation-based ingredients reflects broader trends toward sustainability and clean-label products. As consumer awareness grows and regulatory frameworks increasingly favor natural over synthetic additives, the demand for these ingredients is expected to expand. Continued investment in research to optimize production processes and scale up microbial systems will be crucial to overcoming current cost barriers and unlocking the full potential of these versatile compounds (Table 5.1).

The commercial success of fermentation-derived pigments depends on several critical factors: consumer acceptance, regulatory approval, and the significant capital investment required to bring these products to market (Furubayashi et al. 2015; Heider et al. 2014a, b). Approximately 25 years ago, prominent figures in the food industry expressed doubts about the commercial feasibility of pigments produced through algal and microbial fermentation. Their skepticism stemmed from the high costs associated with establishing necessary infrastructure, such as open pond systems, photobioreactors, and large-scale fermentation facilities. Furthermore, the lengthy and stringent toxicity evaluations required by regulatory authorities added to the challenges, increasing both time and financial demands. Compounding these issues was the unfavorable public perception of fungal-derived products for food use, which presented a significant obstacle to market adoption. These concerns highlighted the complex interplay of economic, regulatory, and social hurdles that needed to be addressed to achieve the successful commercialization of such pigments.

Currently, numerous food-grade pigments derived from microbial and microalgal fermentation have successfully penetrated global markets, demonstrating the growing viability of these bioprocesses. Notable examples include Monascus pigments produced by fungi, Arpink Red™ (rebranded as Natural Red™) from *Penicillium oxalicum*, and phycocyanin extracted from the cyanobacterium *Arthrospira* (*Spirulina*) *platensis*. Additional commercially significant pigments include riboflavin from the mold *Ashbya gossypii*, lycopene and carotene from the tropical mold *Blakeslea trispora* (Chen et al. 2016), β-carotene from the microalga *Dunaliella salina*, and astaxanthin sourced from both the bacterium *Paracoccus carotinifaciens* and the microalga *Haematococcus pluvialis* (Grassi et al. 2016). For example, *Blakeslea trispora* cultures have achieved impressive β-carotene yields of up to 17 g/L (Dufossé et al. 2014; Ramamurthy et al. 2024), highlighting the efficiency of optimized microbial systems.

The food industry and academic researchers are increasingly captivated by the potential of microbial and microalgal pigments, driving exploration through diverse approaches ranging from straightforward screening techniques (Lebeau et al. 2017; Fouillaud et al. 2017) to advanced molecular biology methods. The pivotal role of microorganisms as microbial cell factories in the industrial production of food-grade pigments is expanding rapidly. Their value is rooted in several key

Table 5.1 Production of pigments, colorants, and dyes by microbes and microalgae

Name of the coloring molecule	Color	Microorganism (bacteria/fungus/yeast/algae)
Astaxanthin	Pink-red	*Agrobacterium aurantiacum*
Phycocyanin	Blue	*Arthrospira* sp.
Riboflavin	Yellow	*Ashbya gossypii*, *Candida famata*, *Bacillus subtilis*
β-carotene	Yellow-orange	*Blakeslea trispora*
Lycopene	Red	*Blakeslea trispora*
Canthaxanthin	Dark red	*Bradyrhizobium* sp.
Isorenieratene and OH derivatives(aryl carotenoids)	Orange	*Brevibacterium linens*
Lutein	Yellow	*Chlorella* and many other microalgae
Violacein	Deep violet	*Chromobacterium violaceum*, *Janthinobacterium lividum*
Naphtoquinone	Deep blood-red	*Cordyceps unilateralis*
Melanin	Black	*Cryptococcus neoformans* var. *nigricans*
β-carotene	Yellow-orange	*Dunaliella salina*
Indigoidine	Blue	*Erwinia chrysanthemi*, *Streptomyces lavendulae*
Zeaxanthin	Yellow	*Flavobacterium* sp.
β-carotene	Yellow-orange	*Fusarium sporotrichioides*
Lycopene	Red	*Fusarium sporotrichioides*
Indigo	Blue	Genetically modified *Escherichia coli*
Indigoid tyrian purple (6,6'dibromoindigo)	Purple	Genetically modified *Escherichia coli*
Astaxanthin	Pink-red	*Haematococcus pluvialis*
Azulenes	Blue	*Lactarius* sp.
Chlorophylls	Green	Microalgae
Ankaflavin (azaphilone)	Yellow	*Monascus* sp.
Monascorubramin (azaphilone)	Red	*Monascus* sp.
Rubropunctatin (azaphilone)	Orange	*Monascus* sp.
β-carotene	Yellow-orange	*Mucor circinelloides*
β-carotene	Yellow-orange	*Neurospora crassa*
Astaxanthin	Pink-red	*Paracoccus carotinifaciens*
Zeaxanthin	Yellow	*Paracoccus zeaxanthinifaciens*
Anthraquinones	Red and other hues	*Penicillium oxalicum*
β-carotene	Yellow-orange	*Phycomyces blakesleeanus*
Phycoerythrin	Red	*Porphyridium cruentum*
Phenazines	Broad range of hues	*Pseudomonas* and *Streptomyces* sp.
Torularhodin	Orange-red	*Rhodotorula* sp.

(continued)

Table 5.1 (continued)

Name of the coloring molecule	Color	Microorganism (bacteria/fungus/yeast/algae)
Prodigiosins	Red	*Serratia marcescens*
Rubrolone	Red	*Streptomyces echinoruber*
Azaphilones	Red	*Talaromyces albobiverticillius*
Azaphilones	Red	*Talaromyces atroroseus*
Astaxanthin	Pink-red	*Xanthophyllomyces dendrorhous*

advantages: a broad spectrum of pigment colors and chemical compositions (as illustrated in Fig. 5.1), adaptability to scalable cultivation systems, and the use of well-characterized, optimized production strains with a proven track record in industrial applications (as shown in Fig. 5.2). These attributes make microbial and microalgal systems highly attractive for sustainable and efficient pigment production, meeting the rising demand for natural colorants in food and other industries. The combination of biodiversity, scalability, and established industrial processes underscores the transformative potential of these bioprocesses in modern food science and technology.

5.2 Carotenoids

5.2.1 Zeaxanthin

Zeaxanthin, chemically known as 3,3′-dihydroxy-β-carotene, is a vibrant yellow xanthophyll pigment widely utilized in poultry feed to intensify the yellow hue of broiler skin and egg yolks. Beyond its application as a feed and food colorant, zeaxanthin offers significant health benefits, notably in preventing age-related macular degeneration (AMD), which has led to its incorporation into nutraceuticals and dietary supplements (Chen et al. 2021). In the mid-1960s, researchers at Hoffmann-La Roche identified marine bacteria capable of synthesizing zeaxanthin. One prominent strain, originally classified as *Flavobacterium* sp. (ATCC 21588) under the taxonomic standards of the time, was found to produce up to 190 mg of zeaxanthin per liter of culture when cultivated in a nutrient medium containing glucose or sucrose, sulfur-containing amino acids (such as methionine, cystine, or cysteine), pyridoxine, and divalent metal ions like Fe^{2+}, Co^{2+}, Mo^{2+}, or Mn^{2+}. This strain yielded a specific pigment concentration of 16 mg/g of dried cell mass (US Patent 5,827,652). Another strain, *Flavobacterium multivorum* (ATCC 55238), has been extensively studied for zeaxanthin production (Thawornwiriyanun et al. 2012). Taxonomic reclassification later reassigned these zeaxanthin-producing bacteria to the genus *Paracoccus*, specifically *Paracoccus zeaxanthinifaciens*. This species shows significant industrial potential, comparable to *Paracoccus carotinifaciens*, which is used to produce Panaferd-AX™, a commercial astaxanthin-based pigment for animal feed developed by JX Nippon Oil and Energy (US Patent 8,853,460; US Patent Application 20130012594).

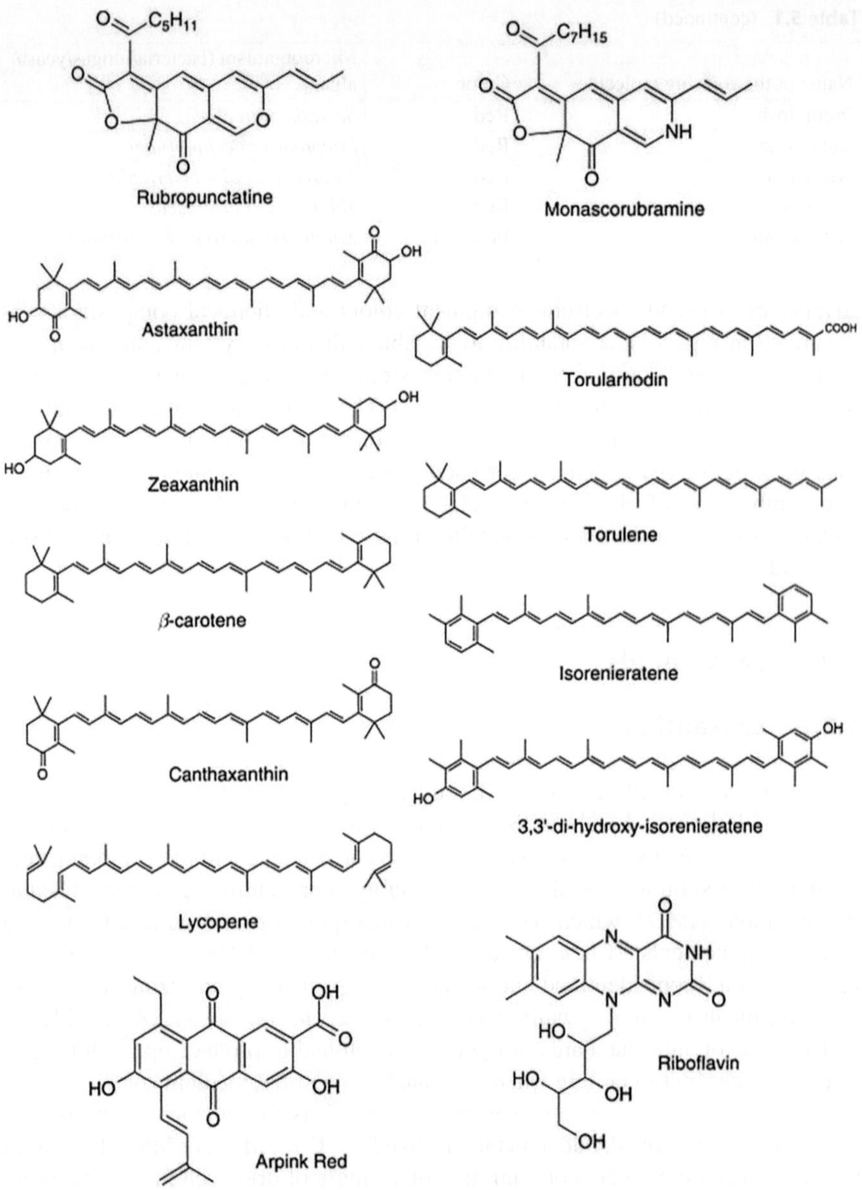

Fig. 5.1 Chemical formulae of some microbial food-grade pigments

Additionally, a zeaxanthin-producing strain isolated from a microbial mat in a French Polynesian atoll is noteworthy for its simultaneous production of exopolysaccharides, which could enhance its industrial versatility (Raguenes et al. 2004). Another bacterium, *Sphingobacterium multivorum*, has been identified as a zeaxanthin producer utilizing the deoxyxylulose-5-phosphate (DXP) pathway, and a

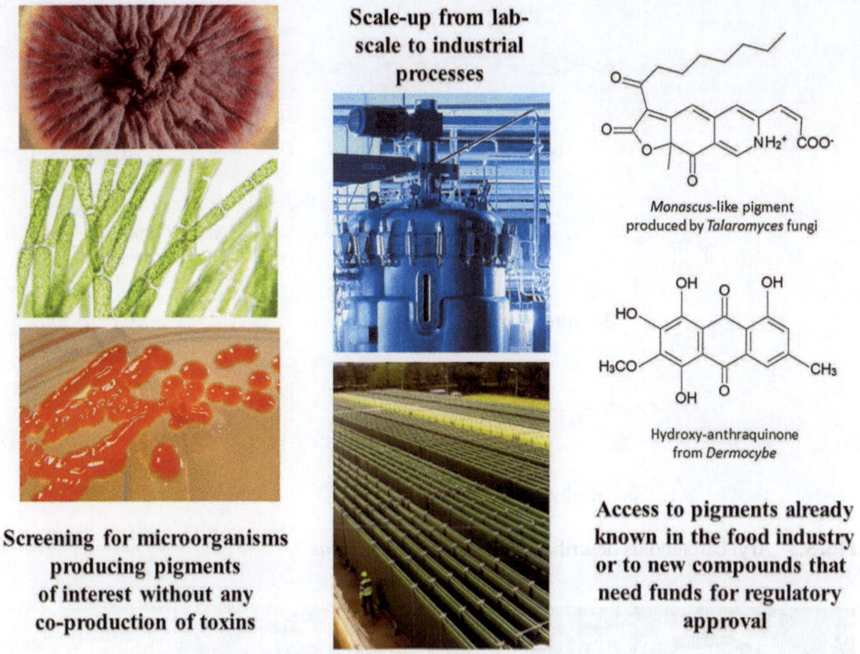

Fig. 5.2 Using microalgae, fungi, yeasts, and bacteria as cell factories for the production of natural food pigments

genetically modified strain has been developed for enhanced zeaxanthin production (US Patent 5,308,759). Advances in understanding the two primary isopentenyl pyrophosphate (IPP) biosynthesis pathways—the mevalonate and DXP pathways—have improved our knowledge of their distribution across microorganisms (Lange et al. 2000; Calegari-Santos et al. 2016). The identification of genes involved in IPP biosynthesis has paved the way for metabolic engineering, enabling the optimization of microbial production of valuable isoprenoids, including carotenoids like zeaxanthin. These developments underscore the potential for scalable, efficient production of zeaxanthin for food, feed, and pharmaceutical applications.

5.2.2 Aryl Carotenoids: Unique Orange Pigments with Potent Antioxidant Properties

Aryl carotenoids, such as isorenieratene and its hydroxylated derivatives (as depicted in Fig. 5.3), are rare orange pigments produced by a limited group of microorganisms. These compounds are notable for their strong antioxidant properties, making them valuable for both industrial and health-related applications. Among the producers, certain anaerobic, photosynthetic bacteria, including members of the *Chlorobiaceae* and *Chromatiaceae* families, are capable of synthesizing

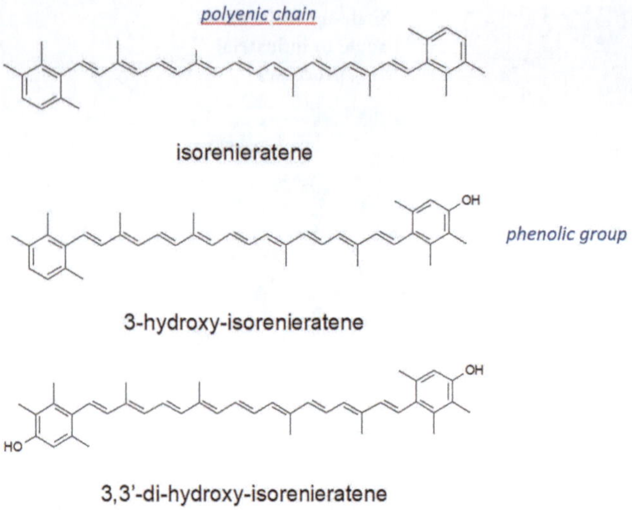

Fig. 5.3 Aryl carotenoids described in *Brevibacterium linens*

Fig. 5.4 Production of aryl carotenoids by *Brevibacterium linens* on a (**a**) Petri dish and on (**b**) Livarot cheeses

these pigments. The restricted distribution of aryl carotenoid production among microorganisms highlights their specialized biosynthetic pathways and underscores the need for further research to explore their potential for commercial applications, such as in food colorants, cosmetics, or nutraceuticals.

Nonphotosynthetic bacteria that produce aryl carotenoids, such as Brevibacterium linens, Streptomyces mediolani, and Mycobacterium aurum, are of significant interest due to their relative ease of cultivation compared to photosynthetic microorganisms (Fig. 5.4). These species, all belonging to the order Actinomycetales within the Actinobacteria phylum, offer promising platforms for industrial pigment production (Dufossé and de Echanove 2005). Their nonphotosynthetic nature simplifies cultivation processes, as they do not require light-dependent systems like photobioreactors, reducing infrastructure costs and complexity. The aryl carotenoids they produce, characterized by their vibrant orange hues and potent antioxidant properties, hold potential for applications in food colorants, cosmetics, and nutraceuticals. The robust metabolic capabilities and established industrial use of Actinomycetales further enhance their appeal for scalable production of these valuable pigments, driving interest in their optimization through biotechnological approaches such as metabolic engineering.

Brevibacterium linens, widely recognized for its role in the ripening of smear cheeses, is a promising candidate for producing aryl carotenoids due to its established use in food production, which may streamline regulatory approval for its biomass-derived pigments as food additives. However, its commercial application is hindered by low pigment yields in liquid culture, ranging from only 0.5 to 1.0 mg/L, rendering it economically unviable for large-scale industrial production (Guyomarc'h et al. 2000). Among carotenoids evaluated for antioxidant activity, including astaxanthin, β-cryptoxanthin, zeaxanthin, lycopene, and lutein, the aryl carotenoid 3,3′-dihydroxyisorenieratene (DHIR) stands out for its exceptional radical scavenging and singlet oxygen quenching capabilities. These properties are attributed to its unique structure, combining a polyenic backbone with phenolic groups, which enhance its antioxidant efficiency (Wagener et al. 2012; Martin et al. 2009). Research on aryl carotenoids began decades ago, with the first related patent issued in 1971 (US Patent 3,598,702). Despite the food-grade status of *B. linens* and its safety for pigment production, its low yield limits cost-effectiveness, making chemical synthesis advanced by companies like BASF and various research groups (Valla et al. 2003) the more practical short- to mid-term solution for producing carotenoids like isorenieratene and DHIR. However, the limited large-scale availability of these compounds remains a challenge. Genetic engineering offers a promising alternative by enhancing the biosynthesis of polyenic-phenolic carotenoids through modified microbial pathways, potentially enabling cost-effective production of these high-value antioxidants for food and nutraceutical applications (US Patent 7,186,523; WO Patent Application 2008009675).

5.2.3 C50 Carotenoids: Unique Microbial Pigments

C50 carotenoids, such as sarcinaxanthin and decaprenoxanthin, are exclusively produced by bacteria within the Actinomycetales order and are not found in plants (Sandmann 2015, 2021). Three distinct biosynthetic pathways for C50 carotenoids have been identified: (1) the β-cyclic C.p. 450 pathway, initially discovered in *Corynebacterium poinsettiae* and later studied in Dietzia sp. CQ4; (2) the γ-cyclic sarcinaxanthin pathway in *Micrococcus luteus* NCTC2665; and (3) the ε-cyclic decaprenoxanthin pathway in *Corynebacterium glutamicum*. Unlike the vast majority of carotenoids, which are derived from the symmetrical C40 phytoene backbone, C50 carotenoids, along with even rarer C30 carotenoids, are distinguished by their unique structures, featuring multiple conjugated double bonds and at least one hydroxyl group. These structural characteristics contribute to their potent antioxidant properties, making them highly promising for applications in food coloring, nutraceuticals, and pharmaceuticals (Heider et al. 2014a, b). The microbial exclusivity and strong oxidative defense mechanisms of C50 carotenoids position them as valuable candidates for industrial development, particularly as consumer demand for natural and functional ingredients grows.

5.2.4 Directed Evolution and Molecular Breeding for Novel Carotenoids

Carotenoids are among the most thoroughly studied natural pigments, with significant advancements in enhancing their production and developing novel variants with improved color strength, stability, and antioxidant properties (Berman et al. 2014; Guedes et al. 2011; Silva et al. 2016; Du et al. 2025). Directed evolution, a powerful biotechnological tool, has been instrumental in these efforts by accelerating mutation and selection of target genes to produce enzymes with optimized traits. For example, the directed evolution of geranylgeranyl diphosphate (GGPP) synthase from Archaeoglobus fulgidus, a critical enzyme in carotenoid biosynthesis, was applied in engineered *Escherichia coli* strains. Screening over 10,000 transformants for color intensity identified eight mutants that nearly doubled lycopene yields, while also revealing key enzyme regions affecting activity and expression (Jin et al. 2015; Schmidt-Dannert et al. 2000). Another innovative approach involves engineering membrane-bound carotenoid synthases to accept longer-chain precursors, enabling the synthesis of novel C45 and C50 carotenoid backbones from C20 and C25 isoprenyl diphosphates, respectively. Previous studies even demonstrated the production of a C35 carotenoid backbone, which could be further modified by downstream enzymes like cyclases and desaturases from existing C30 and C40 pathways. These enzymes, whether natural or engineered, exhibit sufficient substrate flexibility to process new backbones, potentially creating entirely new carotenoid families. With approximately 750 known carotenoid structures, such innovations could double or triple this number, vastly expanding their structural diversity.

Further advancements involve constructing novel biosynthetic pathways by recombining genes from diverse organisms and evolving new enzyme functionalities through random mutagenesis, DNA shuffling, and selective screening (Wang et al. 2007). A critical factor for success is the functional compatibility of carotenogenic Crt enzymes from different microbial sources within a single heterologous host like E. coli, where most cloned carotenoid biosynthetic genes, except for rare cases like zeaxanthin C(5,6) epoxidase, have been successfully expressed (Dufossé 2009). For instance, engineering a chimeric desaturase variant extended desaturation reactions, producing the fully conjugated carotenoid 3,4,3′,4′-tetradehydrolycopene with six double bonds instead of four. Combining a four-step desaturase with a five-step desaturase enabled the biosynthesis of 3,4-didehydrolycopene in *E. coli* (Zhang et al. 2012). Additional modifications with enzymes like 1,2-hydratase and C-3,4 desaturase generated novel hydroxylated carotenoids, such as

1-hydroxy-3,4,3′,4′-tetrahydrolycopene and 1,1′-dihydroxy-3,4-didehydrolycopen e, which exhibit strong antioxidant properties. Recent metabolic engineering efforts in *E. coli* and yeast have significantly increased carotenoid yields by enhancing flux through isoprenoid precursor pathways (Kim et al. 2016; Li et al. 2015; Takemura et al. 2015). Despite over three decades of research since the cloning of the carotenogenic gene cluster from *Erwinia herbicola* in 1991, commercial availability of engineered carotenoids remains limited, reflecting challenges in translating research advancements into market-ready products.

5.3 Azaphilones: Mycotoxin-Free Monascus Red

For over a millennium, *Monascus* species have been used in Asia to produce natural food colorants and health-promoting supplements, with more than a billion people incorporating *Monascus*-fermented products into their daily diets (Zhang et al. 2021). Historical records, such as a red pot-roast lamb recipe from AD 965 in the Qing Yilu, document the use of "hong qu" (red rice koji) produced by Monascus purpureus. These fungi synthesize six primary azaphilone pigments: yellow monascin and ankaflavin, orange monascorubrin and rubropunctatin, and red monascorubramine and rubropunctamine. Over 50 azaphilone compounds have been identified, owing to their reactivity with nitrogen-containing compounds, forming diverse adducts (Yang et al. 2015).

Recent genomic advancements, including the full sequencing of the *Monascus purpureus* YY-1 industrial strain, have revealed a 24.1-megabase genome with eight chromosomes and 7491 genes, facilitated by next-generation sequencing and optical mapping (Yang et al. 2015). As a member of the *Aspergillaceae* family, alongside *Aspergillus* and *Penicillium*, *M. purpureus* exhibits a smaller genome (13.6–40% smaller than related fungi), likely due to gene loss during adaptation to starch-rich environments. The genome includes highly expressed polyketide synthase (PKS) genes under pigment-inducing conditions, as well as a gene cluster for the mycotoxin citrinin (PKS C6.123). Disrupting this citrinin PKS gene does not affect red pigment production, confirming independent biosynthetic pathways (Fu et al. 2007). Transcriptomic studies indicate that carbon starvation, often triggered by low-quality carbon sources, suppresses central carbon metabolism while increasing acetyl-CoA availability, a key precursor for pigment synthesis. Similar investigations into *Penicillium marneffei*, a thermally dimorphic fungus, highlight its potential for producing black, yellow, and red azaphilone pigments. Its faster growth rate and fully sequenced genome make it an attractive alternative to Monascus. The red pigment of *P. marneffei* comprises over 16 compounds, primarily amino acid conjugates of monascorubrin and rubropunctatin, formed via nonenzymatic Schiff base reactions (Woo et al. 2014). These advancements pave the way for engineering mycotoxin-free strains and optimizing azaphilone production for safe, scalable use in food and nutraceutical applications (Fig. 5.5).

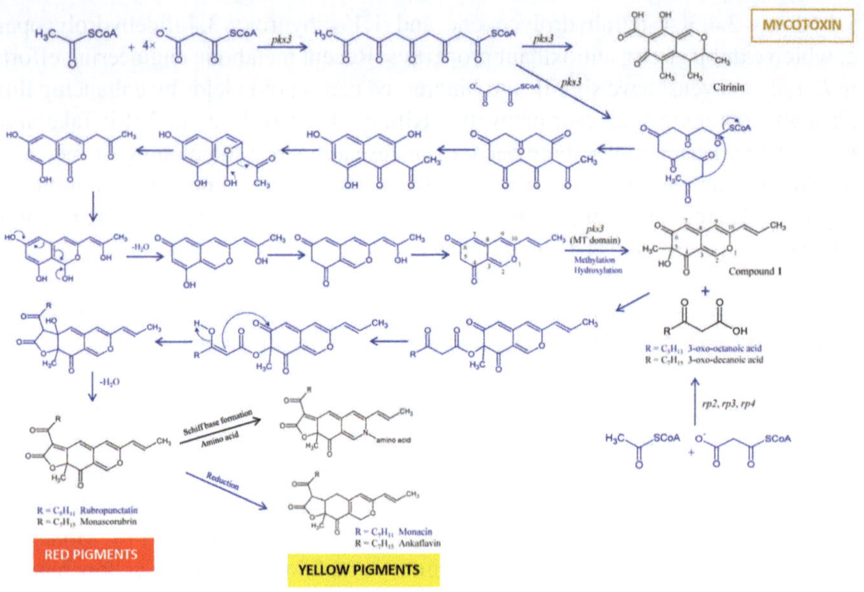

Fig. 5.5 Hypothetical pathway of monascorubrin, ankaflavin, and citrinin biosynthesis in *P. marneffei*. (Adapted from Woo et al. 2014)

The polyketide synthase (PKS) gene cluster in *Penicillium marneffei* is responsible for synthesizing red pigments, as well as ankaflavin and the nephrotoxic mycotoxin citrinin, which is harmful to mammalian kidney function (Kumar et al. 2014). A comprehensive genomic study by Woo et al. (2014) identified 23 putative PKS genes and two PKS-nonribosomal peptide synthase (NRPS) hybrid genes within the *P. marneffei* genome. To elucidate their roles, systematic knockouts of all 25 PKS-related genes were performed, alongside disruptions of genes upstream and downstream of the PKS gene linked to red pigment synthesis, to map the biosynthetic pathway. Despite these efforts, a significant hurdle persists: the challenge of producing red pigments free of citrinin and mevinolin/lovastatin, a cholesterol-lowering compound that may be undesirable in food products. Evidence indicates that citrinin is produced early in the biosynthetic pathway, raising concerns about whether pigment production can be fully decoupled from mycotoxin synthesis in *P. marneffei*. This interdependence poses a critical barrier to developing safe, food-grade pigments from this fungus, necessitating further research into genetic engineering or alternative strains to eliminate mycotoxin coproduction while preserving pigment yield.

5.3.1 Nontoxigenic Talaromyces Strains Pigments

Certain species within the *Talaromyces* genus, such as *Talaromyces purpurogenus*, *T. marneffei*, and *T. minioluteus* (previously classified under *Penicillium*), are recognized for their ability to produce substantial quantities of red pigments, particularly azaphilones. However, not all strains offer the pigment stability required for consistent, large-scale industrial production, limiting their commercial applicability. *T. purpurogenus* has garnered attention for its capacity to produce extracellular enzymes alongside red pigments, enhancing its biotechnological potential. Unfortunately, some strains also synthesize toxic secondary metabolites, including mycotoxins such as rubratoxin A and B, luteoskyrin, and other harmful extrolites like spiculisporic acid and rugulovasine A and B, which have demonstrated toxicity in animal studies (e.g., in cats via intraperitoneal or intravenous administration) (Frisvad et al. 2004). Due to these safety concerns, Frisvad et al. (2013) have advised against the use of *T. purpurogenus* for industrial pigment production, highlighting the need for safer alternatives.

In contrast, the newly described species *Talaromyces atroroseus* sp. nov. offers a promising solution for safe pigment production. This species produces azaphilone pigments, including mitorubrins and *Monascus*-type pigments, without coproducing known mycotoxins, as supported by patent applications (WO2012022765, US 20110250656). Another notable strain, *Talaromyces albobiverticillius* 30548, isolated from a lagoon on Réunion Island, is under investigation by researchers at the University of Réunion (Venkatachalam et al. 2018, 2020, 2021). Like *Monascus* species, *Talaromyces* strains produce azaphilone pigments that react with amino group-containing compounds, forming vibrant dark red pigments through nonenzymatic Schiff base reactions (Mapari et al. 2010; Gao et al. 2013). This reactivity, which gives azaphilones their name, enhances their potential as natural colorants for food, cosmetics, and other biotechnological applications. The identification of nontoxigenic *Talaromyces* strains, combined with their chemical versatility and pigment production capabilities, positions them as viable candidates for sustainable, safe, and scalable production of natural red pigments, addressing the limitations of toxigenic strains and advancing the development of food-grade colorants (Fig. 5.6).

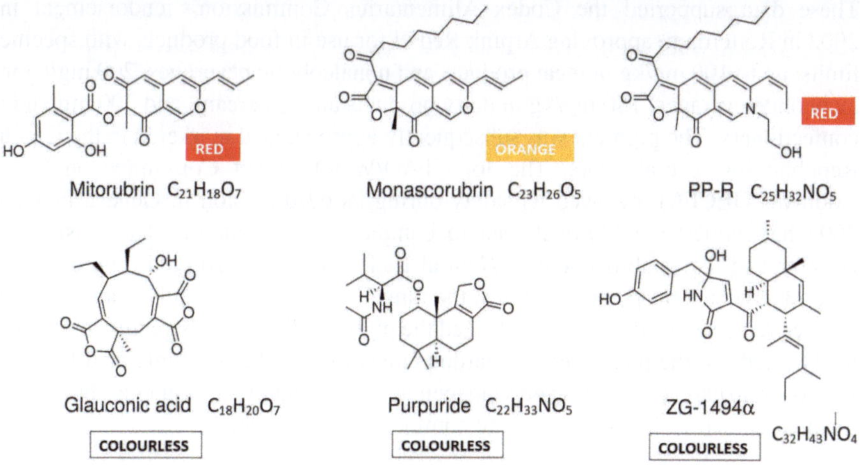

Fig. 5.6 Structures of some of the most characteristic compounds produced by *Talaromyces atroroseus*. (Adapted from Frisvad et al. 2013)

5.4 Anthraquinones: Fungal-Derived Pigments for Food and Beyond

5.4.1 Fungal Natural Red™: A Pioneering Anthraquinone Pigment

Arpink Red™, rebranded as Natural Red™, represents the first commercially successful anthraquinone pigment derived from fungal fermentation. Developed initially by Ascolor Biotech s.r.o. and later by Natural Red™ in the Czech Republic, this pigment is produced using *Penicillium oxalicum* var. *armeniaca* CCM 8242, a strain originally isolated from soil. The production process, detailed in several patents (WO 9,950,434, CZ 285,721, EP 1,070,136, US 6,340,586; cited in Sardaryan et al. 2004), involves fermenting the fungus in a liquid medium enriched with carbohydrate sources (e.g., sucrose or molasses), nitrogen sources (e.g., corn extract or yeast autolysate), and mineral salts such as zinc sulfate and magnesium sulfate. Optimal pigment production is achieved at a pH of 5.6–6.2 and temperatures between 27 °C and 29 °C. The red pigment is secreted into the culture broth by day two, reaching yields of 1.5–2.0 g/L within 3–4 days. Postfermentation, the broth is filtered or centrifuged to separate the biomass, and the supernatant is acidified to pH 2.5–3.0 to precipitate the pigment. This precipitate is dissolved in ethyl alcohol, filtered, and dried to produce a dark red microcrystalline powder. The resulting pigment imparts a vibrant raspberry-red hue in aqueous solutions, maintaining stability at pH levels above 3.5 and temperatures from 4 °C to 60 °C, with solutions at pH 7.0 remaining stable even after 30 minutes of boiling.

Extensive toxicological studies, documented in the aforementioned patents, have evaluated the pigment's safety through acute oral toxicity tests in mice, 90-day subchronic toxicity studies, skin and eye irritation tests, antitumor assays, micronucleus and Ames tests, and screenings for antibiotic activity and five major mycotoxins. These data supported the Codex Alimentarius Commission's endorsement in 2002 in Rotterdam, approving Arpink Red™ for use in food products with specific limits: up to 100 mg/kg in meat products and nonalcoholic beverages, 200 mg/kg in alcoholic beverages, 150 mg/kg in dairy products and ice cream, and 300 mg/kg in confectionery. The pigment was subsequently approved and marketed in the Czech Republic for several years. The Joint FAO/WHO Expert Committee on Food Additives (JECFA) reviewed its safety during its 63rd session in Geneva in June 2004 but requested additional data to complete the evaluation. The closure of Ascolor and the establishment of Natural Red™ in 2012 disrupted progress, and unresolved concerns persist. Notably, the fungal strain is not publicly available, and no peer-reviewed studies have confirmed the pigment's chemical structure or definitively ruled out the presence of hazardous mycotoxins like secalonic acid D. This lack of transparency complicates independent risk-benefit assessments, hindering the pigment's broader regulatory and commercial acceptance.

5.4.2 Diverse Fungal Anthraquinones: A Versatile Pigment Class

Anthraquinones are widely produced by various fungal genera, including *Aspergillus*, *Eurotium*, *Fusarium*, *Dreschlera*, *Penicillium*, *Emericella purpurea*, *Curvularia lunata*, *Mycosphaerella rubella*, and *Microsporum*, among others (Caro et al. 2012; Gessler et al. 2013). Their prevalence across fungal taxa makes them a compelling alternative to plant- or animal-derived pigments, as fungal production is less dependent on climatic or agricultural constraints, offering greater flexibility and scalability. Fungal anthraquinones exhibit a broad range of biological activities, including antibacterial, antifungal, antiviral, herbicidal, and insecticidal properties, which enhance their potential for applications beyond food coloring, such as in pharmaceuticals and agriculture (Gessler et al. 2013). In natural ecosystems, these secondary metabolites likely play ecological roles, such as mediating interspecies interactions. For example, anthraquinones produced by endophytic fungi may protect host plants from microbial pathogens or insect predation, highlighting their ecological significance.

The biosynthesis of fungal anthraquinones is highly complex, influenced by a multitude of environmental and nutritional factors. Their chemical diversity, as illustrated in Fig. 5.7, results in a wide array of structural variants, with production modulated by factors such as the fungus's habitat, light exposure, culture pH, temperature, oxygen availability, growth medium (liquid or solid), carbon and nitrogen sources, C:N ratio, organic acids, minerals, and inoculum density (Caro et al. 2012). This variability allows for tailored optimization of pigment production but also poses challenges in achieving consistent yields and compositions for industrial applications. The versatility of fungal anthraquinones, combined with their potent bioactivities and independence from traditional agricultural limitations, positions them as a promising resource for developing sustainable, natural colorants and functional compounds for food, cosmetics, and other industries.

Current research prioritizes a select group of fungal species that meet stringent criteria for safe and efficient anthraquinone production, as outlined by Mapari et al. (2009). These criteria, established through a chemotaxonomic approach to identify safe fungal cell factories for polyketide-based natural food colorants, include: (1) the fungus must be nonpathogenic to humans, ensuring safety for food applications; (2) it must be nontoxigenic across a wide range of production conditions to prevent the formation of harmful secondary metabolites; and (3) it must be capable of producing pigments in liquid media, which is essential for scalable industrial fermentation. These requirements guide the selection of fungal strains to ensure both safety and practicality in large-scale pigment production.

5.5 Phycobiliproteins: Vibrant and Functional Microalgal Pigments

Phycobiliproteins (PBPs) are water-soluble chromoproteins that serve as accessory light-harvesting pigments in cyanobacteria, red algae (*Rhodophyceae*), and cryptophytes (*Cryptophyceae*). Characterized by bilin chromophores, open-chain

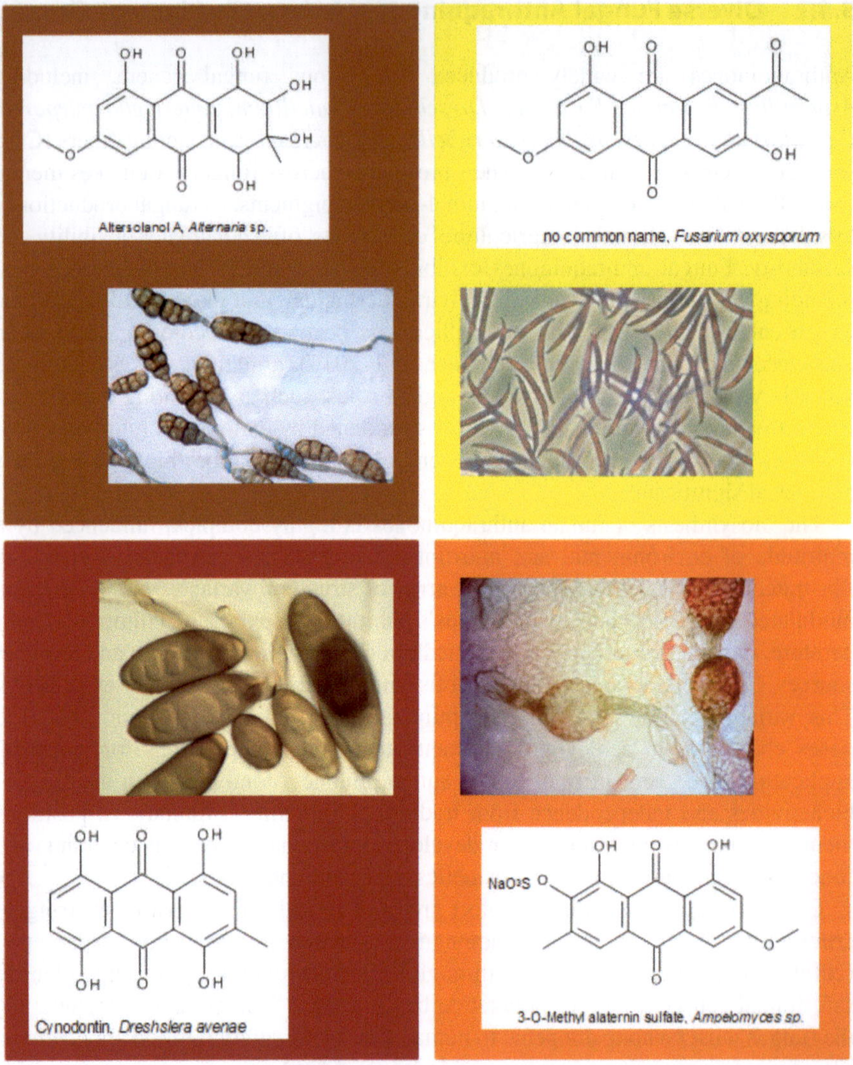

Fig. 5.7 Diversity of fungal anthraquinones

tetrapyrrole groups covalently bound to the protein backbone, PBPs are classified into four main types based on their spectral properties: phycoerythrins (PE, pink to purple, λmax 540–570 nm), phycocyanins (PC, blue, λmax 610–620 nm), phycoerythrocyanins (PEC, orange, λmax 560–600 nm), and allophycocyanins (APC, bluish-green, λmax 650–655 nm) (Kuddus et al. 2013). These pigments form phycobilisomes, complex structures on thylakoid membranes that extend the light absorption range beyond chlorophyll a (430 nm), enhancing photosynthetic efficiency (Eriksen 2008). Typically organized as αβ heterodimers, PBPs assemble into trimers (~120 kDa) or hexamers (~240 kDa), contributing to their intense fluorescence, particularly in phycoerythrin, which makes it a promising natural colorant for food and beverages, such as UV-reactive cocktail dyes.

Commercial phycoerythrin is primarily extracted from mesophilic red algae like *Porphyridium* species, but its low thermal stability limits its use in heat-intensive processes. To overcome this, researchers have explored thermophilic cyanobacteria and extremophiles. For instance, *Leptolyngbya* sp., isolated from hot springs, produces phycoerythrin that retains over 80% stability after 30 minutes at 60 °C (Pumas et al. 2012). Similarly, *Thermosynechococcus elongatus* TA-1, sourced from Taiwanese hot springs, yields C-phycocyanin stable across 4–60 °C and pH 4–9 for at least 4 hours (Leu et al. 2013). *Synechococcus lividus* PCC 6715, from Yellowstone National Park hot springs, produces one of the most thermally stable phycocyanins, maintaining 90% stability after 5 hours at 50 °C and 70% after 2 weeks, with robust pH tolerance (pH 4–8) (Liang et al. 2018). The red alga *Cyanidioschyzon merolae* produces phycocyanin with a midpoint thermal denaturation at 83 °C (pH 5) and an average fluorescence lifetime of 40 minutes (Rahman et al. 2017). However, PBP fluorescence depends more on the tetrapyrrole microenvironment than on aggregation state or thermophilicity (Seibert and Connolly 1984). Stability is assessed through spectrophotometry, fluorescence assays, and antioxidant activity tests (e.g., DPPH assays). Genomic and structural analyses of thermostable PBPs are driving genetic engineering efforts to enhance pigment robustness for industrial applications, such as heat-processed food products.

5.6 Conclusion: The Future of Microbial Pigments

Rising consumer demand for natural, organic, and clean-label products is driving interest in microbial pigments as sustainable alternatives to synthetic colorants. These pigments, produced through advanced biotechnological processes, offer a diverse palette of hues and functional benefits, such as antioxidant properties. Well-known pigments like β-carotene and lycopene are being optimized for higher yields, while unique microbial-derived compounds, such as aryl carotenoids, C50 carotenoids, and azaphilones from *Monascus* and *Talaromyces* species, provide novel options not found in plants. The success of these pigments in food, beverage, and feed applications hinges on their stability under heat and light, vibrant coloration, and regulatory approval. Many microbial pigments already exhibit these desirable traits, and ongoing advancements in metabolic engineering, directed evolution, and genomic analysis are enabling the development of tailored, robust colorants. As start-ups and research efforts continue to innovate, microbial and microalgal pigments are poised to deliver customized, sustainable solutions for a wide range of industrial applications, transforming the food, textile, and nutraceutical sectors.

References

Berman J, Zorrilla-López U, Farré G, Zhu C, Sandmann G, Twyman RM, Capell T, Christou P (2014) Nutritionally important carotenoids as consumer products. Phytochem Rev 14(5):727–743

Calegari-Santos R, Diogo RA, Fontana JD, Bonfim TMB (2016) Carotenoid production by halophilic archaea under different culture conditions. Curr Microbiol 72(5):641–651

Caro Y, Anamale L, Fouillaud M, Laurent P, Petit T, Dufossé L (2012) Natural hydroxyanthraquinoid pigments as potent food-grade colorants: an overview. Nat Prod Bioprospect 2:174–193

Chen Y, Xiao W, Wang Y, Liu H, Li X, Yuan Y (2016) Lycopene overproduction in *Saccharomyces cerevisiae* through combining pathway engineering with host engineering. Microb Cell Factories 15(1):113

Chen X, Lim X, Bouin A, Lautier T, Zhang C (2021) High-level *de novo* biosynthesis of glycosylated zeaxanthin and astaxanthin in *Escherichia coli*. Bioresour Bioprocess 8(1):67

Du J, Bao Y, Zhu J, Pang X, Ren D, Yin X, Zhou P (2025) Engineering of β-carotene hydroxylase for enhanced astaxanthin production in *Saccharomyces cerevisiae*. Biochem Eng J 219:109722. https://doi.org/10.1016/j.bej.2025.109722

Dufossé L (2006) Microbial production of food-grade pigments. Food Technol Biotechnol 44:313–321

Dufossé L (2009) Pigments, microbial. In: Schaechter M (ed) Encyclopedia of microbiology, 3rd edn. Elsevier/Academic Press, New-York, pp 457–471

Dufossé L, de Echanove C (2005) The last step in the biosynthesis of aryl carotenoids in the cheese ripening bacteria *Brevibacterium linens* ATCC 9175 (*Brevibacterium aurantiacum* sp. nov.) involves a cytochrome P450-dependent monooxygenase. Food Res Int 38:967–973

Dufossé L, Galaup P, Yaron A, Arad SM, Blanc P, Chidambara Murthy KN, Ravishankar GA (2005) Microorganisms and microalgae as sources of pigments for food use: a scientific oddity or an industrial reality? Trends Food Sci Technol 16:389–406

Dufossé L, Fouillaud M, Caro Y, Mapari SAS, Sutthiwong N (2014) Filamentous fungi are large-scale producers of pigments and colorants for the food industry. Curr Opin Biotechnol 26:56–61

Eriksen NT (2008) Production of phycocyanin - a pigment with applications in biology, biotechnology, foods and medicine. Appl Microbiol Biotechnol 80:1–14

Fouillaud M, Venkatachalam M, Llorente M, Magalon H, Cuet P, Dufossé L (2017) Biodiversity of pigmented fungi isolated from marine environment in Reunion Island, Indian Ocean: new resources for colored metabolites. J Fungi 3(3):36. (22 pages). https://doi.org/10.3390/jof3030036

Frisvad JC, Smedsgaard J, Larsen TO, Samson RA (2004) Mycotoxins, drugs and other extrolites produced by species in *Penicillium* subgenus *Penicillium*. Stud Mycol 49:201–241

Frisvad JC, Yilmaz N, Thrane U, Rasmussen KB, Houbraken J, Samson RA (2013) *Talaromyces atroroseus*, a new species efficiently producing industrially relevant red pigments. PLoS One 8(12):e84102

Fu G, Xu Y, Li Y, Tan W (2007) Construction of a replacement vector to disrupt pksCT gene for the mycotoxin citrinin biosynthesis in *Monascus aurantiacus* and maintain food red pigment production. Asia Pac J Clin Nutr 16(Suppl. 1):137–142

Furubayashi M, Ikezumi M, Takaichi S, Maoka T, Hemmi H, Ogawa T, Saito K, Tobias AV, Umeno D (2015) A highly selective biosynthetic pathway to non-natural C 50 carotenoids assembled from moderately selective enzymes. Nat Commun 6:7534

Gao JM, Yang SX, Qin JC (2013) Azaphilones: chemistry and biology. Chem Rev 113:4755–4811

Gessler NN, Egorova AS, Belozerskaya TA (2013) Fungal anthraquinones. Appl Biochem Microbiol 49:85–99

Grassi TLM, do Espírito Santo EF, de Siqueira Marcos MT, Cavazzana JF, Oliveira DL, Bossolani ILC, Ponsano EHG (2016) Bacterial pigment for Nile tilapia feeding. Aquac Int 24(2):647–660

Guedes AC, Amaro HM, Malcata FX (2011) Microalgae as sources of carotenoids. Mar Drugs 9:625–644

Guyomarc'h F, Binet A, Dufossé L (2000) Production of carotenoids by *Brevibacterium linens* : variation among strains, kinetic aspects and HPLC profiles. J Ind Microbiol Biotechnol 24:64–70

Heider SA, Peters-Wendisch P, Wendisch VF, Beekwilder J, Brautaset T (2014a) Metabolic engineering for the microbial production of carotenoids and related products with a focus on the rare C50 carotenoids. Appl Microbiol Biotechnol 98:4355–4368

Heider SAE, Peters-Wendisch P, Netzer R, Stafnes M, Brautaset T, Wendisch VF (2014b) Production and glucosylation of C50 and C40 carotenoids by metabolically engineered *Corynebacterium glutamicum*. Appl Microbiol Biotechnol 98(3):1223–1235

Jaswir I, Noviendri D, Hasrini RF, Octavianti F (2011) Carotenoids: sources, medicinal properties and their application in food and nutraceutical industry. J Med Plant Res 5(33):7119–7131

Jin W, Xu X, Jiang L, Zhang Z, Li S, Huang H (2015) Putative carotenoid genes expressed under the regulation of Shine–Dalgarno regions in *Escherichia coli* for efficient lycopene production. Biotechnol Lett 37(11):2303–2310

Kim SH, Kim MS, Lee BY, Lee PC (2016) Generation of structurally novel short carotenoids and study of their biological activity. Sci Rep 6:21987

Kim K, Shim K, Wang YW, Yang D (2025) Synthetic biology strategies for the production of natural colorants and their non-natural derivatives. ACS Synth Biol 14(3):662–676. https://doi.org/10.1021/acssynbio.4c00799

Kuddus M, Singh P, Thomas G, Al-Hazimi A (2013) Recent developments in production and biotechnological applications of C-phycocyanin. Biomed Res Int 2013. https://doi.org/10.1155/2013/742859

Kumar M, Dwivedi P, Sharma AK, Sankar M, Patil RD, Singh ND (2014) Apoptosis and lipid peroxidation in ochratoxin A- and citrinin-induced nephrotoxicity in rabbits. Toxicol Ind Health 30:90–98

Lange BM, Rujan T, Martin W, Croteau R (2000) Isoprenoid biosynthesis: the evolution of two ancient and distinct pathways across genomes. Proc Natl Acad Sci USA 97:13172–13177

Lebeau J, Venkatachalam M, Fouillaud M, Petit T, Vinale F, Dufossé L, Caro Y (2017) Production and new extraction method of polyketide red pigments produced by ascomycetous fungi from terrestrial and marine habitats. J Fungi 3(3):34. (21 pages). https://doi.org/10.3390/jof3030034

Leu J, Lin T, Selvamani MJP, Chen H, Liang J, Pan K (2013) Characterization of a novel thermophilic cyanobacterial strain from Taian hot springs in Taiwan for high CO_2 mitigation and C-phycocyanin extraction. Process Biochem 48(1):41–48. https://doi.org/10.1016/j.procbio.2012.09.019

Li X-R, Tian G-Q, Shen H-J, Liu J-Z (2015) Metabolic engineering of *Escherichia coli* to produce zeaxanthin. J Ind Microbiol Biotechnol 42(4):627–636

Liang Y, Kaczmarek MB, Kasprzak AK, Tang J, Shah MMR, Jin P, Daroch M (2018) Thermosynechococcaceae as a source of thermostable C-phycocyanins: properties and molecular insights. Algal Res 35:223–235. https://doi.org/10.1016/j.algal.2018.08.037

Mannazzu I, Landolfo S, da Silva TL, Buzzini P (2015) Red yeasts and carotenoid production: outlining a future for non-conventional yeasts of biotechnological interest. World J Microbiol Biotechnol 31(11):1665–1673

Mapari SAS, Meyer AS, Thrane U, Frisvad JC (2009) Identification of potentially safe promising fungal cell factories for the production of polyketide natural food colorants using chemotaxonomic rationale. Microb Cell Factories 8:24

Mapari SA, Thrane U, Meyer AS (2010) Fungal polyketide azaphilone pigments as future natural food colorants? Trends Biotechnol 28:300–307

Martin HD, Kock S, Scherrers R, Lutter K, Wagener T, Hundsdörfer C, Frixel S, Schaper K, Ernst H, Schrader W, Görner H, Stahl W (2009) 3,3′-dihydroxyisorenieratene, a natural carotenoid with superior antioxidant and photoprotective properties. Angew Chem Int Ed 48:400–403

Pumas C, Peerapornpisal Y, Vacharapiyasophon P, Leelapornpisid P, Boonchum W, Ishii M, Khanongnuch C (2012) Purification and characterization of a thermostable phycoerythrin from hot spring cyanobacterium *Leptolyngbya* sp. KC45. Int J Agric Biol 14:121–125

Raguenes G, Moppert X, Richert L, Ratiskol J, Payri C, Costa B, Guezennec J (2004) A novel exopolymer-producing bacterium, *Paracoccus zeaxanthinifaciens* subsp *payriae*, isolated from a "kopara" mat located in Rangiroa, an atoll of French Polynesia. Curr Microbiol 49:145–151

Rahman DY, Sarian FD, van Wijk A, Martinez-Garcia M, van der Maarel MJEC (2017) Thermostable phycocyanin from the red microalga *Cyanidioschyzon merolae*, a new natural blue food colorant. J Appl Phycol 29(3):1233–1239. https://doi.org/10.1007/s10811-016-1007-0

Ramamurthy K, Priya PS, Murugan R, Arockiaraj J (2024) Hues of risk: investigating genotoxicity and environmental impacts of azo textile dyes. Environ Sci Pollut Res 31. https://doi.org/10.1007/s11356-024-33444-1

Sandmann G (2015) Carotenoids of biotechnological importance. Adv Biochem Eng Biotechnol 148:449–467

Sandmann G (2021) Carotenoid biosynthesis in the phylum Actinobacteria. Adv Exp Med Biol 1261:175–181

Sardaryan E, Zihlova H, Strnad R, Cermakova Z (2004) Arpink red—meet a new natural red food colorant of microbial origin. In: Dufossé L (ed) Pigments in food, more than Colours…. Université de Bretagne Occidentale Publ, Quimper, pp 207–208

Schmidt-Dannert C, Umeno D, Arnold FH (2000) Molecular breeding of carotenoid biosynthetic pathways. Nat Biotechnol 18:750–753

Seibert M, Connolly JS (1984) Fluorescence properties of C-phycocyanin isolated from a thermophilic cyanobacterium. Photochem Photobiol 40(2):267–271. https://doi.org/10.1111/j.1751-1097.1984.tb04585.x

Silva TP, Paixão SM, Alves L (2016) Ability of *Gordonia alkanivorans* strain 1B for high added value carotenoids production. RSC Adv 6(63):58055–58063

Takemura M, Maoka T, Osawa A, Higashinaka H, Shimada H, Shindo K, Misawa N (2015) (6E) and (6Z)-9′-Aporhodoxanthinone, novel carotenoids produced in zeaxanthin-synthesizing-*Escherichia coli* by redox stress. Tetrahedron Lett 56(44):6063–6065

Thawornwiriyanun P, Tanasupawat S, Dechsakulwatana C, Techkarnjanaruk S, Suntornsuk W (2012) Identification of newly zeaxanthin-producing bacteria isolated from sponges in the Gulf of Thailand and their zeaxanthin production. Appl Biochem Biotechnol 167:2357–2368

Valla A, Cartier D, Valla B, Le Guillou R, Andriamialisoa Z, Labia R, Breithaupt DE, Savy S, Binet A, Dufossé L (2003) New syntheses of natural carotene isorenieratene (φ,φ-carotene) and its 3,3′-dimethoxy analog. Helv Chim Acta 86:3314–3319

Venkatachalam M, Zelena M, Cacciola F et al (2018) Partial characterization of the pigments produced by the marine-derived fungus *Talaromyces albobiverticillius* 30548. Towards a new fungal red colorant for the food industry. J Food Compos Anal 67:38–47

Venkatachalam M, Shum-Chéong-Sing A, Dufossé L, Fouillaud M (2020) Statistical optimization of the physico-chemical parameters for pigment production in submerged fermentation of *Talaromyces albobiverticillius* 30548. Microorganisms 8(5):711

Venkatachalam M, Shum-Chéong-Sing A, Caro Y, Dufossé L, Fouillaud M (2021) OVAT analysis and response surface methodology based on nutrient sources for optimization of pigment production in the marine-derived fungus Talaromyces albobiverticillius 30548 submerged fermentation. Mar Drugs 19(5):248

Wagener S, Völker T, De Spirt S, Ernst H, Stahl W (2012) 3,3′-dihydroxyisorenieratene and isorenieratene prevent UV-induced DNA damage in human skin fibroblasts. Free Radic Biol Med 53:457–463

Wang F, Jiang JG, Chen Q (2007) Progress on molecular breeding and metabolic engineering of biosynthesis pathways of C-30, C-35, C-40, C-45, C-50 carotenoids. Biotechnol Adv 25:211–222

Woo PC, Lam CW, Tam EW, Lee KC, Yung KK, Leung CK, Sze KH, Lau SK, Yuen KY (2014) The biosynthetic pathway for a thousand-year-old natural food colorant and citrinin in *Penicillium marneffei*. Sci Rep 4:6728. https://doi.org/10.1038/srep06728

Yang Y, Liu B, Du X, Li P, Liang B, Cheng X, Du L, Huang D, Wang L, Wang S (2015) Complete genome sequence and transcriptomics analyses reveal pigment biosynthesis and regulatory mechanisms in an industrial strain, *Monascus purpureus* YY-1. Sci Rep 5:8331. https://doi.org/10.1038/srep08331

Zhang JH, Lu LL, Yin LJ, Xie S, Xiao M (2012) Carotenogenesis gene cluster and phytoene desaturase catalyzing both three- and four-step desaturations from *Rhodobacter azotoformans*. FEMS Microbiol Lett 333:138–145

Zhang J, Gao J, Li M, Shao Y, Chen F (2021) MrGcn5 is required for the mycotoxin production, sexual and asexual development in *Monascus ruber*. Food Biosci 43:101304

Zhou P, Ye L, Xie W, Lv X, Yu H (2015) Highly efficient biosynthesis of astaxanthin in *Saccharomyces cerevisiae* by integration and tuning of algal crtZ and bkt. Appl Microbiol Biotechnol 99(20):8419–8428

Insect and Animal-Derived Biocolours

6

Maithilee Chaudhary, Preeti Solanki, Rashmi Sharma, Seema Joshi, Basanti Brar, Abhishek Walia, and Chayanika Putatunda

Abstract

Man has always been attracted by the aesthetics of nature, which is only possible with wonderful, attractive, bright and enchanting colours. Since ancient times humans have tried to add colour in his life, be in the form of design of walls, paintings, fabrics, or even food items, body art, and cosmetics. Colour is one of the basic attributes of any material, which catches the eye. The ancient civilizations used plants, animals, minerals, etc. as the source of colours. Some animal origin biocolours which have been used across the globe since ancient times include Cochineal dye, Kermes dyes, lac dyes (all insect origin dyes) and Tyrian Purple (from mollusc). The advent of synthetic dyes, revolutionized the colour

M. Chaudhary
GN Ramachandran Protein Centre, CSIR-IMTECH, Chandigarh, India

P. Solanki
University Institute of Biotechnology, Chandigarh University, Mohali, Punjab, India

R. Sharma
VRDL, Department of Microbiology, Shri Lal Bahadur Shastri Govt. Medical College, Mandi, Himachal Pradesh, India

S. Joshi
Department of Environmental Science, Om Sterling Global University, Hisar, Haryana, India

B. Brar
Department of Biotechnology, Om Sterling Global University, Hisar, Haryana, India

A. Walia
Department of Microbiology, College of Basic Sciences, CSKHPKV, Palampur, Himachal Pradesh, India

C. Putatunda (✉)
Department of Biotechnology, Om Sterling Global University, Hisar, Haryana, India

Department of Microbiology, Om Sterling Global University, Haryana, Hisar, India

R. K. Bhatia, A. Walia (eds.), *Biocolours*,
https://doi.org/10.1007/978-981-95-1738-1_6

industries by providing an economical and easy to use source of colour as opposed to plant or animal based colours and hence, the animal origin dyes fell out of use. However, with the pace of time, we are becoming more aware about the detrimental effects of synthetic dyes on the environment as well as the human health. So, the interest is again shifting towards the biological sources of colours. This chapter focuses on the various common animal based biocolours including their structure, source, and applications along with highlighting some recent advancements in the field of animal-based biocolours.

Keywords

Animal biocolour · Insect biocolour · Cochineal dye · Kermes dyes · Lac dyes · Tyrian purple

6.1 Introduction

Colours have fascinated humans from time immemorial. The colourful flora and fauna occupying the planet earth have always been a source of intrigue and awe for the human mind. With evolution, humans started to add colour to their lives, as is evident from the cave paintings of prehistoric man (Karagiannidou 2014). The colours have become an integral part of lives, be it for textiles, food, cosmetics, buildings, furniture, pharmaceuticals, etc. or for arts and crafts. Initially, the sources of colours were mostly of mineral origin or biological origin. The biological origin colours, or the biocolours, were predominantly derived from plant sources or animal sources. The evidence of animal-origin biocolours dates back to very ancient times. Sotiropoulou et al. (2021) reported the use of purple pigment of molluscan origin in the wall paintings of early Late Bronze Age Aegean regions. The "true purple" or "royal purple" colour obtained from sea molluscs of the Muricidae family was very widely used for dyeing fabrics in the regions near the Mediterranean Sea since the Bronze Age (Cardon 2007). Depending on the process and conditions, the dye presented purplish-red to violet-blue shades in the final product. The "royal purple," as is evident from the name itself, was associated with the royalty and the persons of upper strata of the society, like the high priests (Mac Donald 2019).

The dye was produced primarily from three species of mollucks, viz., *Hexaplex trunculus* (*Murex trunculus*), *Bolinus brandaris* (*Murex brandaris*), and *Stramonita haemastoma* (*Thais haemastoma*) (Sukenik et al. 2016). Sukenik et al. (2021) reported on the evidence of royal purple dyed textile from the Timna Valley of Israel. Apart from such mollusc-origin colourants, certain insects were also used as sources of biocolours by the earlier human societies. Some of the classical examples are the lac dyes, kermes dyes, and cochineal dyes (scale insects). The evidence of lac insect (*Kerria lacca*) cultivation in India dates back thousands of years. In fact, one of the ancient scriptures of Hinduism, the Atharva Veda (approximately 1500 BCE), mentions the medical application of lac dye (Shamim et al. 2014). Lac dye and shellac are obtained from the lac insect *Kerria lacca* (Cooksey 2018). The lac dye gives a pink, red, or purple colouration and has been used in ancient times for textiles, silk, colouring, paintings, etc. (Dyer et al. 2018; Berbers et al. 2019). Another insect-based colourant is the Kermes dye, which has been found in the dyed fabrics obtained from ancient archaeological sites in France (Shamim et al. 2014). Kermes dye was obtained from dried female coccid *Kermes* insects found in Turkey as well as the Mediterranean region of southern Europe. These insects were meticulously harvested and processed to extract the vibrant red dye, which was highly valued for its richness and durability. The use of Kermes dye not only showcases the ingenuity of ancient textile artisans but also reflects the trade networks that existed across regions, as the demand for such colourants influenced cultural exchanges and economic relationships. It was a highly prized and expensive dye, as is clear from the name "King's Red" associated with this dye (Cooksey 2018; El Sayed and Ibrahim 2019). In fact, in 1464, the Pope directed scarlet colour from Kermes dye to be used for colouring the garbs of high priests (Cooksey 2018).

Similarly, another very important insect-based colourant is the cochineal dye. This dye is obtained from the female *Dactylopius coccus* insects, which live on *Opuntia* cactus plants. The cochineal dyes had been used by the ancient Aztec civilization in Mexico (South America) and later travelled to Europe and the rest of the world, where it became a coveted product and was used for dyeing the garments of royalty and religious heads. Even the famous artists like Tintoretto and Van Gogh have used the dye in their painting (Achim 2021). The insect- and animal-based dyes, although they played a very crucial role in the development of human history, fell out of use with the advent of chemical dyes. The unpredictability in the supply of raw materials, tedious and lengthy processing steps, and the requirement of skilled labour (along with historical and political factors) made such colourants quite costly. The chemical dyes, on the other hand, were much cheaper, easy to handle, more colourfast, and also presented a wide array of shades (Bhatia et al. 2017). Hence, the chemical dyes almost completely replaced the animal- and insect-based colouring agents. However, people are now becoming more aware of the inherent disadvantages of the chemical dyes. Since the chemical dyes are more stable, they tend to remain in the environment for a very long period of time. In fact, the effluents from the dyeing industries have wreaked havoc on several water bodies. Apart from affecting the environment, especially the aquatic ecosystems, the chemical dyes tend to have an adverse effect on human health also, as several of the commonly used chemicals have also been found to be carcinogenic and deleterious for human health (Bhatia et al. 2017; Gurav et al. 2011). The health issues become even more important in respect of the food colorants. All these factors have led to a renewed interest in the natural colour sources, including the animal and insect-based biocolours, because not only are these biodegradable, but they also offer several advantages in the form of medicinal properties and hence can be used for the production of bioactive textile materials also (Repon et al. 2023). Hence, scientists are working on improving the production and properties of these colourants.

6.2 Insect Based Biocolourants

The main sources of natural dyes are insect secretions and dried insect body parts (Desai et al. 2023). These insects feed on plants and trees, secreting protective substances that also contain pigments composed of compounds such as carminic acid, kermesic acid, and laccaic acid. Extraction, purification, and manipulation of these compounds provide natural, vivid dyes.

6.2.1 Cochineal Dyes

Cochineal dye, an insect-based biocolour, has been widely used for centuries. In ancient times, Mexican dyers discovered and used the red hue generated from cochineal "*Dactylopius coccus Costa*." By the sixteenth century, American cochineal had grown to become the principal source of cochineal dye due to its great production (Deveoglu 2020). Later in the nineteenth century, Peru and Mexico became key trading centres for cochineal insects throughout the colonial era. In the same era, cochineal-containing bio-colours were also being utilized by prominent artists such as Vincent Van Gogh in their works (Galappaththi and Patabendige 2021).

6.2.1.1 Source and Extraction

Cochineal dye is obtained from the dried bodies of female Dactylopius coccus insects, which live on Opuntia cacti (prickly pear cactus), aka Nopalea coccinellifera. It is also known as nopal, opuntia, or Indian fig that is mainly found in Mexico, Central America, and Brazil in South America (Cooksey 2018; Valipour et al. 2013). However, nowadays, the main sources of such dye are Peru and the Canary Islands. The cochineal insects are soft-bodied, flat, and oval-shaped. Female insects are distinguished from males as wingless, 5 mm in length, and form clusters on cactus pads. However, the adult males are smaller, have wings, and survive longer mainly for mating. The juvenile females rely for nourishment on cactus juices, remaining immobile until they reproduce. After mating, females enlarge in size and produce nymphs also known as crawlers. These nymphs release a waxy white coating for their protection from external factors like heat and water, which makes them appear white and grey; however, inside they produce red pigment, making their insides look dark purple. The nymph generates long wax filaments, and as they move to the edge of the cactus pad, the filaments are carried away with the wind, aiding in the dispersal of cochineals to new host cacti (Nejad and Nejad 2013).

6.2.1.2 Chemical Composition and Structure

Carminic acid is a hydroxy anthraquinone derivative present in cochineal dye along with trace amounts of kermesic and flavokermesic acids. Cochineal dye contains carminic acid (C22H20O13, Natural Red 4, C.I. 75470) as the main pigment, having a molecular weight of 492 g/mol (Mahltig 2024; Valipour et al. 2013). The chemical structure of carmine exhibits an anthraquinone chromophore, the carboxylic acid group along with a glucose unit (Fig. 6.1).

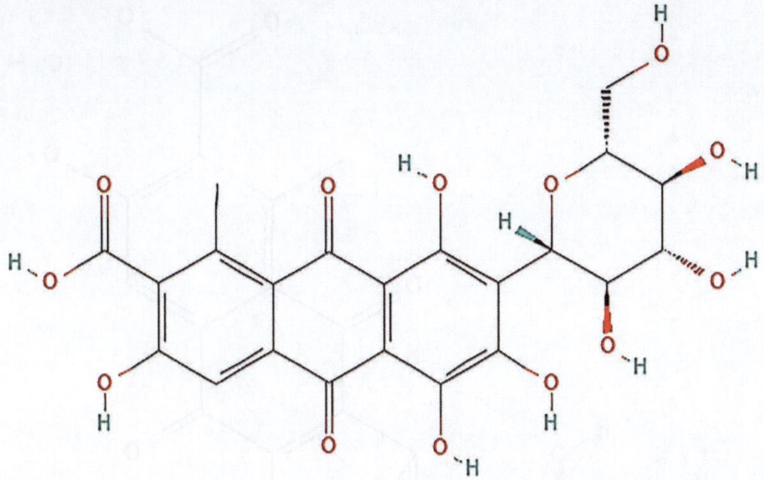

Fig. 6.1 Carminic Acid. (Source: National Centre for Biotechnology Information (2025). PubChem Compound Summary for CID 10255083, Carminic acid. Retrieved January 13, 2025 from https://pubchem.ncbi.nlm.nih.gov/compound/Carminic-acid)

This dye deprotonates in water and imparts a negative charge, showing its hydrophilic nature, but is insoluble in petroleum, ether, benzene, and chloroform. There are 14 distinct tautomeric forms of carminic acid, which depend upon varying pH values. In a similar manner, at different pH values, a range of colours can be obtained from carminic acid; for example, at acidic pH 4.5 or below, a pale orange colour is observed; light red and red are visible at neutral pH 7–7.7, and at basic pH above 12, the colour obtained is magenta red (Müller-Maatsch and Gras 2016). The orange pigment carminic acid is a hydroxy anthraquinone derivative that is produced by aqueous alkali or ethanol extraction of the female cochineal cactus insect. Harvesting and processing the insects is not an easy task, as it determines the final quality of the colour. Crude orange extracts of carminic acid are thus purified before being put into colour formulation production. Carmine acid can be transformed into carmine by chelating the carmine acid with various metal ions; however, a calcium-aluminium chelate form is most common, where calcium is used to enhance the precipitation process. Carmine likely consists of two carminic acid molecules co-ordinately bound to a single aluminium atom, forming carbonyl-hydroxyl pairs (Fig. 6.2).

In the presence of light, carmine maintains stability and also resists oxidation. Unlike carminic acid, carmine colour is pH-independent, as somewhere below pH 7, it imparts a red colour; however, at acidic pH 3.5 and below, the dye starts precipitating; hence, the usage of carmine for acidic products is avoided. Carmine provides a range of red shades like crimson, scarlet, and blue, depending on the mordants used. Using mordants like alum, chromium, iron, and copper, a range of colours from purple to grey is obtained (Gupta 2020). The appearance of carminic acid and cochineal extract at low pH is quite challenging due to the instability of the

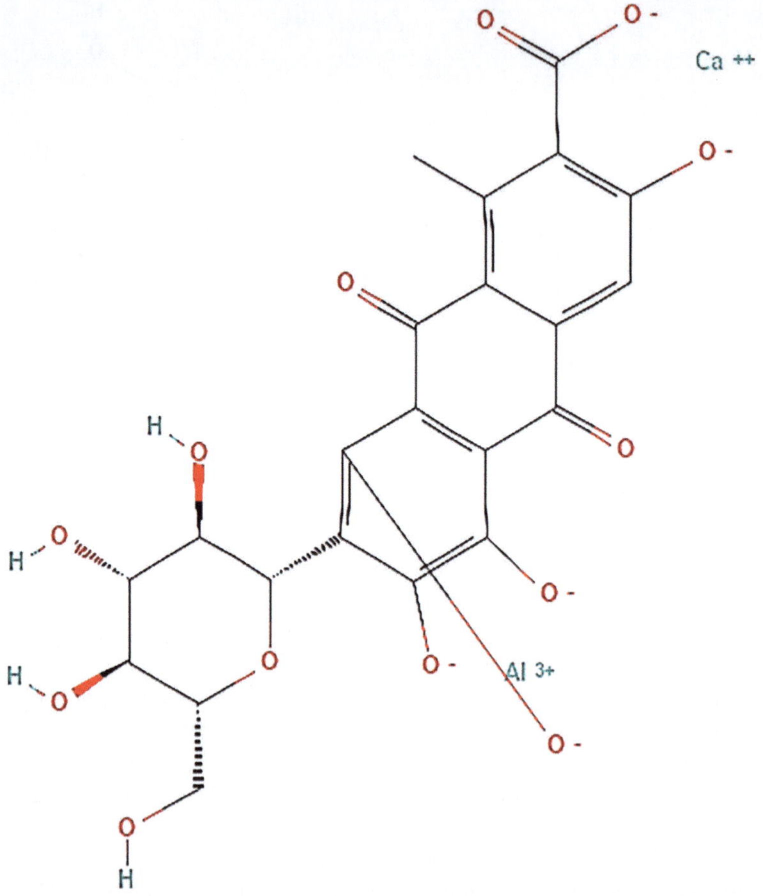

Fig. 6.2 Carmine red. (Source: National Center for Biotechnology Information (2025). PubChem Compound Summary for CID 102004931, Carmine Red. Retrieved January 13, 2025 from https://pubchem.ncbi.nlm.nih.gov/compound/Carmine-Red)

structure; a shift from a rich red hue to orange is observed. Therefore, an acid-stable derivative like 4-aminocarminic acid (Fig. 6.3) was prepared by reacting carminic acid with a hot ammonia solution, which is also widely used in the food industry (Dapson 2007).

6.2.1.3 Applications

The cochineal dye has been in use since the Renaissance period because of its vivid hues and organic appeal, which adds uniqueness and sustainability to artistic creations. These dyes are evidently used in the historical and contemporary paintings of Caravaggio, Sebastián López de Arteaga, and others (Akçay 2024). Applications of cochineal dye are expanded to several sectors; in the food industry, this dye is used quite often and is regarded as E 12. In textiles, natural fibres like silk, wool,

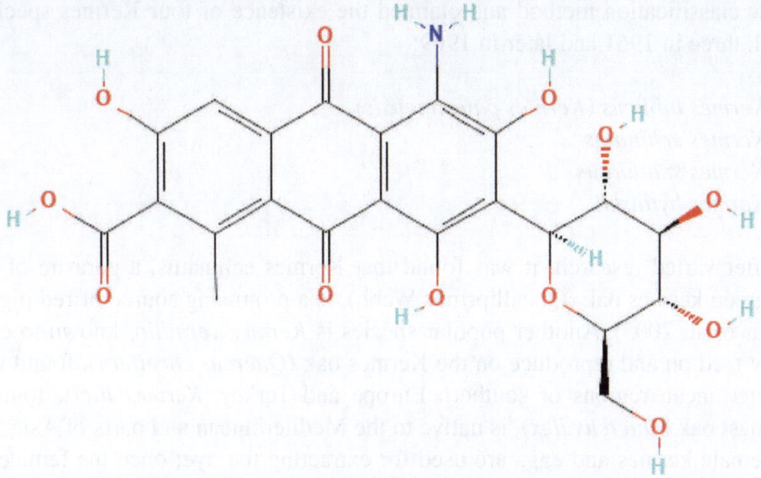

Fig. 6.3 4-Aminocarminic acid. (Source-National Center for Biotechnology Information (2025). PubChem Compound Summary for CID 102023189, 4-Aminocarminic Acid, >85%. Retrieved January 13, 2025 from https://pubchem.ncbi.nlm.nih.gov/compound/4-Aminocarminic-Acid_-_85)

cotton, and synthetic fabrics are dyed with cochineal dye; moreover, wool when dyed with carminic acid in amalgamation with metal mordants provides the fabric with antimicrobial activity (Mansour 2018). At present, it is extensively used in cosmetics and beverages as well. There are numerous applications of the dye and several benefits, but the low extraction yield and poor fastness have become a limiting factor. However, recent advancements, such as microwave-assisted extraction of the dye and the use of bio-mordants, have improved the extraction yield and fastness of the dye, ensuring its relevance as a sustainable dye (Amin et al. 2019).

6.2.2 Kermes Dyes

Kermes insects are identified as small, round, and resembling vegetable seeds, often referred to as "kermes grains" or "grana." These insects live on the oak tree and feed on the sap from leaves and stems. The female insect of this genus contains the red pigment and is sourced by drying them. Historical evidence shows the usage of kermes dyes in ancient times in regions of England, South France, Spain, Turkey, and Scotland for textile dyeing (Adeel et al. 2018).

6.2.2.1 Source and Extraction

Bodenheimer was the first researcher to conduct a study on coccoid scales in Israel and identified three Kermes species; later, another researcher, Balachowsky, renamed these species in their work. Finally, Sternlicht in his dissertation developed a new classification method and claimed the existence of four Kermes species in Israel, three in 1961 and later in 1969:

(a) *Kermes biblicus* (*Kermes palestinensis*)
(b) *Kermes echinatus*
(c) *Kermes spatulatus*
(d) *Kermes bytinskii*

After varied research, it was found that Kermes echinatus, a parasite of local evergreen kermes oak (Q. calliprinos Webb), is a promising source of red pigment (Amar et al. 2005). Another popular species is *Kermes vermilio,* known to exclusively feed on and reproduce on the Kermes oak (*Quercus coccifera*), found in the Mediterranean regions of southern Europe and Turkey. *Kermes ilicis,* found on Durmast oak (*Quercus ilex*), is native to the Mediterranean and parts of Asia. Both the female kermes and eggs are used for extracting the dye; once the females lay eggs, they are killed using vinegar fumes or by submerging them into it, and then they are dried and crushed to obtain the pigment (Seyhan et al. 2019).

6.2.2.2 Chemical Composition and Structure

The colourant obtained from the female kermes is water soluble and gives a yellowish-red colour due to the presence of Kermesic acid (CI 75460, also known as CI Natural Red 3), which is similar to carminic acid but lacks a carbohydrate moiety (Dapson 2007). Kermesic acid's structure is composed of an anthraquinone core with attached carboxylic acid and four hydroxy groups (Fig. 6.4).

During protonation, the acid group gives up hydrogen to become a negatively charged molecule exhibiting a hydrophilic nature to the compound. Because of this anionic property, they show a strong affinity for protein fibres like wool and silk.

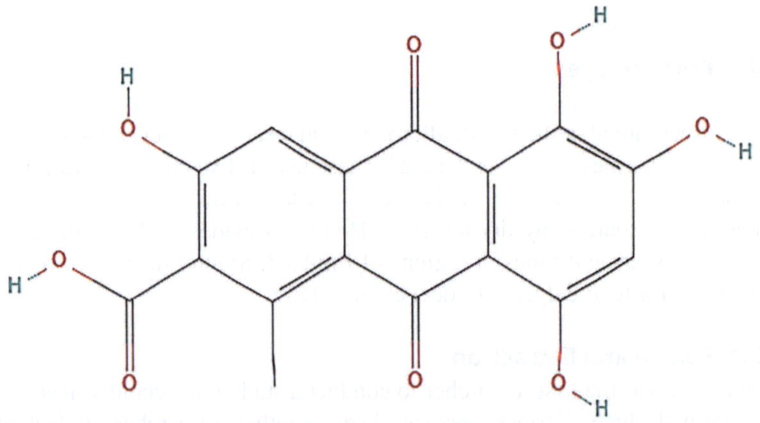

Fig. 6.4 Kermesic acid. (Source-National Center for Biotechnology Information (2025). PubChem Compound Summary for CID 11727234, Kermesic Acid. Retrieved January 13, 2025 from https://pubchem.ncbi.nlm.nih.gov/compound/Kermesic-Acid)

Another notable dye constituent in kermes is flavokermesic acid, which is similar to kermesic acid but contains one less hydroxy group (Mahltig 2024). The colour range of kermes dyes is varied depending upon the extraction process and use of mordant. Alum is the most commonly used mordant along with this dye, which produces a crimson red. However, using chrome, iron, and tin as mordants would produce purple and scarlet, respectively (Abdel-Kareem 2012). Mustafa et al. (2020) underlined the potential application of Kermes dye as a histological stain, comparing its staining properties to eosin. Furthermore, the study reported that Kermes dye exhibits a specific affinity towards collagen and muscle fibres, proposing a potential scope of this dye in studying collagen and muscle-related disorders. Overharvesting has reduced the insect's population, and its use has declined due to the availability of cochineal and synthetic alternatives.

6.2.3 Lac Dyes

Lac dye, one of the most ancient insect-derived dyes, has been utilized since prehistoric times. The term "lac" is derived from the Indian word "lakh," which refers to the small quantity of dye produced by countless insects. This dye is obtained from the resinous secretion of the lac insects, *Laccifer lacca* and *Kerria* lacca (Adeel et al. 2018). They are mainly found on tree twigs in Southeast Asia, India being the largest lac producer, with an annual production exceeding 20,000 tons, out of which approximately 75% is exported to over 100 countries. The major contributing states include (the top producer) Chhattisgarh, (second) Jharkhand, and (third) Madhya Pradesh, West Bengal, Maharashtra, Odisha, and others (Boruah and Konwar 2018).

6.2.3.1 Source and Extraction

Lac dye is derived from the insect *Coccus laccae* (*Laccifer lacca* or *Kerria lacca)* and belongs to the family Lacciferidae (Chungkrang and Bhuyan 2020). They are soft-bodied, round, and tiny parasitic insects that complete their lifecycle on host plants within 6 months. During the life cycle, these insects secrete reddish-brown resinous substances around the branches of host plants for their protection. The extraction of lac dye begins with crushing and washing sticklac to produce seedlac. During this process, water-dissolved dye is acidified with 0.1% (v/v) hydrochloric acid or sulfuric acid, which causes the dye to precipitate and settle in tanks. The extracted sludge is treated with boiling water thrice to recover the dye. The combined filtrate is treated with calcium carbonate or quicklime, precipitating the calcium salt of the dye. This is filtered, washed, suspended in cold water, acidified with 10% aqueous acid, and left at room temperature for 7–10 days to crystallize. To use the dye in food products, the previously filtered dye is dissolved in a mild alkali to filter out insoluble components, then acidified, and left to crystallize for 8 days. The final product is bright red in colour; however, in aqueous solutions (beverages and cold drinks), it produces a deep orange-red hue (Sharma et al. 2020).

6.2.3.2 Chemical Composition and Structure

Lac dye (CI Natural Red 25; CI 75450) is derived from sticklac, a resin secreted by these insects while feeding on specific host trees. It is a compound of at least five anthraquinone derivatives, i.e. Laccaic acids A, B, C, D, and E, out of which Laccaic Acid A, Laccaic Acid B, and Laccaic Acid C are two major components (Fig. 6.5) (Bashir et al. 2022).

By nature, it is acidic and soluble in water and other solvents, including methyl alcohol, amyl alcohol, acetone, acetic acid, and formic acid; however, it is insoluble in solvents like ether, chloroform, and benzene. This dye imparts a range of colours like orange-red at acidic pH and reddish-violet at alkaline pH; however, it rapidly degrades in alkaline solutions (Sharma 2023).

For centuries, lac dye has been in use and is evident through its traditional application as *alakta* or *Alta*, a liquid dye prepared by dissolving lac in water. This solution was frequently employed by women for staining their hands and feet during cultural ceremonies. In ancient India, lac dye was commonly employed for wood staining, polishing, and dyeing wool and silk. Moreover, it holds ornamental significance, being applied to gold and silver ornaments and in the crafting of traditional bangles. Lac dye is regarded as a traditional dye used for dyeing leather in China. It is also being used in the food industry for producing colours in candies, carbonated drinks, juices, jams, etc. It has also been used as varnish in composition with alcohol for wooden instruments like violins, etc (Reshma et al. 2018). Lac dye has received regulatory approval from the United States Food and Drug Administration (FDA) and the European Union (EU) due to its non-toxic and non-allergenic properties,

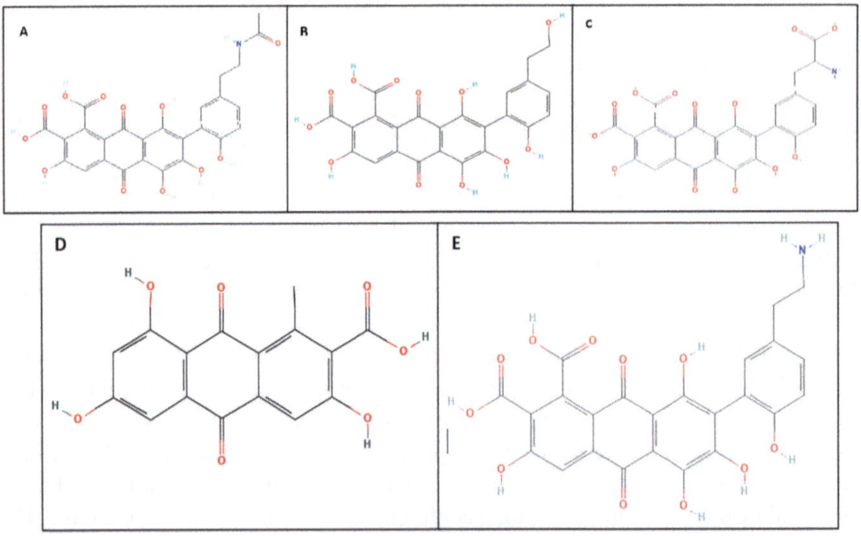

Fig. 6.5 (**a**) Laccaic acid A, (**b**) Laccaic acid B, (**c**) Laccaic acid C, (**d**) Laccaic acid D, and (**e**) Laccaic acid E. (Source: National Center for Biotechnology Information (2025). PubChem Compound Summary for CID 133109320, Natural Red 25. Retrieved January 13, 2025 from https://pubchem.ncbi.nlm.nih.gov/compound/Natural-Red-25)

making it suitable for use as a natural colourant in pharmaceutical products like capsules, tablets, and syrups. In addition to its application as a safe colourant, it also exhibits remarkable bioactive properties, including radical scavenging activity, anti-oxidative potential, anti-inflammatory effects, and antimicrobial efficacy. These attributes of lac dye promise it as a valuable natural resource and not only affirm its utilizations in current applications but also pave the way for innovative research and future advancements in diverse fields (Dagur and Ghosh 2025).

6.3 Animal-Derived Biocolours

Dyes that are based on animals' sources can be derived from various parts of the animals. The scales, bones, wings or other skin appendages, and secretions of the animals can be used for deriving colourants that can be useful in imparting colours to textiles, ceramics, pottery, and other such materials. Animal-based dyes have historical, economic, and environmental significance. They have been pivotal in traditional practices of trade and culture, bringing various shades of hues and tones to the dye industry, while posing environmental concerns like that for over-harvesting of Murex snails. The sources of colourants from animals besides cochineal include sea snails (Tyrian purple), cuttlefish (Sepia), various other molluscs (shellfish purple), fish scales (guanine), cattle bones (bone black), and squids (squid ink) (Table 6.1, Alegbe and Uthman 2024).

6.3.1 Mollusc-Derived Biocolours

6.3.1.1 Tyrian Purple

The dye derived from Atlantic and Mediterranean seasnails (Families Muricidae and Thaisidae) contains primarily 6,6'-Dibromoindigo (DBI) and is highly

Table 6.1 Animal dyes and their applications

Animal source	Name of the dye	Colour of the dye	Application
Dried body of female insect—*Dactylopius coccus*	Cochineal dye	Red	Food and Textile
Secretions of sea snails—*Murex brandaris* and *M. Trunculus*	Tyrian Purple or royal purple	Blue to purple or reddish-purple	Textile
Molluscs—*Hexaplextrunculus*	Shellfish Purple	Purple	Textile
Ink sac from cuttlefish—*Sepia officinalis*	Sepia dye	Reddish-brown	Ink, painting, dye, pharmacology
Ink sac of Squids—*Loligo* sp., *Sepia* sp.	Squid ink	Black	Food colour and flavouring
Scales of fishes—herring and sturgeon	Guanine	Silver	Cosmetics, colourant
Animal (Cattle) Bones	Bone black	Black	Paintings, inks, dyes

expensive due to its minute quantities present in the snails. This oldest and best-known dye is known by many names, such as "Royal purple," "Purple of the ancients," "True purple," "Imperial purple," "Tyrian Lila," or "Shellfish purple" (Cooksey 2001; Karapanagiotis 2019). With time, however, the production and hence the use of this dye have decreased due to the diminished number of molluscs in the seas as a result of increased pollution in water bodies and the aftereffects of global warming. Many molluscs like *T. haemastoma* have even gone extinct, while the numbers of *M. trunculus* have tremendously dwindled as per recent surveys (Karapanagiotis 2019). Precursor chromogens are present in the hypobranchial glands of the molluscs that can undergo spontaneous photo-conversion and oxidative dimerization in the presence of air and sunlight to the coloured product—DBI. Other analogous chemical variants produced include 6,6′–dibromoindirubin, 6′–bromoindirubin, 6–bromoindigotin, 6–bromoindirubin, Indigotin, and Indirubin (Karapanagiotis et al. 2006). Today it is possible to produce these chemicals in pure form and large amounts synthetically. Archaeologically, shellfish purple has been documented to be used in textiles, pottery, wall paintings, and such artworks (Karapanagiotis 2019). In the recent decades, the medicinal and nutraceutical importance of these chemicals is also being explored (Ngandjui et al. 2024). The species generally employed in the production of Tyrian purple are Bolinus *brandaris* (purple dye murex), *Murex* or *Hexaplex* or *Phyllonotus trunculus* (banded dye murex), *Stramonita* haematoma (Florida rock snail), and *Dolabella auricularia* sea slug. Tekhelet dye, or hyacinthine purple, is another violet-blue dye like Tyrian purple derived from molluscs (Florida rock snail) and was used historically for the painting of chalets by Jews (Al Ashkar and Hassabo 2021).

6.3.1.2 Inks from Squids, Cuttlefish, and Octopuses

Cephalopods are a class of marine molluscs consisting of squids, cuttlefishes, and octopuses, generally found across different oceans. All cephalopods of order Coleoidea, except a few like suborder Cirrata, possess ink sacs developed since hatching containing ink produced by these organisms. This ink comprises two component secretions, the black ink made of melanin coming from the ink gland and the mucus coming from the funnel organ (Derby 2014). Some species also secrete luminescent particles or inks (Robison et al. 2003; Otjacques et al. 2023). The ink sac and secretion of ink are under neural control, and the most ecological way of collecting ink is from the sea water when it is secreted by live organisms naturally, although it has the disadvantage of dilution and contamination due to sea water. A syringe technique of extracting ink from the ink sac from freshly killed or dead cephalopods, with or without squeezing or milking the ink gland, or else homogenizing the ink sacs and glands are other methods of collecting inks for various purposes (Wood et al. 2010). Dark-brown-coloured eumelanin is the ink found in cephalopods, and every ink sac of *Sepia* has ~1 g of eumelanin (Prota et al. 1981; Palumbo 2003). Chemically, tyrosine-derived 5,6-dihydroxyindole and 5,6-dihydro xyindole-2-carboxylic acid constitute the polymer eumelanin in the ratio 5:1. For the melanogenesis, dihydroxyphenylalanine (DOPA) and tyrosine are converted by tyrosinases to dopachrome, which in turn yields 5,6-dihydroxyindole and

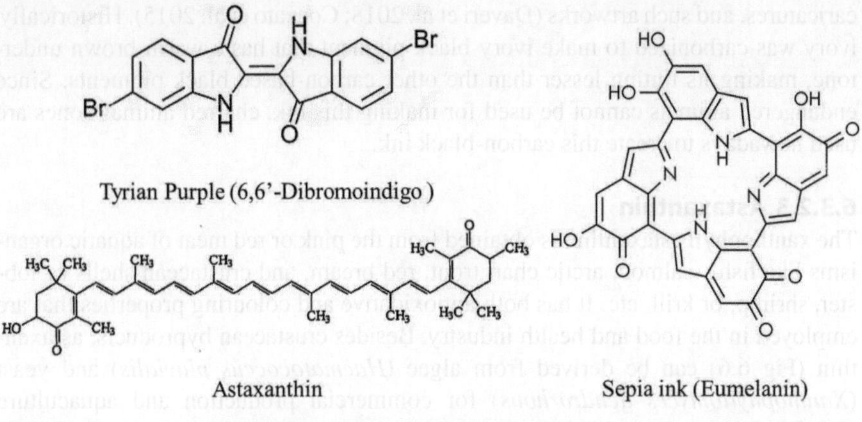

Fig. 6.6 Structures of some selected animal-derived biocolours

5,6-dihydroxyindole-2-carboxylic acids that are in turn enzymatically peroxidized to eumelanin (Prota et al. 1981) (Fig 6.6).

The sepia ink has been used since the nineteenth century for writing, making artwork like paintings, and in cosmetics (Derby 2014). Cephalopod ink, especially from cuttlefish, is used in the food industry as a flavouring agent for making black rice, sauces, soups, and caviar (Marquinet and Inaki 2003). The applications of cephalopod ink for human use include applications beyond being used as an ink, like in the fields of medicine: anti-cancer, anti-hypertensive, anti-inflammatory, anti-microbial, anti-oxidant, anti-retroviral, anti-ulcerogenic, and haematopoietic (Derby 2014).

6.3.2 Biocolours from Other Animal Sources

6.3.2.1 Pearl Essence or Guanine Silver

It is a silvery, iridescent crystalline dye present in the scales and epidermis of fishes, helping them camouflage, that was used in various cosmetics and inks or paints. It is produced commercially in many developed countries like the USA, Japan, etc. Pearl essence is found in 0.125–0.290 g/100 g of body weight and 0.07–0.2 g/g of scales in various fishes, with the highest being present in ribbon fish. It is used mostly in making artificial pearls, besides its application in the form of spray or dip paint for various items, including cosmetics and textiles, imparting them a pearly shine. Since fish scales are waste products from food industries, they can be used for making pearl essence (Mhatre et al. 2022) (Fig 6.6).

6.3.2.2 Ivory and Bone Black

The chemical composition of animal inks, like the ones derived from burnt ivory or bones, is primarily hydroxyapatite, besides carbon, magnesium, calcium, and phosphorus. They have been used extensively in writings, oil paintings, mural paintings, caricatures, and such artworks (Daveri et al. 2018; Coccato et al. 2015). Historically, ivory was carbonized to make ivory black pigment that has a warm-brown undertone, making its tinting lesser than the other carbon-based black pigments. Since endangered animals cannot be used for making this ink, charred animal bones are used nowadays to create this carbon-black ink.

6.3.2.3 Astaxanthin

The xanthophyll astaxanthin is obtained from the pink or red meat of aquatic organisms like fish—salmon, arctic char, trout, red bream, and crustacean shells of lobster, shrimp, or krill, etc. It has both antioxidative and colouring properties that are employed in the food and health industry. Besides crustacean byproducts, astaxanthin (Fig 6.6) can be derived from algae (*Haematococcus pluvialis*) and yeast (*Xanthophyllomyces dendrorhous*) for commercial production and aquaculture (Stachowiak and Szulc 2021).

6.4 Large-Scale Production of Insect- and Animal-Derived Biocolours

Red animal biocolours (pigments) are formed when lactic corrosive (Lac colour), carminic acid (cochineal), Tyrian purple, and kermes corrosive (kermes) are released from dried bodies of insects and molluscs, respectively. These potentials are popular for the pigments employed in ancient times (Yusuf et al. 2017). The cochineal insect, native to South Africa, is a rich source of vibrant red pigments. Carminic acid, the primary anthraquinone colourant, exhibits water solubility and striking colour-shifting properties. As acidity levels decrease, the colour transforms from red to violet, showcasing the unique characteristics of this natural pigment (Bowers and Sobeck 2016). Carmine, a natural colour additive, is predominantly used in the food and beverage industry, where these insects' vibrant pigments are in high demand. The production of carminic acid, also known as cochineal dye, begins with the harvesting of cochineal insects that feed on specific cactus species. As a defence mechanism, these insects produce carminic acid, which yields the distinctive red colour. Large-scale carmine production involves a multi-step process that transforms this natural pigment into a versatile colouring agent. These steps involve:

6.4.1 Cultivation and Harvesting

Cochineal insect harvesting involves cultivating these insects on cactus plants in controlled environments. Skilled labourers carefully handpick the insects from the cactus plants, primarily in countries such as Mexico and those in South America, where the ideal climate and conditions support their growth.

6.4.2 Drying and Crushing

Following collection, the cochineal insects undergo a drying process, where they are exposed to sunlight, transforming into a dense, dark red mass. This dried material is then pulverized into a fine, uniform powder, preparing it for further processing.

6.4.3 Extraction

For large-scale extraction, the dry powder of cochineal insect bodies is immersed in a preheated solvent for an optimal duration of several minutes to a few hours, with tens of minutes being the preferred timeframe. The solvent's pH level is flexible, but a neutral to alkaline range of 6–8 is recommended, with a pH of 7–8 being ideal and 7.5 being the most suitable. To achieve the desired pH, food-grade alkalizing agents such as sodium hydroxide, sodium carbonate, sodium hydrogen carbonate, or sodium citrate can be selected and added to the solvent.

6.4.4 Purification and Filtration

The filtration and purification process typically employs membrane filters to remove impurities and achieve a concentrated carmine solution. This crucial step ensures the elimination of unwanted substances, resulting in a refined and purified carmine solution that is ready for further processing.

6.4.5 Precipitation

In the precipitation stage, the carmine solution is combined with a metal salt, typically alum or calcium, triggering the carmine to settle out of the solution and form vibrant red pigment particles, effectively concentrating the colourant.

6.4.6 Drying and Milling

The final step involves collecting the precipitated carmine, drying it to remove excess moisture, and then milling it into a fine, uniform powder. This powdered pigment is now ready for incorporation into a wide range of products, from food and cosmetics to art supplies and textiles. Hence, to produce high-quality colour compositions, crude orange extracts of carminic acid undergo purification. A widely used form of carminic acid (cochineal) is carmine, which is obtained by reacting cochineal extract with aluminium sulphate in an acidic environment. This reaction results in the formation of a compound between the metal and carminic acid, shifting its absorption peak and yielding an insoluble red lake pigment through sol–gel polymerization. Further processing converts this pigment into water-dispersible powders, with the option of using calcium ions to facilitate lake precipitation.

6.5 Applications of Biocolours

Industries such as pharmacy, nutraceuticals, dairy, printing, textiles, and many more used natural colourants as quality indicators and colouring agents. A discussion of each of these uses is provided below. There are various applications of biocolours in different industries as described below.

6.5.1 Pharmaceutical Industry

From ancient times, numerous drugs have been prepared and used all over the globe to reduce deaths and injuries from severe and chronic conditions such as cancer, heart and kidney diseases, respiratory infections, and brain disorders, among others. Biocolours have several benefits to human health because they contain some physiologically active compounds having pharmacological properties such as powerful antimutagenic antioxidants, anti-inflammatory effects, and anti-arthritic properties (Hewlings and Kalman 2017). Carmine is used to colour tablets, capsules, cough syrup, and ointments in pharmaceuticals (Wang et al. 2019). These dyes are used to cure cardiac disease, cancer chemoprevention, Alzheimer's disease, and other related problems. A pigment known as 6-bromoindirubin-3'oxime (BIO) obtained from molluscs is a powerful kinase inhibitor that can be utilized as an anticancer, intrabrain, and neuroprotective inhibitor (Liu et al. 2020).

6.5.2 Food Industry

Synthetic dyes and pigments, mainly sodium nitrite, are associated with significant health issues by increasing food toxicity, leading to problems like indigestion and many more. To resolve these environmental and health-related issues, there is an urgent need for natural food colourants over artificial synthetic dyes. Nowadays scientific communities have verified that natural colours derived from animals are effective and safe for humans as food additives. In one study, lac dye was used as a food colourant from a natural source in meat and their products and imparted a vibrant red hue and also had slight antibacterial properties. Lac dye is incorporated into fruit drinks, vegetable juices, flavoured fruit items, and chocolate-containing goods due to its nontoxic property (de Vargas et al. 2017a, b). Similarly, carmine, obtained from carminic acid calcium aluminium, is non-soluble in water and broadly utilized in sausages, toppings, meat, dairy products, desserts, and also used in canned fruits like berries, cherries, jams, and pulps (de Vargas et al. 2017a, b). Thus, natural colours based on animals are widely utilized in the food industry as safer and trustworthy alternatives to synthetic pigments and dyes.

6.5.3 Fish Industry

Fish farming is an industry that is growing rapidly all over the world, which involves the cultivation of several species of fishes such as finfish, shellfish, molluscs, and ornamental fish. Consumers are gaining interest in pigmented aquatic animals, but fishes are unable to synthesize their own colouring pigments de novo, so the coloured pigments derived from some plants, animals, and other natural sources need to be incorporated into the diet. Astaxanthin has been proven as a promising pigment to impart colour to the skin of fish. Some commercially available astaxanthin (carotenoid) and their products, Phaffia Rhodozyma, and fermentation product of Xanthophyllomyces dendrorhous, have also been extensively used. Microalgae *Chlorella vulgaris* gives yellow-blue hues, yielding both muscle and skin pigmentation effects (Empis 1997).

6.5.4 Textile Industry

The textile industry directly or indirectly discharges their wastewater in various environmental components that mainly consist of synthetic and artificial dyes. These dyes have been widely used in the clothing industries due to their ease and cost-effectiveness in synthesis and high stability towards light and temperature. But these dyes are mostly hazardous, mutagenic, and carcinogenic and lead to problems such as cancer and allergic reactions in humans (Gurav et al. 2011). Because of several problems related to the usage of synthetic dyes, scientific communities advise people to promote eco-friendly dyes of natural origin that not only protect them from several health issues but also provide them with health and economic benefits. Natural colours derived from animals are very significant in several fields, mainly textiles. Lac, cochineal, and kermes are the most commonly used dyes in textile industries (Swamy et al. 2016). Scientists like Chairat et al. (2008) utilized natural colour and man-made materials with insect-based dyes. Thus, the worldwide demand for the dyes of natural origin is increasing rapidly in the textile industry.

6.6 Recent Advances in Animal and Insect-Based Biocolours

Biocolours derived from biological sources, particularly insects and animals, are gaining attention due to their biodegradability, non-toxicity, and potential health benefits. Insect-derived dyes, such as carminic acid from cochineal (Dactylopius coccus), kermesic acid from kermes (Kermes vermilio), and laccaic acid from lac (Kerria lacca), produce vibrant red and purple pigments. Similarly, marine animals like molluscs and crustaceans yield astaxanthin, Tyrian purple, and carotenoids, widely used in aquaculture, food, textiles, and cosmetics (Pereira et al., 2014; Karapanagiotis 2019; Karadag 2023; Bashir et al. 2024). Traditional extraction techniques—including aqueous and solvent-based extraction, thermal and alkaline treatments—often result in low yield, pigment degradation, and environmental concerns (Borges et al. 2012; Usman et al. 2023). To address these issues, recent innovations emphasize green and efficient extraction strategies such as enzyme-assisted, microwave-assisted, and supercritical fluid extraction (Ferreyra-Suarez et al. 2024; Al Ashkar and Hassabo 2021). Enzyme-assisted methods utilize cellulases, chitinases, and pectinases to disrupt cell walls, enhancing pigment release while preserving stability (Shen et al. 2023).

Solvent extraction remains central but has evolved to incorporate greener solvents like ethanol, methanol, and deep eutectic solvents, improving selectivity and reducing environmental impact (Mussagy et al. 2020; Zannou et al. 2024). Adjusting extraction parameters—such as temperature, time, and solvent ratio—has increased yields of compounds like carminic acid and astaxanthin (Feizi et al. 2024; Venkatachalam and Dufossé 2024). Supercritical CO_2 (SC-CO_2) extraction offers benefits like low solvent residues and thermolabile pigment protection while delivering higher yields and shorter extraction times (Muilu-Mäkelä et al. 2024). Radiation-based methods—UV, plasma, gamma, ultrasound, and microwave—enhance dye uptake and colour strength while minimizing chemical use. For example, microwave and ultrasound methods improve cell wall rupture and energy efficiency, benefiting the application of dyes like lac and cochineal in textile dyeing (Adeel et al. 2018; Ghosh et al. 2025; Deveoglu 2020). Bio-mordants have emerged as eco-friendly alternatives to conventional chemical mordants. Natural options like Ludhra and biopolymers such as chitosan enhance dye uptake and fastness while minimizing toxicity (Mohamed et al. 2022; Jimtaisong 2020). Cationic agents like polyethyleneimine (PEI) also improve bonding with insect-derived dyes (Vuthiganond et al. 2022).

Ongoing research is exploring genetic and biochemical pathways for pigment biosynthesis, offering opportunities for synthetic biology applications and tailored dye production. Additionally, valorizing waste and by-products from insect and marine industries as biocolour sources could enhance sustainability and economic viability (Karadag 2023; Sotiropoulou et al. 2021).

6.7 Conclusions and Future Perspectives

The animal-origin biocolours have been an integral part of the development and evolution of human civilization. The evidence of usage of animal-based biocolours like mollusc-derived Tyrian purple dye, insect-origin dyes like cochineal dyes, kermes dyes, and lac dyes have been found in the form of paintings, garments, and pottery at various archaeological sites apart from several written records going back thousands of years. However, owing to certain inherent properties of these dyes, like tedious extraction and processing steps, issues related to the availability of raw materials, colour-fastness, etc. these biocolours were outcompeted by the cheaper chemical dyes. But, with greater awareness related to the adverse impact of chemical colourants on health as well as the environment, there is a renewed interest in the animal-origin biocolours. Although several scientists are carrying out efforts to improve the properties and availability of these biocolours, still a lot of research efforts are needed to make the use of these dyes more economical and improve their binding properties to different fabrics or strata. Another direction that calls for more research is to make use of their other functional attributes, like medicinal properties, along with their colouring properties.

References

Abdel-Kareem O (2012) History of dyes used in different historical periods of Egypt. Res J Text Appar 16(4):79–92. https://doi.org/10.1108/rjta-16-04-2012-b009

Achim M (2021) Cochineal. Book Section, 177

Adeel S, Rafi S, Mustaan MA, Salman M, Ghaffar A (2018) Animal based natural dyes: a short review. In: Handbook of renewable materials for coloration and finishing, pp 41–74. https://doi.org/10.1002/9781119407850.ch4

Akçay AA (2024) The use of insect pigment in art-works. Insects 15(7):519. https://doi.org/10.3390/insects15070519

Al Ashkar A, Hassabo A (2021) Recent use of natural animal dyes in various field. J Text Color Poly Sci 18(2):191–210. https://doi.org/10.21608/jtcps.2021.79791.1067

Alegbe EO, Uthman TO (2024) A review of history, properties, classification, applications and challenges of natural and synthetic dyes. Heliyon 10(13):e33646. https://doi.org/10.1016/j.heliyon.2024.e33646

Amar Z, Gottlieb H, Varshavsky L, Iluz D (2005) The scarlet dye of the Holy Land. Bioscience 55(12):1080. https://doi.org/10.1641/0006-3568(2005)055

Amin N, Rehman F, Adeel S, Ahamd T, Muneer M, Haji A (2019) Sustainable application of cochineal-based anthraquinone dye for the coloration of bio-mordanted silk fabric. Environ Sci Pollut Res 27(7):6851–6860. https://doi.org/10.1007/s11356-019-06868-3

Bashir NH, Chen H, Munir S, Wang W, Chen H, Sima Y, An J (2022) Unravelling the role of lac insects in providing natural industrial products. Insects 13(12):1117. https://doi.org/10.3390/insects13121117

Bashir I, Pandey VK, Dar AH, Dash KK, Shams R, Mir SA, Fayaz U, Khan SA, Singh R, Zahoor I (2024) Exploring sources, extraction techniques and food applications: a review on biocolors as next-generation colorants. Phytochem Rev 23(4):1–26. https://doi.org/10.1007/s11101-023-09908-6

Berbers SVJ, Tamburini D, van Bommel MR, Dyer J (2019) Historical formulations of lake pigments and dyes derived from lac: a study of compositional variability. Dyes Pigments 170:107579. https://doi.org/10.1016/j.dyepig.2019.107579

Bhatia D, Sharma NR, Singh J, Kanwar RS (2017) Biological methods for textile dye removal from wastewater: a review. Crit Rev Environ Sci Technol 47(19):1836–1876

Borges ME, Tejera RL, Díaz L, Esparza P, Ibáñez E (2012) Natural dyes extraction from cochineal (*Dactylopius coccus*). New extraction methods. Food Chem 132(4):1855–1860. https://doi.org/10.1016/j.foodchem.2011.12.018

Boruah RR, Konwar M (2018) A review paper on lac and lac dye. Int J Res Appl, Nat Soc Sci 6:61–66

Bowers LMR, Sobeck SJS (2016) Impact of medium and ambient environment on the photodegradation of carmine in solution and paints. Dyes Pigments 127:18–24

Cardon D (2007) Natural dyes, sources, tradition, technology and science. Archetype, London

Chairat M, Rattanaphani S, Bremner JB, Rattanaphani V (2008) Adsorption kinetic study of lac dyeing on cotton. Dyes Pigments 76(2):435–439

Chungkrang L, Bhuyan S (2020) Natural dye sources and its applications in textiles: a brief review. Int J Curr Microbiol App Sci 9(10):261–269. https://doi.org/10.20546/ijcmas.2020.910.034

Coccato A, Jehlicka J, Moens L, Vandenabeele P (2015) Raman spectroscopy for the investigation of carbon-based black pigments. J Raman Spectrosc 46(10):1003–1015. https://doi.org/10.1002/jrs.4715

Cooksey CJ (2001) Tyrian purple: 6,6'-Dibromoindigo and related compounds. Molecules 6(9):736–769. https://doi.org/10.3390/60900736

Cooksey CJ (2018) The red insect dyes: carminic, kermesic and laccaic acids and their derivatives. Biotech Histochem 94(2):100–107. https://doi.org/10.1080/10520295.2018.1511065

Dagur P, Ghosh M (2025) Lac dye: beyond textiles and tradition, a comprehensive review of its bioactive potential. Pharmacogn Mag. https://doi.org/10.1177/09731296241305858

Dapson R (2007) The history, chemistry and modes of action of carmine and related dyes. Biotech Histochem 82(4–5):173–187. https://doi.org/10.1080/10520290701704188

Daveri A, Malagodi M, Vagnini M (2018) The bone black pigment identification by noninvasive, in situ infrared reflection spectroscopy. J Anal Methods Chem:1–8. https://doi.org/10.1155/2018/6595643

Derby C (2014) Cephalopod ink: production, chemistry, functions and applications. Mar Drugs 12(5):2700–2730. https://doi.org/10.3390/md12052700

Desai J, Chauhan J, Mankad A, Maitreya B (2023) Natural colourants: a review. Int Assoc Biol Comput Digest 2(1):261–270. https://doi.org/10.56588/iabcd.v2i1.121

Deveoglu O (2020) A review on Cochineal (*Dactylopius coccus* Costa) Dye. Res J Recent Sci 9(3):37–43

Dyer J, Tamburini D, Sotiropoulou S (2018) The identification of lac as a pigment in ancient Greek polychromy–the case of a Hellenistic oinochoe from Canosa di Puglia. Dyes Pigments 149:122–132

El Sayed MA, Ibrahim MA (2019) Natural kermes dye as an effective additive for electrochemical deposition of nickel from watts-type nickel bath. Int J Electrochem Sci 14(6):4957–4973

Empis J (1997) Use of *Chlorella vulgaris* in rainbow trout, *Oncorhynchus mykiss*, diets to enhance muscle pigmentation. J Appl Aquac 7:2

Feizi P, Maghsoudlou Y, Shahiri Tabarestani H, Jafari SM, Bahri A (2024) Comparison of ultrasound with soaking method for extracting astaxanthin pigment from shrimp (*Fenneropena eusmerguiensis*) and gammarus (*Pontogammarus maeoticus*) using ionic liquid microemulsion in water. Food Process Preserv J 16(2):53–66. https://doi.org/10.22069/fppj.2024.21715.1778

Ferreyra-Suarez D, Paredes-Vargas L, Jafari SM, García-Depraect O, Castro-Muñoz R (2024) Extraction pathways and purification strategies towards carminic acid as natural-based food colorant: a comprehensive review. Adv Colloid Interf Sci 323:103052. https://doi.org/10.1016/j.cis.2023.103052

Galappaththi M, Patabendige N (2021) Cochineal chemistry, related applications and problems: a mini review. Academia Letters. https://doi.org/10.20935/al1792

Ghosh J, Repon MR, Pranta AD, Rupanty NS, Khan F, Noor T (2025) Bioactive component integrated textiles: a promising source of medicine and healthcare. J Eng Fibers Fabrics 20:15589250241308561. https://doi.org/10.1177/15589250241308561

Gupta VK (2020) Fundamentals of natural dyes and its application on textile substrates. IntechOpen eBooks. https://doi.org/10.5772/intechopen.89964

Hewlings SJ, Kalman DS (2017) Curcumin: a review of its effects on human health. Foods 6(10):92

Jimtaisong A (2020) Aluminium and calcium lake pigments of lac natural dye. Braz J Pharm Sci 56:e18140. https://doi.org/10.1590/s2175-97902019000418140

Karadag R (2023) Establishing a new international standard for natural dyed textile goods [natural organic dye standard (NODS)]. J Nat Fibers 20(1):2162187. https://doi.org/10.1080/15440478.2022.2162187

Karagiannidou EG (2014) Colors in the prehistoric and archaic era. Chem Chronica/Association Greek Chem 76:18–25

Karapanagiotis I (2019) A review on the archaeological chemistry of shellfish purple. Sustainability 11(13):3595. https://doi.org/10.3390/su11133595

Karapanagiotis I, De Villemereuil V, Magiatis P, Polychronopoulos P, Vougogiannopoulou K, Skaltsounis A (2006) Identification of the coloring constituents of four natu-

ral indigoid dyes. J Liq Chromatogr Relat Technol 29(10):1491–1502. https://doi.org/10.1080/10826070600674935
Liu X, Tang H, Liu J, Wang X (2020) hsa_circRNA_101237: a novel diagnostic and prognostic biomarker and potential therapeutic target for multiple myeloma. Cancer Manag Res:2109–2118
Mac Donald N (2019) The priestly vestments. In: Berner C, Schäfer M, Schott M, Schulz S, Weingärtner M (eds) Clothing and nudity in the Hebrew bible and the ancient near east. T&T Clark, London, pp 435–448
Mahltig B (2024) Overview on natural dyes and their IR-spectra - part IV: dyes originated from insects. Commun Develop Assembl Text Prod 5(1):107–119. https://doi.org/10.25367/cdatp.2024.5.p107-119
Mansour R (2018) Natural dyes and pigments: extraction and applications. In: Handbook of renewable materials for coloration and finishing, pp 75–102. https://doi.org/10.1002/9781119407850.ch5
Marquinet A, Inaki J (2003) Process for producing a food colorant, colorant thus obtained and uses Thereof (Patent No. 6,329,010). U.S. Patent
Mhatre NM, Muniv NY, Thakre NR (2022) The extraction of pearl essence from different types of fishes. Int J Adv Res Sci Commun Technol:237–239. https://doi.org/10.48175/ijarsct-3094
Mohamed A, Shaarawy S, Elshemy N, Hebeish A, Hassabo A (2022) Treatment of cotton fabrics using polyamines for improved coloration with natural dyes extracted from plant and insect sources. Egypt J Chem. https://doi.org/10.21608/ejchem.2022.137464.6053
Muilu-Mäkelä R, Brännström H, Weckroth M, Kohl J, Da Silva Viana G, Diaz M, Ghalibaf M, Hagner M, Hiidenhovi J, Ilvesniemi J, Jyske T, Järvenpää E, Katajajuuri J-M, Kilpeläinen P, Korpinen R, Kumar A, Kunnas S, Linnakoski R, Marnila P et al (2024) Valuable biochemicals of the future: the outlook for bio-based value-added chemicals and their growing markets. Natural Resources Institute Finland, Luonnonvarakeskus. https://jukuri.luke.fi/handle/10024/555418
Müller-Maatsch J, Gras C (2016) The "Carmine Problem" and potential alternatives. Elsevier eBooks, pp 385–428. https://doi.org/10.1016/b978-0-08-100371-8.00018-x
Mussagy CU, Santos-Ebinuma VC, Kurnia KA, Dias ACRV, Carvalho P, Coutinho JAP, Pereira JFB (2020) Integrative platform for the selective recovery of intracellular carotenoids and lipids from Rhodotorulaglutinis CCT-2186 yeast using mixtures of bio-based solvents. Green Chem 22(23):8478–8494. https://doi.org/10.1039/D0GC02992K
Mustafa NE, Elmahi OM, Waggiallah HA, Eltayeb LB (2020) Kermes dye extract from coccus ilicis insect as an alternative counter stain instead of eosin in various tissue constituents: an experimental study. Pharmacophore 11(5):14–19
Nejad HE, Nejad AE (2013) Cochineal (Dactylopius coccus) as one of the most important insects in industrial dyeing. Int J Adv Biol Biomed Res 1(11):1302–1308
Ngandjui YAT, Kereeditse TT, Kamika I, Madikizela LM, Msagati TAM (2024) Nutraceutical and medicinal importance of marine molluscs. Mar Drugs 22(5):201. https://doi.org/10.3390/md22050201
Otjacques E, Pissarra V, Bolstad K, Xavier JC, McFall-Ngai M, Rosa R (2023) Bioluminescence in cephalopods: biodiversity, biogeography and research trends. Front Mar Sci 10. https://doi.org/10.3389/fmars.2023.1161049
Palumbo A (2003) Melanogenesis in the ink gland of *Sepia officinalis*. Pigment Cell Res 16(5):517–522. https://doi.org/10.1034/j.1600-0749.2003.00080.x
Pereira D, Valentão P, Andrade P (2014) Marine natural pigments: chemistry, distribution and analysis. Dyes Pigments 111:124–134. https://doi.org/10.1016/j.dyepig.2014.06.011
Prota G, Ortonne JP, Voulot C, Khatchadourian C, Nardi G, Palumbo A (1981) Occurrence and properties of tyrosinase in the ejected ink of cephalopods. Comp Biochem Physiol Part B Comp Biochem 68(3):415–419. https://doi.org/10.1016/0305-0491(81)90149-8
Repon M, Islam T, Islam T, Ghorab AE, Rahman MM (2023) Cleaner pathway for developing bioactive textile materials using natural dyes: a review. Environ Sci Pollut Res 30:48793–48823. https://doi.org/10.1007/s11356-023-26131-0
Reshma BV, Manohar RN, Anaha VI (2018) A review on *Laccifer lacca*. World J Pharma Res 7(10). https://doi.org/10.20959/wjpr201810-12282

Robison BH, Reisenbichler KR, Hunt JC, Haddock SHD (2003) Light production by the arm tips of the deep-sea cephalopod *Vampyroteuthis infernalis*. Biol Bull 205(2):102–109. https://doi.org/10.2307/1543231

Seyhan SA, Demirbağ C, Dölen E (2019) Identification of the main dyestuffs obtained from Kermes (Kermes vermilio) in the Northwest of Turkey. AnaleleUniversităţii "Ovidius" Constanţa. Seria Chimie/"Ovidius" Univ Ann Chem 30(1):1–4. https://doi.org/10.2478/auoc-2019-0001

Shamim G, Ranjan SK, Pandey DM, Ramani R (2014) Biochemistry and biosynthesis of insect pigments. Eur J Entomol 111:149–164

Sharma KK (2023) Advances in LAC insect culture research and scope for innovative entrepreneurship using natural resins and gums. Indian J Entomol:40–51. https://doi.org/10.55446/ije.2022.913

Sharma KK, Chowdhury AR, Srivastava S (2020) Chemistry and applications of LAC and its byproduct. Springer eBooks, pp 21–37. https://doi.org/10.1007/978-3-030-36610-0_2

Shen N, Ren J, Liu Y, Sun W, Li Y, Xin H, Cui Y (2023) Natural edible pigments: a comprehensive review of resource, chemical classification, biosynthesis pathway, separated methods and application. Food Chem 403:134422. https://doi.org/10.1016/j.foodchem.2022.134422

Sotiropoulou S, Karapanagiotis I, Andrikopoulos KS, Marketou T, Birtacha K, Marthari M (2021) Review and new evidence on the molluscan purple pigment used in the early late bronze age Aegean Wall paintings. Heritage 4(1):171–187. https://doi.org/10.3390/heritage4010010

Stachowiak B, Szulc P (2021) Astaxanthin for the food industry. Molecules 26(9):2666. https://doi.org/10.3390/molecules26092666

Sukenik N, Iluz D, Varvak A, Amar Z (2016) A re-evaluation of the textile dyes in the cave of letters. In: Ortiz J, Alfaro C, Turell L, Martínez MJ (eds) Ancient textiles and dyes in the ancient mediterranean world. Publicacions de la Universitat de València, Valencia, pp 263–273

Sukenik N, Iluz D, Amar Z, Varvak A, Shamir O, Ben-Yosef E (2021) Early evidence of royal purple dyed textile from Timna Valley (Israel). PLoS One 16(1):e0245897. https://doi.org/10.1371/journal.pone.0245897

Swamy VN, Gowda KN, Sudhakar R (2016) Dyeing of silk using *Madhuca longifolia* as natural dye source. Ind J Fibre Text Res (IJFTR) 40(4):419–424

Usman M, Rehman F, Afzal M, Javed M, Ibrahim M, Amin N, Adeel S, Imran M, Mansour R (2023) Sustainable appraisal of lac (*Kerria Lacca*) based anthraquinone natural dye for chemical and bio-mordanted viscose and silk dyeing. Sci Prog 106(4):00368504231215944. https://doi.org/10.1177/00368504231215944

Valipour P, Ekrami E, Nateri AS (2013) Colorimetric properties of wool dyed with cochineal: effect of Dye-bath pH. Prog Colour Color Coat 7(2):129–138

de Vargas EF, Jablonski A, Flôres SH, Rios ADO (2017a) Obtention of natural dyes from industrial blackberry pulp residues (Rubus sp). J Food Process Preserv 41(1):e12777

de Vargas EF, Jablonski A, Flôres SH, Rios ADO (2017b) A stepwise approach for the detection of carminic acid in saffron with regard to religious food certification. Food Chem 267:410–419

Venkatachalam M, Dufossé L (2024) Red biotechnological colorants for application in plant-based foods. https://doi.org/10.13140/RG.2.2.35639.94881

Vuthiganond N, Chitichotpanya P, Nakpathom M, andMongkholrattanasit, R. (2022) Ecological dyeing of acrylic yarn with colorant derived from natural lac dye. J Metals Mater Minerals 32(4):71–78. https://doi.org/10.55713/jmmm.v32i4.1529

Wang M, Firrman J, Liu L, Yam K (2019) A review on flavonoid apigenin: dietary intake, ADME, antimicrobial effects, and interactions with human gut microbiota. Biomed Res Int 2019(1):7010467

Wood JB, Maynard AE, Lawlor AG, Sawyer EK, Simmons DM, Pennoyer KE, Derby CD (2010) Caribbean reef squid, *Sepioteuthis sepioidea*, use ink as a defense against predatory French grunts, *Haemulon flavolineatum*. J Exp Mar Biol Ecol 388(1–2):20–27. https://doi.org/10.1016/j.jembe.2010.03.010

Yusuf M, Shabbir M, Mohammad F (2017) Natural colorants: historical, processing and sustainable prospects. Nat Prod Bioprospect 7:123–145

Zannou O, Oussou KF, Mohammed S, Chabi IB, Kpoclou YE, Tekgüler B, Dabadé DS, Koca I, Esatbeyoglu T (2024) Application of emerging technologies for extraction of pigments using green solvents: case of deep eutectic solvents combined with ultrasound-assisted extraction technique. Biomass Convers Biorefinery. https://doi.org/10.1007/s13399-024-05674-3

Gurav AA, Ghosh JS, Kulkarni GS (2011) Decolorization of anthroquinone based dye Vat Red 10 by Pseudomonas desmolyticum NCIM 2112 and Galactomyces geotrichum MTCC 1360. Inter J Biotechnol Mol Biol Res 2(6):93–97

Analytical Techniques in Biocolor Analysis

7

Poonam Jyoti, Jyoti Bhardwaj, and Sudesh Kumar Yadav

Abstract

With this increasing demand of sustainable, eco-friendly, and non-toxic colors, there is an increase in the use of bio pigments in various industries such as food, cosmetics, pharmaceuticals, and many more. Bio pigments have the ability to provide vibrant colors at the same time prioritizing safety or sustainability. However, their use is limited because of the challenges associated with their stability, complexity, and regulatory concerns. Advancements in analytical methods help in overcoming some of these challenges for example improving production efficiency, detecting impurities and optimizing the storage conditions. This chapter explores the key analytical techniques used to address the issues related to biocolor production and analysis. It also delves into how future studies and advance techniques could revolutionize the biocolor analysis without damaging or altering the biocolor properties and enhancing their production and scalability.

Keywords

Biocolors · Analytical techniques · Pigment quantification · Sustainability

---

Author Sudesh Kumar Yadav is the co-corresponding author for this chapter.

P. Jyoti (✉) · J. Bhardwaj · S. K. Yadav
Department of Biotechnology, CSIR-IHBT, Palampur, India
e-mail: poonamjyoti@ihbt.res.in

R. K. Bhatia, A. Walia (eds.), *Biocolours*,
https://doi.org/10.1007/978-981-95-1738-1_7

7.1 Introduction to Biocolor Analysis

Color of any product is an important visual character that plays a vital role in its consumption. For example, the color of food, beside its flavor has huge impact in stimulating our appetite and making overall experience more desirable. This is how the use of dyes and pigments from ancient times evolved into the development of synthetic colorants. Due to their easy availability and lower cost, synthetic colors quickly became the preferred choice, primarily for their stability compared to the natural colors (Yang et al. 2011). However, these synthetic colorants come with numerous harmful effects (Dey and Nagababu 2022). As awareness of environmentally friendly and safer alternatives grows, there is an increasing demand for natural or biocolors in several industries such as food, cosmetics, pharmaceuticals, and textiles (Mohammad et al. 2022; Renita et al. 2023; Bashir et al. 2024). The global demand for the bio-based food colors in 2023 was worth 1.86 billion U.S. dollars, and is expected to reach approximately 2.61 billion dollars by 2028 (Shahbandeh 2024 Statista report).

However, despite the promising potential of biocolors, there are several challenges like inconsistent composition, stability of biocolor, co-elution, ethical concerns, and regulations which hamper their large-scale production. Analytical methods have important role in the reliable production of these pigments (Fig. 7.1) right from the assessment of raw material used for the biocolor production, exact composition, characterization, purity, concentration measurements to stability, storage, and safety assessment (Parmar and Singh 2018; Meenakshi and Chauhan 2018; Kaur et al. 2019; Hagos et al. 2022).

Potential analytical techniques used in biocolor analysis include spectroscopic methods, chromatographic techniques and mass spectrometry which provide the detailed information about the biocolor including molecular composition, purity, and stability (Custodio et al. 2024). Recent advancements in the analytical methods, such as hyperspectral imaging and multivariate statistical analysis have a huge impact on addressing the challenges associated with the correct assessment of biocolors (Zahra et al. 2024; Guerri et al. 2024). These analytical advancements not only ensure the quality of biocolors but also help optimize the production processes for enhanced yields and sustainability. In this chapter, we have compiled a range of analytical techniques (Table 7.1) that are commonly used in biocolor production, providing insights to improve the overall production and application of biocolors.

7.2 Spectroscopic Techniques for Biocolor Identification and Quantification

For the characterization and analysis of biocolors, the spectroscopic techniques are UV-Visible, Fluorescence Spectroscopy, Fourier Transform Infrared (FTIR) Spectroscopy, and Raman Spectroscopy which are widely used. These analytical methods provide information related to the molecular structure of the biocolor as well as also give information of functional groups along with detailed qualitative and quantitative insights.

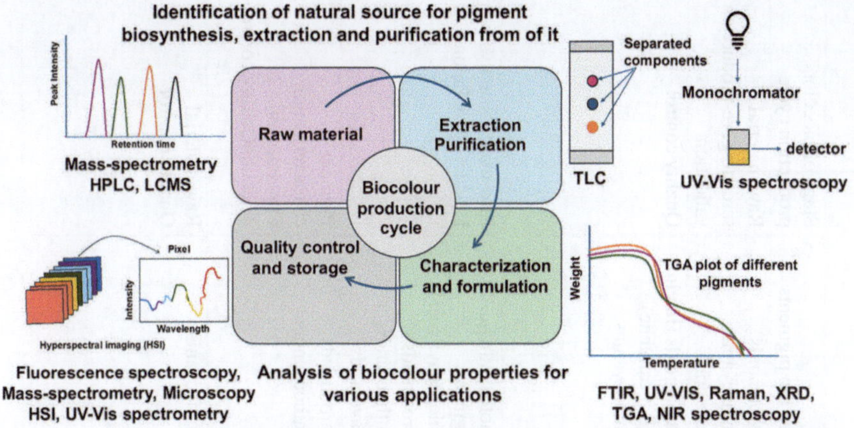

Fig. 7.1 Various analytical platforms employed throughout the biocolor production cycle. Mass spectrometry is used for raw material analysis, while Thin-Layer Chromatography (TLC) and UV-Vis spectroscopy are commonly applied during the extraction of biocolor from biological sources. Fourier Transform Infrared (FTIR) spectroscopy, Thermogravimetric Analysis (TGA), and X-ray Diffraction (XRD) are utilized for the characterization of biocolor. Fluorescence spectroscopy and Hyperspectral Imaging (HSI) are used for monitoring the degradation and ensuring the quality control of the biocolor. These analytical methods are integral at different stages throughout the entire biocolor production lifecycle

7.2.1 UV-Visible Spectroscopy

This method is widely used for primary identification of the bio pigments (Narendrababu and Shishupala 2017; Sharma and Ghoshal 2020), where pigment molecules absorb the light in the ultraviolet (UV, 100–380 nm) and visible (Vis, up to 800 nm) regions and emit to generate the structure of the pigment molecules (Rocha et al. 2018). This method is widely used for quick assessment of concentration of the ingredients. For the quantification of molecules, the Beer-Lambert Law is often applied in UV-Visible Spectroscopy, which states that absorption of light by the sample is directly proportional to the concentration of the sample and path length (Verma and Mishra 2018).

$$A = abc$$

where

- A is the absorbance
- a is the absorptivity
- c is the concentration of the pigment
- b is the path length

Near UV-Vis spectroscopy is used for the various bio pigment analysis, typically the pigments like carotenoids and chlorophylls. For example, in extra virgin olive

Table 7.1 Various analytical platforms and their application in biocolor analysis

S. No.	Analytical platform	Application	Industrial use	Key bio pigments	Stage of biocolor production cycle
1.	Thin-layer chromatography	Primary screening Quick identification and separation	Research & Development (initial pigment profiling)	Carotenoids Anthocyanins Flavonoids	Raw material screeningPre-production validation
2.	UV-visible spectroscopy	Quick assessment of biocolor concentration Composition analysis Monitoring color stability	Food & beverage Cosmetics (color composition) Textiles (color consistency)	Carotenoids (lutein and β-carotene) Anthocyanins	Quality control
3.	High-performance liquid Chromato-graphy (HPLC)	Identification Separation Quantification Impurity detection	Food & Beverage (Colorant quality and stability) Cosmetics (colorant stability) Pharmaceuticals (ingredient identification)	Photosynthetic pigments Chlorophylls and carotenoids Anthocyanins Flavonoids	Raw material analysis Product formulation
4.	Mass spectrometry	Detailed profiling Identification of trace components Detection of contaminants	Food & Beverage (Authenticity testing) Pharmaceuticals (Composition analysis and quality check)	Carotenoids Anthocyanins	Pigment characterization Contaminant detectionQuality control
5.	Matrix-assisted laser desorption ionization (MALDI)	High-throughput analysis Structural information Interaction of bio pigments with other proteins	To check stability, Degradation and color modification in various products like food, cosmetics, textile, and pharmaceuticals	Anthocyanins	Raw material Quality control

6.	Microscopy	Localization and distribution of bio pigments Monitoring minor colorants with high sensitivity Properties of biocolor	Nanoparticles loaded with bio pigments Insights of sources of biocolor production	Carotenoids Anthocyanins	Raw data analysis Real-time production monitoringquality control
7.	X-ray diffraction (XRD)	Crystalline and amorphous properties of the biocolor, Stability Solubility	Cosmetics (analyze contaminants) Color consistency Food adulteration	Anthocyanins	Help characterizing the changes in the physical properties Efficiency of encapsulation processes
8.	Thermogravimetric analysis (TGA)	Thermal stability Moisture content	Food industry (color property change with temperature) Pharmaceutical (effect of temperature on ingredient color)	Betalains Anthocyanins	Quality control
9.	Hyperspectral imaging (HSI)	Real-time analysis Non-invasive More accuracy	Food industry (analyze contaminants) Color consistencyAgriculture (detection of ripening) Pharmaceutical (analyze contaminants)	Photosynthetic pigments Anthocyanins	Real-time color MonitoringFinal product inspection

oils, to quantify the pigments the absorption spectrum between 390 and 720 nm were used and four pigments were observed which were carotenoids (lutein and β-carotene) and two chlorophylls (pheophytin-a and pheophytin-b) (Domenici et al. 2014; Borello and Domenici 2019). For curcumin analysis, curcumin ethanol solution was used and the absorption was measured at 421nm while for β-carotene, hexane solution was used and the absorption was measured at 454 nm and for lycopene the absorption wavelength was 503 nm in UV-Vis spectrometer (Kabir et al. 2024).

7.2.2 Fluorescence Spectroscopy

Some biocolors absorb light in the UV range and they emit light of a longer wavelength and this emitted light intensity is proportional to the concentration of the biocolor which is used in fluorescence spectroscopy method. This method is more sensitive compared to UV-Vis spectroscopy and widely used for phenolics detections. In the skin of berries, flavanols and hydroxycinnamic acids were observed to be the main phenolics after fluorescence microscopy (Kolb et al. 2003; Cerovic et al. 2008; Roca et al. 2016).

7.2.3 Fourier Transform Infrared (FTIR) Spectroscopy

FTIR Spectroscopy is an excellent spectroscopy method to identify the functional groups present in the bio pigment sample and also to characterize the chemical structure or secondary structural changes (Rocha et al. 2023). FTIR rely on an interferometer (a signal generated by the interference between two IR beams), to collect the spectrum of infrared light absorbance (Valand et al. 2020). Attenuated Total Reflectance-Fourier Transform Infrared Spectroscopy (ATR-FTIR) method allow the characterization of the functional groups, which help in the assessment of pigment stability under different temperature and pH conditions. The bands characterized by ATR-FTIR in a study of betalains from amaranth plants were linked to the conservation and degradation processes of betalains (Lopez et al. 2023).

7.2.4 Raman Spectroscopy

It is a widely used, non-destructive method of pigments analysis. When a sample interact with the light, it scattered light of same frequency, but a small fraction of scattered light has different frequency due to inelastic scattering, which generate a Raman spectrum and can be recorded over 4000–10 cm^{-1} (Bumbrah and Sharma 2016). Pigments extracted from microbial cultures have often strong and distinct signals therefore, this technique is extensively used in recent years to identify the microbial pigments such as chlorophylls, carotenoids, and scytonemin (Jehlička et al. 2014).

Raman spectroscopy is important for the identification and quantification of the carotenoids such as lutein, zeaxanthin, and beta-carotene present in the biological samples. Raman spectroscopy helps in overcoming the challenges associated with the differentiation of carotenoids, especially xanthophylls like lutein and zeaxanthin. Carotenoids are similar in structure and the only difference is the length of the conjugated chain but Raman spectroscopy can differentiate them by analyzing the frequency shifts which occurs due to the carbon-to-carbon double bond (C=C) of carotenoids that correlates with the conjugation length (Udensi et al. 2022).

7.3 Chromatographic Techniques for Separation and Purification

Thin-Layer Chromatography (TLC) and High-Performance Liquid Chromatography (HPLC) are important techniques used to identify, separate, quantify, and characterize biocolors. While HPLC is highly detailed and quantitative method for pigment analysis, TLC is a simpler and faster way for preliminary screening.

1. *TLC* involves a flow of liquid (developing solvent) over sorbent which is the dissolved sample on a flat plate already layered with silica gel (SiO_2) (Silver 2020). TLC was used for the separation of the chlorophylls and carotenoids from different algae (green, brown), diatoms, dinoflagellates, chrysomonads, and the plant leaves. Various pigments were separated in this study using TLC which includes chlorophylls (a, b, and c), pheophytins (a and b), carotenes, lutein, neoxanthin, peridinin, neoperidinin, and some other pigments (Jeffrey 1968; Akase et al. 1999).
2. *HPLC* is a good method to separate the components from a mixture or solution. In this method, mobile phase (solvent) is introduced with high pressure into the column (stationary phase), which separate the compounds based on their affinity for the stationary phase (Lozano et al. 2018; Akash and Rehman 2020). Solvent gradients, composition of mobile phase, and choice of stationary phase play important role in optimizing pigment separation. For example, anthocyanins are water-soluble pigments (polar) having the aromatic rings with substituent groups, which require polar mobile phase while carotenoids are fat-soluble colorants and may need more non-polar solvent. HPLC with UV–Vis or photodiode array detectors (PDA) are the most used methods for anthocyanins separation. The reverse phase (RP) HPLC using C18 column helped separating photosynthetic pigments like neoxanthin, lutein, chlorophyll, and carotene (De et al. 1989). RP-HPLC was also used for the identification and quantification of betalain pigments from 37 species Amaranthaceae and a total of 16 betacyanins and 3 betaxanthins were studied using the RP-HPLC method (Cai et al. 2005). HPLC or ultrahigh-performance liquid chromatography (UPLC) are the choice of techniques for carotenoids and chlorophyll analysis (Sousa 2022).
3. *Gas Chromatography (GC) and Liquid Chromatography (LC)* are commonly used method for quantitative analyses owing to their higher separation capacity

and excellent sensitivity (David and Moldoveanu 2022). GC-MS was utilized along with other techniques like HPLC and column chromatography to extract, purify, and analyze the furfural, a brown colored pigment from *T. viride.* (Chitale et al. 2012).

7.4 Mass Spectrometry (MS) for Structural and Molecular Analysis

Mass spectrometry techniques are quite sensitive for the identification and quantification of compounds based on mass to charge ratio. MS include an ion source, mass analyzer, and a detector. For ionization, EI (electron ionization), ESI (electron spray ionization), and MALDI (matrix-assisted laser desorption ionization) are commonly used while time-of-flight (TOF), ion trap, orbitrap, and quadrupole are commonly used for bio pigment analysis (Sousa 2022). Tandem mass spectrometry (tandem MS) or MS/MS are type of MS where two or more than two mass analyzers are involved which are separated by collision cell. Ions get fragmented in the collision cells and resulted fragments associated with the molecular structures of the bicolor (Sousa 2022). MS/MS is an important approach to identify the intermediate products or end products of compounds during degradation. When combined with chromatography techniques like HPLC or GC, MS are ideal for identification of bio pigments, analyzing their stability or degradation under the influence of environmental factors like temperature, and pH (Grauwet et al. 2015).

When GC coupled with mass spectrometry (GC-MS), this combination helps in the identification of molecules based on their mass/charge (m/z) and the fragmentation pattern. This combination is important for identifying small molecules from a mixture of plant extract. GC-MS is used to analyze the biocolor degradation such as flavonoids (Colombini et al. 2007), effect of high pressure and high temperature on carrot pigments (Kebede et al. 2014; Grauwet et al. 2015). LC-MS due to its high specificity and sensitivity is an important tool to detect the compounds present in trace amounts. LC-MS help profiling the biocolors with precise identification and also play important role in detecting the impurities (Lohachoompol et al. 2008). High-resolution Quadrupole Time-of-Flight (QToF) (Quadrupole Time-of-Flight) MS was used to determine the fragmentation pathways of certain dyes, including aminoorceins and hydroxyorceins (Lech and Fornal 2020). Matrix-Assisted Laser Desorption/Ionization (MALDI) is an ideal technique to generate detailed mass profiles of complex mixtures and structural identification by generating high-throughput analysis (Wingerath et al. 1996; Singhal et al. 2015). MALDI is a useful technique to find the interaction of these bio pigments with other proteins (Ferrer et al. 2015).

7.5 Electrophoretic Techniques for Charge-Based Separation

Capillary Electrophoresis (CE) and Gel Electrophoresis are useful techniques which help in studying the charged pigments and also the pigment-protein interactions. TLC, GC, HPLC are excellent methods for pigment analysis but these methods can be challenging for low concentration samples where CE offers high resolution and quick analysis with minimal amount of sample. There are several modes of separation included in CE where capillary zone electrophoresis (CZE) and micellar electrokinetic chromatography (MEKC) are suitable for small molecules. CZE is widely used approach owing to its ability to separate both anions and cations (Watanabe and Terabe 2000). Here the separation of components is based on their electrophoretic mobility which is directly proportional to the charge and inversely to the viscosity of solvent and atom radius. Various bio-pigments such as anthocyanins, betalains, Caramel Classes, isobetanin and their aglycones are separated by CE (Stuppner and Egger 1996; Acunha et al. 2016; Da Costa et al. 2000; Delgado et al. 2000; Watson et al. 2004).

Gel Electrophoresis, such as agarose gel electrophoresis and polyacrylamide gel electrophoresis (PAGE), is used to study pigment-associated proteins or pigment-protein complexes, this understanding is crucial to know how pigments are stabilized in the plant systems or their function in photosystems (Gong et al. 2022).

7.6 Structural Analysis Techniques for Biocolors

For comprehensive analysis of biocolors (Fig. 7.2), analytical techniques such as Nuclear Magnetic Resonance (NMR) Spectroscopy, X-Ray Diffraction (XRD), and X-Ray Fluorescence (XRF) play important role. NMR provides the molecular structure information along with the purity of the pigments, XRD provides the information related to the crystallinity and physical properties, which help deciphering the stability of the compound while XRF plays significant role in the detection of metal ions and the contaminants present in the biocolors.

1. *NMR spectroscopy* is a non-invasive method which provide information about the chemical environment of hydrogen and carbon atoms with the application of strong magnetic field. For NMR, sample of either liquid or solid-state is placed in a strong magnetic field and provided with radiofrequency radiation. These radiations resonate the nuclei of certain atoms at specific frequencies and result in chemical shifts, which provide information of the molecular structure (Gerothanassis et al. 2002).

   NMR is used for extensive characterization of bio pigments such as betacyanin, anthocyanin-flavanol pigments (1D and 2D NMR), ethyl-linked anthocyanin–flavanol (Stintzing et al. 2004; Mateus et al. 2002; Lee et al. 2004). The interaction of anthocyanins with other phenolic compounds play crucial role in the stability and can lead to color alterations, these changes can be monitored

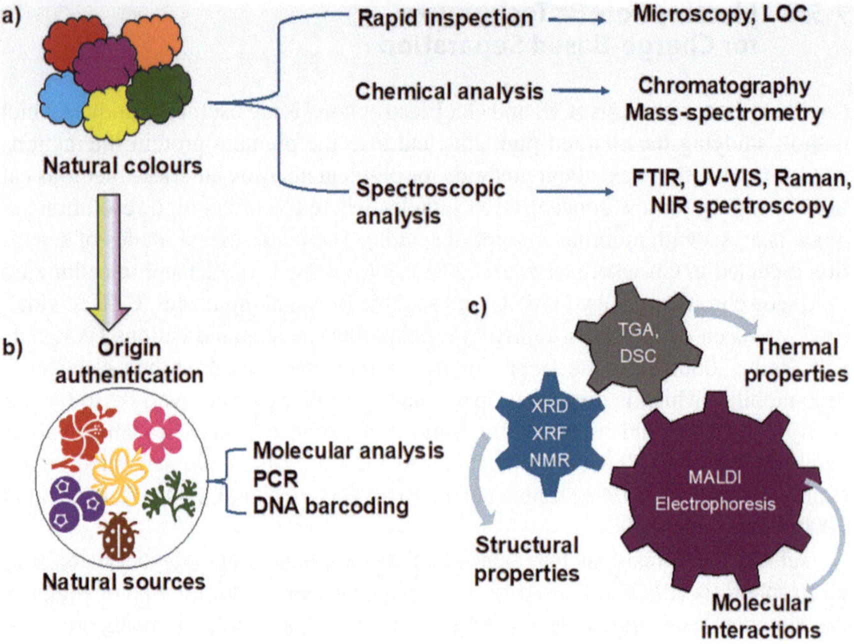

Fig. 7.2 Analytical methods are crucial for detecting the contamination or adulteration in biocolor. This ensures the purity and authenticity of both raw materials and final products. (**a**) Microbial contamination in the raw materials can be identified using microscopy or Lab-on-a-Chip (LOC) technology which offers high-throughput screening for detecting specific contaminants while the presence of foreign substances can be detected using mass spectrometry, or spectroscopy-based techniques. (**b**) In the final bio-color product, molecular methods like PCR-based methods or DNA barcoding can be employed for detecting the origin source, ensuring the authentication and confirming that biocolor is free from unauthorized adulteration. (**c**) To detect other impurities or alterations in the biocolor composition advanced methods such as Thermogravimetric Analysis (TGA), X-ray Diffraction (XRD), or Nuclear Magnetic Resonance (NMR) etc., can be used to ensure purity and safety

through NMR and are believed to be responsible for alterations in color (Mateus et al. 2002). Acylated anthocyanins possess great stability by enhancing self-association of anthocyanins. The findings from NMR NOESY showed that malvidin-3-O-(6-coumaroylglucoside) form intra molecular association with the coumaric acid on the anthocyanidin which contributed to aggregation phenomenon. These findings highlight the role of acylation for enhancing stability of anthocyanins, which could be a beneficial biocoloring application (Fernandes et al. 2015).

2. *XRD* is useful to analyze the physical properties of the sample, stability, solubility, and color properties (West et al. 2013; Wang et al. 2019). It utilizes a monochromatic beam of X-rays and generates the diffraction patterns by constructive interference of scattered X-rays (Bunaciu et al. 2015). The demand of biocolors is increasing in various industries but their stability is a major concern. Several

attempts are being done to enhance stability by immobilization. For example, anthocyanin extracted from purple sweet potato was immobilized on different inorganic solids such as smectite clays and a natural montmorillonite (Ogawa et al. 2017), black carrot anthocyanins and halloysite hybrid (Ghavami et al. 2024). XRD help characterizing these hybrids by providing information about crystalline and amorphous properties. Changes in the physical properties indicate the efficiency of encapsulation processes (Sarkar et al. 2021; Ghavami et al. 2024; Lee et al. 2013).

3. *XRF* is widely used as non-invasive, sensitive, and high-resolution technique for characterization and quantification of mineral pigments (Popelka et al. 2016). Other advantages of XRF are cost effectiveness, easy to use, high accuracy, and quick and precise analysis within few minutes. Its portable form can be used in the field also (Panebianco et al. 2022). In XRF, the object is bombarded with X-rays, which result in emission of X-rays through the object. By analyzing the intensity of emitted rays, XRF identifies the elemental composition (Pinto et al. 2018). XRF is mainly used for the analysis of impurities, metal ions, contaminants, assessment of quality, and elemental distribution in the plant tissues (Margui et al. 2007; Sarret et al. 2013; Sharma et al. 2018; Panebianco et al. 2022; Bisht and Garg 2022).

7.7 Thermal Analysis Techniques for Stability Testing

Stability of pigments is a crucial component for industrial applications. Due to their natural origin they tend to degrade easily with temperature variations during processing and storage. Thermal analysis techniques, including Differential Scanning Calorimetry (DSC) and Thermogravimetric Analysis (TGA), are used to check their thermal stability of bio pigments (Nguyen and Bechtold 2021; Oancea 2021) DSC measures the heat flow and provides phase transitions information, melting points, and thermal stability while TGA measures the change in mass when sample is heated at a controlled rate and provide information on volatility, moisture content, and degradation temperature (Müller et al. 2020). These methods are helpful to assess the temperature ranges for the biocolors when they undergo structural changes or degrade, for example, anthocyanin extracted from red onion cultivars were studied through DSC and found that their degradation starts when temperature goes over 45° (Oancea et al. 2013), and betalains from beet root (20–27°C) (Hernández et al. 2021). Pigment hybrids are created by combining the natural pigments with more stable compounds, for example, through encapsulation. This hybrid method enhance the stability of natural pigments. TGA method is widely used to characterize these hybrids and assess their thermal stability. For example, anthocyanin/montmorillonite hybrids have been studied using TGA (Li et al. 2021), and purpurin combined with different clay minerals also characterized using TGA (Szadkowski et al. 2023). Similarly, encapsulated bioactive compounds, including phenolics and carotenoids, have been analyzed for their stability using TGA (Bodbodak et al. 2024.

These methods help finalize the processing or storage conditions for the products which include biocolors. For example, microencapsulated and spray dried powder of anthocyanins from *Bactris guineensis* fruit is stable up to 100°C and the optimal storage conditions are below 37°C and under 76% relative humidity (Osorio et al. 2010).

7.8 Microscopy and Imaging Techniques for Morphological Analysis

Microscopic techniques such as Scanning electron microscopy (SEM), Transmission electron microscopy (TEM), Confocal Microscopy, Fluorescent Microscopy, and Atomic Force Microscopy (AFM) are also important for the characterization of biocolors. SEM provides the high-resolution images examining the morphology and surface texture of powdered pigments. SEM is also helpful to understand the morphological changes in the pigment after processing or encapsulation (Milea et al. 2019; Begum and Deka 2017; Da Fonseca et al. 2018). TEM provides detailed structural information as it provides internal structure of pigment or nanoparticle loaded biocolor or encapsulated biocolors and help in understanding the pigment properties and stability (Ge et al. 2018; Tong et al. 2020).

Confocal microscopy uses laser to scan the samples and provide high-resolution, 3D images which are useful for investigating the localization and distribution of bio pigments inside the biological sample (Marangoni et al. 1996; Vermaas et al. 2008). This technique is also useful in studying the biological interactions of the pigments. Fluorescent microscopy is similar to confocal microscopy but this technique is useful in studying fluorescent molecules like chlorophyll or anthocyanins, which exhibit intrinsic fluorescence (Oxborough 2004; Dima et al. 2006). AFM is a scanning probe type of microscopy, used for scanning biological samples with a sharp tip which scan across the sample. This method is widely used for the characterization of surface properties of the samples during preservation, processing, and storage mechanisms. This method is also applicable for knowing the morphological and structural properties of nanoparticles, bioactive compounds and nanoencapsulations (Raliya et al. 2014; He et al. 2019; Romero et al. 2020; Mezacasa et al. 2020; Huang et al. 2024).

7.9 Colorimetric and Visual Analysis Techniques

Color consistency is an important parameter of quality control (Fig. 7.2) for natural pigments, which can be analyzed by analytical platforms like Tristimulus Colorimetry and digital image analysis software, both these methods complement each other and provide precise information of color. Tristimulus colorimetry utilize three color channels red (R), green (G), and blue (B) and generate XYZ values in response to light emitted by the sample. A calorimeter or spectrometer measures the emitted or reflected light across different wavelengths and analyze the color that human eye can perceive (Gardner 2007). Widely used RGB systems are CIE XYZ, Hunter L a b, CIE L*u*v*, CIE Yxy, and CIE LCH (Pathare et al. 2013). Tristimulus Colorimetry is widely used to characterize bio pigment contents under different conditions. For example, the changes in carotenoids content in oranges (Meléndez et al. 2003); betacyanin content in Amaranth (Arunachalam et al. 2016); anthocyanins from Jaboticaba (Montes et al. 2005); anthocyanins from purple corn (Yang et al. 2008); and carotenoids from pumpkin (Kinsman et al. 2024). Digital image analysis software uses high resolution cameras and sensors to generate detailed image of bio pigments and the strains producing bio pigments (Neuman et al. 1989; Kamath et al. 2005; Pathare et al. 2013; Xia et al. 2016; Alves et al. 2024).

7.10 Molecular Biology Techniques for Source Verification

Molecular techniques such as Polymerase Chain Reaction (PCR) are useful for identification of the biological sources or finding pigment producing genes in various organisms (Fig. 7.2; Lyu et al. 2022). For example, characterization of filamentous fungi *Aspergillus niger*, *Aspergillus tamarii* for natural pigments (Toma et al. 2021; Saravanan et al. 2020); characterization of bixin biosynthetic pathway in *Bixa orellana* seedlings (Sankari et al. 2019); *Nigrospora aurantiaca* analysis for red pigment (Suwannarach et al. 2019); characterization of orange fruit for Anthocyanin Biosynthetic Genes (Cotroneo et al. 2006); and purple potato tubers for anthocyanin (Wu et al. 2024). PCR is also an excellent approach for checking adulteration and ensuring the authenticity and traceability of biocolors. Oh and Jang (2020) developed real-time PCR based markers to check the adulteration of turmeric powder with corn powder (Oh and Jand 2020). A specific genome region like ribosomal RNA (rRNA) gene or the internal transcribed spacer (ITS) region are Fpen amplified and sequenced in DNA barcoding technique to identify the biological sources (Anantha and Johnson 2019). This helps in tracing the origin of biocolor and identifying the plant species or authentication of the biological sources (Poovitha et al. 2016; Nehal et al. 2021; Salo et al. 2021).

7.11 Emerging and Advanced Techniques in Biocolor Analysis

Non-destructive analysis is an emerging way of analysis without damaging or altering the properties of the sample, which mainly utilize spectroscopy, hyperspectral imaging, and fluorescence sensing along with machine learning and AI applications (Manzoor et al. 2022).

Hyperspectral imaging (HSI) is an advanced technology with imaging and spectroscopy enabling both spectral and spatial data analysis simultaneously. HSI records spectral information of wavelength bands from visible to infrared light (Plaza et al. 2009). HSI is widely used non-destructive method for natural pigment analysis, e.g., photosynthetic pigments like phycoerythrin, phycocyanin, and allophycocyanin in *Neopyropia yezoensis* (Che et al. 2023); Pigments from *Chromobacterium violaceum* (Gallardo et al. 2014); and anthocyanin content analysis in mulberry (Li et al. 2023). HSI has revolutionized food industries by making the analysis of biocolors more accurate and real time, for example, predicting color in fruits like strawberry, analyzing the quality of fruits like apples, predicting sugar content, moisture content, acidic content, and so on (ElMasry et al. 2007; Dai et al. 2015). This method is also helpful to check the pigment after processing like detection of bioactive compounds in Dancong tea (semi-fermented oolong) for chlorophyll, carotenoids, and tea polyphenols using visible-near-infrared and short-wave-near-infrared HIS (Long et al. 2024), chlorophyll, and carotenoids content in stacked matcha tea (He et al. 2024), and color analysis of orange peel in Sausages (Feng 2024).

Lab-on-a-chip (LOC) is an emerging potential technique for rapid and low-cost analysis of contaminants including microbial contamination (Javanifar et al. 2025; Lu et al. 2024) and metal ion contamination (Aryal et al. 2024). LOC is suitable for real time analysis with portable sensors and help in quick decision making. Paper LOC (pLOC) is more advantageous with its intrinsic capillary network, which requires low volume of sample and have better sensitivity for food colorants (Kapoor et al. 2024). AI applications, machine learning, computer vision, and deep learning are revolutionizing the pigment analysis making it rapid and accurate. AI models trained to analyze the patterns such as artificial neural network and adaptive neuro fuzzy inference system help finding co-relation and optimizing conditions for better extraction of pigments (fucoxanthin) and models like convolutional neural networks help in quantification (Chong et al. 2023). AI help in HSI data analysis and interpretation enabling more precise and efficient analysis (Khonina et al. 2024). Integrating AI applications with image and spectral analysis enables spatial–temporal detection and efficient management (Ali et al. 2024).

7.12 Conclusion

Biocolor analysis is an important aspect of the natural color industry and various analytical methods play significant role in accurate and reliable detection. Key techniques include spectrophotometric, colorimetric, and fluorescence detection methods while microscopy-based techniques help visualize the change in color. Biocolor analysis, is challenging owing to the complexity of biological sources, variations in pigment distribution, and interference of non-pigment components or contaminants. With the advent of advancements in analytical methods, such as hyperspectral imaging and multivariate analysis, we can address these issues. These methods are reported to improve not only the accuracy and precision, but also for sensitivity which enable them the detection of even the subtle changes in color. Advanced algorithms and machine learning tools enable automated and real-time analysis of biocolors. Integration of biocolor analysis with sensors, smartphone apps, and other analytical techniques will enable consistent quantifications. These developments in biocolor analysis make biocolor detection a powerful tool with broad applications in healthcare research, and environmental sciences.

References

Acunha T, Ibáñez C, García-Cañas V, Simó C, Cifuentes A (2016) Recent advances in the application of capillary electromigration methods for food analysis and Foodomics. Electrophoresis 37:111–141

Akase SI, Yoshikawa T, Sakata T (1999) Pigment analysis of marine microalgae by TLC and HPLC methods. Aquaculture Sci 47:511–518

Akash MS, Rehman K (2020) High performance liquid chromatography. Essentials of pharmaceutical analysis:175–184

Ali F, Razzaq A, Tariq W, Hameed A, Rehman A, Razzaq K, Sarfraz S, Rajput NA, Zai HE, Shahid MS, Ondrasek G (2024) Spectral intelligence: AI-driven hperspectral imaging for agricultural and ecosystem applications. Agronomy 14:260

Alves V, Dos Santos JM, Viegas O, Pinto E, Ferreira IM, Lima VA, Felsner ML (2024) An eco-friendly approach for analysing sugars, minerals, and colour in brown sugar using digital image processing and machine learning. Food Res Int 191:114673

Anantha Narayana DB, Johnson ST (2019) DNA barcoding in authentication of herbal raw materials, extracts and dietary supplements: a perspective. Plant Biotechnol Report 13:201–210

Arunachalam V, Dhargalkar SD, Vainganakar JD, Kevat N (2016) Pigment rich Amaranth by tristimulus colorimetry and progeny test. Nat Acad Sci Letter 39:411–415

Aryal P, Indrianingsih AW, Henry CS (2024) Smartphone-enabled green anthocyanin sensor for Fe (III) sensing on paper using capillary-driven microfluidics. Green Analytic Chem 8:100091

Bashir I, Pandey VK, Dar AH, Dash KK, Shams R, Mir SA, Fayaz U, Khan SA, Singh R, Zahoor I (2024) Exploring sources, extraction techniques and food applications: a review on biocolors as next-generation colorants. Phytochem Rev 12:1–26

Begum YA, Deka SC (2017) Stability of spray-dried microencapsulated anthocyanins extracted from culinary banana bract. Int J Food Prop 20:3135–3148

Bisht A, Garg N (2022) AMF modulated rhizospheric microbial enzyme activities and their impact on sulphur assimilation along with thiol metabolism in pigeonpea under Cd stress. Rhizosphere 21:100478

Bodbodak S, Nejatian M, Ghandehari Yazdi AP, Kamali Rousta L, Rafiee Z, Jalali-Jivan M, Kharazmi MS, Jafari SM (2024) Improving the thermal stability of natural bioactive ingredients via encapsulation technology. Crit Rev Food Sci Nutr 64:2824–2846

Borello E, Domenici V (2019) Determination of pigments in virgin and extra-virgin olive oils: a comparison between two near UV-vis spectroscopic techniques. Foods 8:18

Bumbrah GS, Sharma RM (2016) Raman spectroscopy–basic principle, instrumentation and selected applications for the characterization of drugs of abuse. Egypt J Forensic Sci 6:209–215

Bunaciu AA, UdriŞTioiu EG, Aboul-Enein HY (2015) X-ray diffraction: instrumentation and applications. Crit Rev Anal Chem 45:289–299

Cai Y, Sun M, Corke H (2005) HPLC characterization of betalains from plants in the Amaranthaceae. J Chromatogr Sci 43:454–460

Cerovic ZG, Moise N, Agati G, Latouche G, Ghozlen NB, Meyer S (2008) New portable optical sensors for the assessment of winegrape phenolic maturity based on berry fluorescence. J Food Compos Anal 21:650–654

Che S, Du G, Zhong X, Mo Z, Wang Z, Mao Y (2023) Quantification of photosynthetic pigments in *Neopyropia yezoensis* using hyperspectral imagery. Plant Phenomic 5:0012

Chitale A, Jadhav D, Waghmare S, Ranveer RC (2012) Production & characterization of brown coloured pigment from *Trichoderma viride*. Elec J Env Agricult Food Chem Title 11:529–537

Chong JW, Tang DY, Leong HY, Khoo KS, Show PL, Chew KW (2023) Bridging artificial intelligence and fucoxanthin for the recovery and quantification from microalgae. Bioengineered 14:2244232

Colombini MP, Andreotti A, Baraldi C, Degano I, Łucejko JJ (2007) Colour fading in textiles: a model study on the decomposition of natural dyes. Microchem J 85:174–182

Cotroneo PS, Russo MP, Ciuni M, Recupero GR, Piero AR (2006) Quantitative real-time reverse transcriptase-PCR profiling of anthocyanin biosynthetic genes during orange fruit ripening. J Am Soc Hortic Sci 131:537–543

Custodio-Mendoza JA, Aktaş H, Zalewska M, Wyrwisz J, Kurek MA (2024) A review of quantitative and topical analysis of anthocyanins in food. Molecules 29:1735

Da Costa CT, Horton D, Margolis SA (2000) Analysis of anthocyanins in foods by liquid chromatography, liquid chromatography–mass spectrometry and capillary electrophoresis. J Chromatogr A 881:403–410

Da Fonseca Machado AP, Rezende CA, Rodrigues RA, Barbero GF, Rosa PD, Martínez J (2018) Encapsulation of anthocyanin-rich extract from blackberry residues by spray-drying, freeze-drying and supercritical antisolvent. Powder Technol 340:553–562

Dai Q, Cheng JH, Sun DW, Zeng XA (2015) Advances in feature selection methods for hyperspectral image processing in food industry applications: a review. Crit Rev Food Sci Nutr 55:1368–1382

David V, Moldoveanu SC (2022) Recent applications of gas chromatography in bioanalysis. In: Novel Aspects of Gas Chromatography and Chemometrics, IntechOpen

De Las RJ, Abadia A, Abadia J (1989) A new reversed phase-HPLC method resolving all major higher plant photosynthetic pigments. Plant Physiol 91:190–192

Delgado-Vargas F, Jiménez AR, Paredes-López O (2000) Natural pigments: carotenoids, anthocyanins, and betalains-characteristics, biosynthesis, processing, and stability. Crit Rev Food Sci Nutr 40:173–289

Dey S, Nagababu BH (2022) Applications of food color and bio-preservatives in the food and its effect on the human health. Food Chem Adv 1:100019

Dima E, Manetas Y, Psaras GK (2006) Chlorophyll distribution pattern in inner stem tissues: evidence from epifluorescence microscopy and reflectance measurements in 20 woody species. Trees 20:515–521

Domenici V, Ancora D, Cifelli M, Serani A, Veracini CA, Zandomeneghi M (2014) Extraction of pigment information from near-UV vis absorption spectra of extra virgin olive oils. J Agric Food Chem 62:9317–9325

ElMasry G, Wang N, ElSayed A, Ngadi M (2007) Hyperspectral imaging for nondestructive determination of some quality attributes for strawberry. J Food Eng 1:98–107

Feng CH (2024) Colour analysis of sausages stuffed with modified casings added with citrus peel extracts using hyperspectral imaging combined with multivariate analysis. Sustainability 16:8683

Fernandes A, Brás NF, Mateus N, de Freitas V (2015) A study of anthocyanin self-association by NMR spectroscopy. New J Chem 39:2602–2611

Ferrer-Gallego R, Soares S, Mateus N, Rivas-Gonzalo J, Escribano-Bailón MT, Freitas VD (2015) New anthocyanin–human salivary protein complexes. Langmuir 31:8392–8401

Gallardo MJ, Staforelli JP, Meza P, Bordeu I, Torres S (2014) Characterization of *Chromobacterium violaceum* pigment through a hyperspectral imaging system. AMB Express 4:1–9

Gardner JL (2007) Comparison of calibration methods for tristimulus colorimeters. J Res Natl Inst Stand Technol 112:129

Ge J, Yue P, Chi J, Liang J, Gao X (2018) Formation and stability of anthocyanins-loaded nanocomplexes prepared with chitosan hydrochloride and carboxymethyl chitosan. Food Hydrocoll 74:23–31

Gerothanassis IP, Troganis A, Exarchou V, Barbarossou K (2002) Nuclear magnetic resonance (NMR) spectroscopy: basic principles and phenomena, and their applications to chemistry, biology and medicine. Chem Educ Res Pract 3:229–252

Ghavami A, Zamiri B, Mohebali A (2024) Anthocyanin/halloysite hybrid pigments with enhanced thermal and acid stability. Dyes Pigments 231:112356

Gong J, Zhang H, Zeng Y, Cheng Y, Sun X, Wang P (2022) Combining BN-PAGE and microscopy techniques to investigate pigment-protein complexes and plastid transitions in citrus fruit. Plant Methods 18:124

Grauwet T, Kebede BT, Delgado RM, Lemmens L, Manzoni F, Vervoort L, Hendrickx M, Stephen Elmore J, Van Loey A (2015) Evaluating the potential of high pressure high temperature and thermal processing on volatile compounds, nutritional and structural properties of orange and yellow carrots. Eur Food Res Technol 240:183–198

Guerri MF, Distante C, Spagnolo P, Bougourzi F, Taleb-Ahmed A (2024) Deep learning techniques for hyperspectral image analysis in agriculture: a review. ISPRS 30:100062

Hagos M, Redi-Abshiro M, Chandravanshi BS, Yaya EE (2022) Development of analytical methods for determination of β-carotene in pumpkin (*Cucurbita maxima*) flesh, peel, and seed powder samples. Int J Analytic Chem 1:9363692

He S, Zhang N, Jing P (2019) Insights into interaction of chlorophylls with sodium caseinate in aqueous nanometre-scale dispersion: color stability, spectroscopic, electrostatic, and morphological properties. RSC Adv 9:4530–4538

He Q, Liu Z, Li X, He Y, Lin Z (2024) Detection of the pigment distribution of stacked matcha during processing based on hyperspectral imaging technology. Agriculture 14:2033

Hernández-Aguirre OA, Muro C, Hernández-Acosta E, Alvarado Y, Díaz-Nava MD (2021) Extraction and stabilization of betalains from beetroot (*Beta vulgaris*) wastes using deep eutectic solvents. Molecules 26:6342

Huang W, Hua MZ, Li S, Chen K, Lu X, Wu D (2024) Application of atomic force microscopy in the characterization of fruits and vegetables and associated substances toward improvement in quality, preservation, and processing: nanoscale structure and mechanics perspectives. Crit Rev Food Sci Nutr 64:11672–11700

Javanifar R, Dabagh S, Kaya M, Sengel SB, Ebrahimi A, Ghorbanpoor H, Avci H (2025) A novel green microfiltration approach by developing a lab-on-a-chip system: a case study for *Escherichia coli*. Sep Purif Technol 353:128411

Jeffrey SW (1968) Quantitative thin-layer chromatography of chlorophylls and carotenoids from marine algae. Biochimica et Biophysica Acta (BBA)-Bioenergetics 162:271–285

Jehlička J, Edwards HG, Oren A (2014) Raman spectroscopy of microbial pigments. Appl Environ Microbiol 80:3286–3295

Kabir FN, Palleb SR, Halim MA, Nayan SB, Uddin MB, Aziz MG (2024) Extraction of natural food color from tomato, carrot and turmeric. Eur J Nutr Food Safety 5:175–188

Kamath SB, Chidambar S, Brinda BR, Kumar MA, Sarada R, Ravishankar GA (2005) Digital image processing-an alternate tool for monitoring of pigment levels in cultured cells with special reference to green alga *Haematococcus pluvialis*. Biosens Bioelectron 21:768–773

Kapoor A, Ramamoorthy S, Sundaramurthy A, Vaishampayan V, Sridhar A, Balasubramanian S, Ponnuchamy M (2024) Paper-based lab-on-a-chip devices for detection of agri-food contamination. Trends Food Sci Technol 147:104476

Kaur P, Ghoshal G, Jain A (2019) Bio-utilization of fruits and vegetables waste to produce β-carotene in solid-state fermentation: characterization and antioxidant activity. Process Biochem 76:155–164

Kebede BT, Grauwet T, Palmers S, Vervoort L, Carle R, Hendrickx M, Van Loey A (2014) Effect of high pressure high temperature processing on the volatile fraction of differently coloured carrots. Food Chem 153:340–352

Khonina SN, Kazanskiy NL, Oseledets IV, Nikonorov AV, Butt MA (2024) Synergy between artificial intelligence and hyperspectral imagining—a review. Technologies 12:163

Kinsman ME, Serviss MT, Meru G, Chase CA, Sargent SA, Simonne A, MacIntosh AJ (2024) A greener approach to assess bioactive compounds in tropical pumpkin (*Cucurbita moschata*) using colorimetry. Front Sustain Food Syst 8:1480964

Kolb CA, Kopecký J, Riederer M, Pfündel EE (2003) UV screening by phenolics in berries of grapevine (*Vitis vinifera*). Funct Plant Biol 30:1177–1186

Lech K, Fornal E (2020) A mass spectrometry-based approach for characterization of red, blue, and purple natural dyes. Molecules 25:3223

Lee DF, Swinny EE, Jones GP (2004) NMR identification of ethyl-linked anthocyanin–flavanol pigments formed in model wine ferments. Tetrahedron Lett 45:1671–1674

Lee SV, Halim NA, Arof AK, Abidin ZH (2013) Characterisation of poly (vinyl alcohol) coating mixed with anthocyanin dye extracted from roselle flower with different nitrate salt. Pigm Resin Technol 42:146–151

Li SE, Mu B, Ding JJ, Zhang H, Wang XW, Wang AQ (2021) Fabrication of anthocyanin/montmorillonite hybrid pigments to enhance their environmental stability and application in allochroic composite films. Clay Clay Miner 69:142–151

Li X, Wei Z, Peng F, Liu J, Han G (2023) Non-destructive prediction and visualization of anthocyanin content in mulberry fruits using hyperspectral imaging. Front Plant Sci 14:1137198

Lohachoompol V, Mulholland M, Srzednicki G, Craske J (2008) Determination of anthocyanins in various cultivars of highbush and rabbiteye blueberries. Food Chem 111:249–254

Long T, Tang X, Liang C, Wu B, Huang B, Lan Y, Xu H, Liu S, Long Y (2024) Detecting bioactive compound contents in Dancong tea using VNIR-SWIR hyperspectral imaging and KRR model with a refined feature wavelength method. Food Chem 460:140579

López MA, Luna-Suárez S, Macuil RJ, Cárdenas FD (2023) Simple and efficient protocol for amaranth betalains extraction and stability analysis by ATR-FTIR spectroscopy. J Cereal Sci 113:103745

Lozano-Sánchez J, Borrás-Linares I, Sass-Kiss A, Segura-Carretero A (2018) Chromatographic technique: high-performance liquid chromatography (HPLC). Modern Techniques for Food Authentication, pp 459–526

Lu Y, Zhang J, Lu X, Liu Q (2024) Isothermal nucleic acid amplification based microfluidic "lab-on-a-chip" for the detection of pathogenic bacteria and viruses in agri-foods. Trends Food Sci Technol 7:104482

Lyu X, Lyu Y, Yu H, Chen W, Ye L, Yang R (2022) Biotechnological advances for improving natural pigment production: a state-of-the-art review. Bioresour Bioprocess 9:8

Manzoor MF, Hussain A, Naumovski N, Ranjha MM, Ahmad N, Karrar E, Xu B, Ibrahim SA (2022) A narrative review of recent advances in rapid assessment of anthocyanins in agricultural and food products. Front Nutr 9:901342

Marangoni R, Cubeddu R, Taroni P, Valentini G, Sorbi R, Lorenzini E, Colombetti G (1996) Microspectrofluorometry, fluorescence imaging and confocal microscopy of an endogenous pigment of the marine ciliate *Fabrea salina*. J Photochem Photobiol B Biol 34:183–189

Marguí E, Hidalgo M, Queralt I (2007) X-ray fluorescence spectrometry for trace element analysis of vegetation samples. Spectrosc Eur 19:13–17

Mateus N, Silva AM, Santos-Buelga C, Rivas-Gonzalo JC, de Freitas V (2002) Identification of anthocyanin-flavanol pigments in red wines by NMR and mass spectrometry. J Agric Food Chem 50:2110–2116

Meenakshi NR, Chauhan A (2018) Extraction and characterization of biocolors from bacterial isolates of *Pseudomonas* sp. M1 and MS2. Phytomedicine 7:63–68

Meléndez-Martínez AJ, Vicario IM, Heredia FJ (2003) Application of tristimulus colorimetry to estimate the carotenoids content in ultrafrozen orange juices. J Agric Food Chem 51:7266–7270

Mezacasa AV, Queiroz AM, Graciano DE, Pontes MS, Santiago EF, Oliveira IP, Lopez AJ, Casagrande GA, Scherer MD, dos Reis DD, Oliveira SL (2020) Effects of gold nanoparticles on photophysical behaviour of chlorophyll and pheophytin. J Photochem Photobiol A Chem 389:112252

Milea AȘ, Vasile AM, Cîrciumaru A, Dumitrașcu L, Barbu V, Râpeanu G, Bahrim GE, Stănciuc N (2019) Valorizations of sweet cherries skins phytochemicals by extraction, microencapsulation and development of value-added food products. Foods 8:188

Mohammad Azmin SN, Sulaiman NS, Mat Nor MS, Abdullah PS, Abdul Kari Z, Pati S (2022) A review on recent advances on natural plant pigments in foods: functions, extraction, importance and challenges. Appl Biochem Biotechnol 194:4655–4672

Montes C, Vicario IM, Raymundo M, Fett R, Heredia FJ (2005) Application of tristimulus colorimetry to optimize the extraction of anthocyanins from Jaboticaba (*Myricia Jaboticaba* Berg.). Food Res Int 38:983–988

Müller L, Rubio-Pérez G, Bach A, Muñoz-Rujas N, Aguilar F, Worlitschek J (2020) Consistent DSC and TGA methodology as basis for the measurement and comparison of thermo-physical properties of phase change materials. Materials 13:4486

Narendrababu BN, Shishupala S (2017) Spectrophotometric detection of pigments from *Aspergillus* and *Penicillium* isolates. J Appl Biol Biotechnol 5:53–58

Nehal N, Choudhary B, Nagpure A, Gupta RK (2021) DNA barcoding: a modern age tool for detection of adulteration in food. Crit Rev Biotechnol 41:767–791

Neuman MR, Sapirstein HD, Shwedyk E, Bushuk W (1989) Wheat grain colour analysis by digital image processing I. Methodol J Cereal Sci 10:175–182

Nguyen HL, Bechtold T (2021) Thermal stability of natural dye lakes from Canadian Goldenrod and onion peel as sustainable pigments. J Clean Prod 315:128195

Oancea S (2021) A review of the current knowledge of thermal stability of anthocyanins and approaches to their stabilization to heat. Antioxidants 10:1337

Oancea S, Drăghici O (2013) pH and thermal stability of anthocyanin-based optimised extracts of Romanian red onion cultivars. Czech J Food Sci 31:283–291

Ogawa M, Takee R, Okabe Y, Seki Y (2017) Bio-geo hybrid pigment; clay-anthocyanin complex which changes color depending on the atmosphere. Dyes Pigments 139:561–565

Oh SH, Jang CS (2020) Development and validation of a real-time PCR based assay to detect adulteration with corn in commercial turmeric powder products. Foods 9:882

Osorio C, Acevedo B, Hillebrand S, Carriazo J, Winterhalter P, Morales AL (2010) Microencapsulation by spray-drying of anthocyanin pigments from corozo (*Bactris guineensis*) fruit. J Agric Food Chem 58:6977–6985

Oxborough K (2004) Imaging of chlorophyll a fluorescence: theoretical and practical aspects of an emerging technique for the monitoring of photosynthetic performance. J Exp Bot 55:1195–1205

Panebianco S, Mazzoleni P, Barone G, Musumarra A, Pellegriti MG, Pulvirenti A, Scordino A, Cirvilleri G (2022) Feasibility study of tomato fruit characterization by fast XRF analysis for quality assessment and food traceability. Food Chem 383:132364

Parmar RS, Singh C (2018) A comprehensive study of eco-friendly natural pigment and its applications. Biochem Biophys Report 13:22–26

Pathare PB, Opara UL, Al-Said FA (2013) Colour measurement and analysis in fresh and processed foods: a review. Food Bioprocess Technol 6:36–60

Pinto AH (2018) Portable X-ray fluorescence spectrometry: principles and applications for analysis of mineralogical and environmental materials. Asp Min Miner Sci 1:AMMS.000506

Plaza A, Benediktsson JA, Boardman JW, Brazile J, Bruzzone L, Camps-Valls G, Chanussot J, Fauvel M, Gamba P, Gualtieri A, Marconcini M (2009) Recent advances in techniques for hyperspectral image processing. Remote Sens Environ 113:S110–S122

Poovitha S, Stalin N, Balaji R, Parani M (2016) Multi-locus DNA barcoding identifies matK as a suitable marker for species identification in Hibiscus L. Genome 59:1150–1156

Popelka-Filcoff RS, Lenehan CE, Lombi E, Donner E, Howard DL, de Jonge MD, Paterson D, Walshe K, Pring A (2016) Novel application of X-ray fluorescence microscopy (XFM) for the non-destructive micro-elemental analysis of natural mineral pigments on aboriginal Australian objects. Analyst 141:3657–3667

Raliya R, Tarafdar JC, Singh SK, Gautam R, Choudhary K, Maurino VG, Saharan V (2014) MgO nanoparticles biosynthesis and its effect on chlorophyll contents in the leaves of clusterbean (*Cyamopsis tetragonoloba* L.). Adv Sci Eng Med 6:538–545

Renita AA, Gajaria TK, Sathish S, Kumar JA, Lakshmi DS, Kujawa J, Kujawski W (2023) Progress and prospective of the industrial development and applications of ECO-friendly colorants: an insight into environmental impact and sustainability issues. Foods 12:1521

Roca M, Chen K, Pérez-Gálvez A (2016) Chlorophylls. In: Handbook on natural pigments in food and beverages. Woodhead Publishing, pp 125–158

Rocha FS, Gomes AJ, Lunardi CN, Kaliaguine S, Patience GS (2018) Experimental methods in chemical engineering: ultraviolet visible spectroscopy—UV-vis. Can J Chem Eng 96:2512–2517

Rocha F, de Paula Rezende J, dos Santos Dias MM, Pinto VR, Stringheta PC, dos Santos Pires AC, Vidigal MC (2023) Complexation of anthocyanins, betalains and carotenoids with biopolymers: an approach to complexation techniques and evaluation of binding parameters. Food Res Int 163:112277

Romero N, Visentini FF, Márquez VE, Santiago LG, Castro GR, Gagneten AM (2020) Physiological and morphological responses of green microalgae *Chlorella vulgaris* to silver nanoparticles. Environ Res 189:109857

Salo HM, Nguyen N, Alakärppä E, Klavins L, Hykkerud AL, Karppinen K, Jaakola L, Klavins M, Häggman H (2021) Authentication of berries and berry-based food products. Compr Rev Food Sci Food Saf 20:5197–5225

Sankari M, Hridya H, Sneha P, Doss CG, Christopher JG, Mathew J, Zayed H, Ramamoorthy S (2019) Implication of salt stress induces changes in pigment production, antioxidant enzyme activity, and qRT-PCR expression of genes involved in the biosynthetic pathway of *Bixa orellana* L. Funct Integr Genomics 19:565–574

Saravanan A, Jayasree R, Kumar PS, Varjani S, Hemavathy RV, Jeevanantham S, Yaashikaa PR (2020) Production of pigment using *Aspergillus tamarii*: new potentials for synthesizing natural metabolites. Environ Technol Innov 19:100967

Sarkar R, Dutta A, Patra A, Saha S (2021) Bio-inspired biopolymeric coacervation for entrapment and targeted release of anthocyanin. Cellulose 28:377–388

Sarret G, Smits EP, Michel HC, Isaure MP, Zhao FJ, Tappero R (2013) Use of synchrotron-based techniques to elucidate metal uptake and metabolism in plants. Adv Agron 119:1–82

Shahbandeh M (2024) Global natural food colors market value 2020–2028. Statista. https://www.statista.com/statistics/1003649/global-food-color-market-value/

Sharma R, Ghoshal G (2020) Optimization of carotenoids production by *Rhodotorula mucilaginosa* (MTCC-1403) using agro-industrial waste in bioreactor: a statistical approach. Biotechnol Report 25:e00407

Sharma S, Sarika Bharti A, Tiwari MK, Uttam KN (2018) Effect of manganese stress on the mineral content of the leaves of wheat seedlings by use of X-ray fluorescence excited by synchrotron radiation. Spectrosc Lett 51:302–310

Silver J (2020) Let us teach proper thin layer chromatography technique! J Chem Educ 97:4217–4219

Singhal N, Kumar M, Kanaujia PK, Virdi JS (2015) MALDI-TOF mass spectrometry: an emerging technology for microbial identification and diagnosis. Front Microbiol 6:791

Sousa C (2022) Anthocyanins, carotenoids and chlorophylls in edible plant leaves unveiled by tandem mass spectrometry. Foods 11:1924

Stintzing FC, Conrad J, Klaiber I, Beifuss U, Carle R (2004) Structural investigations on betacyanin pigments by LC NMR and 2D NMR spectroscopy. Phytochemistry 65:415–422

Stuppner H, Egger R (1996) Application of capillary zone electrophoresis to the analysis of betalains from *Beta vulgaris*. J Chromatogr A 735:409–413

Suwannarach N, Kumla J, Nishizaki Y, Sugimoto N, Meerak J, Matsui K, Lumyong S (2019) Optimization and characterization of red pigment production from an endophytic fungus, *Nigrospora aurantiaca* CMU-ZY2045, and its potential source of natural dye for use in textile dyeing. Appl Microbiol Biotechnol 103:6973–6987

Szadkowski B, Kuśmierek M, Kozanecki M, Nowakowska J, Rogowski J, Maniukiewicz W, Marzec A (2023) Ecological hybrid pigments with improved thermal, light, and chemical stability based on purpurin dye and different minerals for applications in polymer materials. Dyes Pigments 217:111430

Toma MA, Nazir KN, Mahmud MM, Mishra P, Ali MK, Kabir A, Shahid MA, Siddique MP, Alim MA (2021) Isolation and identification of natural colorant producing soil-borne aspergillus Niger from Bangladesh and extraction of the pigment. Foods 10:1280

Tong Y, Deng H, Kong Y, Tan C, Chen J, Wan M, Wang M, Yan T, Meng X, Li L (2020) Stability and structural characteristics of amylopectin nanoparticle-binding anthocyanins in *Aronia melanocarpa*. Food Chem 311:125687

Udensi J, Loughman J, Loskutova E, Byrne HJ (2022) Raman spectroscopy of carotenoid compounds for clinical applications-a review. Molecules 27:9017

Valand R, Tanna S, Lawson G, Bengtström L (2020) A review of Fourier transform infrared (FTIR) spectroscopy used in food adulteration and authenticity investigations. Food Addit Contam: Part A 37:19–38

Verma G, Mishra M (2018) Development and optimization of UV-vis spectroscopy-a review. World J Pharm Res 7:1170–1180

Vermaas WF, Timlin JA, Jones HD, Sinclair MB, Nieman LT, Hamad SW, Melgaard DK, Haaland DM (2008) In vivo hyperspectral confocal fluorescence imaging to determine pigment localization and distribution in cyanobacterial cells. Proc Natl Acad Sci 105:4050–4055

Wang X, Mu B, Wang W, Wang Q, Wang A (2019) A comparative study on color properties of different clay minerals/BiVO4 hybrid pigments with excellent thermal stability. Appl Clay Sci 181:105221

Watanabe T, Terabe S (2000) Analysis of natural food pigments by capillary electrophoresis. J Chromatogr A 880:311–322

Watson DJ, Bushway AA, Bushway RJ (2004) Separation of peonidin and cyanidin, two anthocyanidins, in cranberries by capillary electrophoresis. J Liq Chromatogr Relat Technol 27:113–121

West ME, Mauer LJ (2013) Color and chemical stability of a variety of anthocyanins and ascorbic acid in solution and powder forms. J Agric Food Chem 61:4169–4179

Wingerath T, Stahl W, Kirsch D, Kaufmann R, Sies H (1996) Fruit juice carotenol fatty acid esters and carotenoids as identified by matrix-assisted laser desorption ionization (MALDI) mass spectrometry. J Agric Food Chem 44:2006–2013

Wu X, Ma Y, Wang P, Wu J, Li N, Zhang Z, Xie R, Wang D, Nie H (2024) Transcriptome sequencing and screening of anthocyanin related genes in purple potato tubers (*Solanum tuberosum* L.). BMC Genomics 25:1–5

Xia ML, Wang L, Yang ZX, Chen HZ (2016) High-throughput screening of high Monascus pigment-producing strain based on digital image processing. J Ind Microbiol Biotechnol 43:451–461

Yang Z, Fan G, Gu Z, Han Y, Chen Z (2008) Optimization extraction of anthocyanins from purple corn (*Zea mays* L.) cob using tristimulus colorimetry. Eur Food Res Technol 227:409–415

Yang X, Qin H, Gao M, Zhang H (2011) Simultaneous detection of Ponceat 4R and tartrazine in food using adsorptive stripping voltammetry on an acetylene black nanoparticle-modified electrode. J Sci Food Agric 15:2821–2825

Zahra A, Qureshi R, Sajjad M, Sadak F, Nawaz M, Khan HA, Uzair M (2024) Current advances in imaging spectroscopy and its state-of-the-art applications. Expert Syst Appl 238:122172

Bioprocessing of Biocolors for Large-Scale Production

8

Monika Thakur, Divya Tandon, Sakshi Bhardwaj, and Neha Gautam

Abstract

The environmental and health concerns caused by synthetic dyes have driven the demand for natural alternatives, and biocolors have emerged as a green and sustainable choice. In recent years, demand for biocolors has increased globally in the food and pharmaceutical industries due to their multifunctional bioactivities and natural origin. Biocolors are natural coloring agents derived from biological sources, viz., plants, animals, and microbes. Traditional production methods that have been used for many years for biocolor production are source-specific, dependent upon climatic conditions, and subject to small contents of target pigment compounds. Many efforts, viz., optimization of microbial-based bioprocesses and development of in vitro plant tissue culture systems, have been done to increase biocolor production on an industrial scale, but still methodologies and techniques for production, extraction, and purification of biocolors are not up to the mark, which constrains the overall productivity as well as the quality of the final product. The present chapter addresses bioprocessing techniques and technologies to upscale the large-scale production of biocolors.

Keywords

Biocolor · Bioprocess · Submerged fermentation · Downstream processing

---

M. Thakur · S. Bhardwaj · N. Gautam (✉)
Department of Food Science and Technology, Dr. Y. S. Parmar University of Horticulture and Forestry, Nauni, Solan, HP, India

D. Tandon
Department of Microbiology, Shoolini Institute of Life Sciences and Business Management, Solan, HP, India

R. K. Bhatia, A. Walia (eds.), *Biocolours*,
https://doi.org/10.1007/978-981-95-1738-1_8

8.1 Introduction

Color is the first and foremost consideration for consumers while selecting any food product. Consumers equate other crucial attributes like uniqueness, maturity, and safety with color, making it a much more important and primary attribute. The demand for color pigment in a number of industries is growing annually, and the revenue is expected to grow at an annual rate of 4.9% to reach 54.64 billion USD by 2026 (Mohd Said 2023). Therefore, to make any food product more appealing and acceptable for a longer period of time, food, pharmaceutical, and textile coloring is now a common ingredient in many food products for which synthetic colorants have been the major form of colorant to be used (Singh et al. 2023). The low cost of manufacturing and extremely high stability of synthetic dyes have greatly accelerated their use in the food industry. Synthetic colors are being added to adjust and reform the color variance of various natural food ingredients during food processing and to make up for the losses occurring during storage. Despite this, the majority of artificial dyes consisting of azo functional groups and aromatic ring structures may be hazardous to human health since they can cause hyperactivity, allergic reactions, and respiratory conditions (Li et al. 2017). Also, synthetic color is unable to degrade in the natural setting of the environment owing to its chemical composition and structure. Therefore, the prevailing marketing trend is replacing synthetic colorants with less hazardous, cleaner, and safer biocolors with increasing health awareness for quality products among consumers.

Biocolors are sourced from natural and biological sources like plants and microorganisms, and they can also be employed in a variety of end products since they have biological qualities in addition to giving any product a variety of hues. Evidence of human adoption of biocolors dates back at least a hundred thousand years to the ancient era. Embracing pigments sourced from earthly resources, including red ochre, charcoal white from ground calcite, and yellow ochre, early people painted in caves using several traditional protocols like brushing, smearing, dabbing, and spraying methods (Barnett et al. 2006). The artistic, cultural, and environmental significance of biocolors is still acknowledged in contemporary practices, and their long legacy extends from prehistoric artwork in caves of sophisticated historical civilizations to synthetic alternatives that appeared in the nineteenth century. Since biocolors are derived from sustainable and biodegradable resources, they lessen the need for synthetic artificial dyes based on petroleum-based chemicals and the pollution that comes with the production and disposal of chemicals associated with it (Bashir et al. 2024). Even with such significance of biocolors in the food processing industry, the presently available methodologies and techniques for production, extraction, and purification of biocolors extracted from natural sources are not up to the mark, which constrains the overall productivity as well as the quality of the final food product (Murador et al. 2019). Also, stable and secure, massive-scale resources for biocolors are necessary for industrial sectors like food, cosmetics, textiles, and medicines to meet high production levels for consumers. Consequently, recognizing that it allows for constant supply, cost-effectiveness, and wider commercial use, large-scale manufacturing of biocolors is essential for their practical application

(Lyu et al. 2022). This facilitates the growing demand for natural and sustainable colorants in place of synthetic ones. Integrating all these important points together and recognizing the importance of mass production of biocolors to completely restrain the use of synthetic dyes, the present chapter addresses the conventional and novel bioprocessing techniques and technologies to upscale the large-scale production of biocolors has been centered on the sources and production of biocolors using advanced techniques.

8.2 Bioprocessing Techniques

8.2.1 Fermentation Technique

By employing cells, enzymes, or cell components in engineered biological processes (bioprocesses), fermentation biotechnology creates a bioproduct through the metabolism of raw material feedstocks like carbon and nitrogen. Currently, biomanufacturing uses bacteria to create chemicals, enzymes, food ingredients, materials, and medications. Owing to developments in fermentation bioprocess design, food ingredients that can mimic those sourced from animals can now be produced, drawing on insights from the bioindustrial and biopharmaceutical sectors. Fermentation is a bioprocess that turns simple or complicated carbon sources into chemicals of interest (e.g., enzymes, metabolites, proteins, vitamins, or lipids) using microorganisms like yeast, bacteria, and fungi. The microbial transformation of sugar into an organic acid or an alcohol is known as fermentation. According to Paulova et al. (2013), food and beverage fermentation traditionally employs microbes to maintain and enhance the organoleptic qualities of foods like bread, beer, vinegar, yogurt, cheese, and wine. The food industry is currently creating bioprocesses to use microbes and more sophisticated fermentation techniques to produce alternative proteins, lipids, and other nutrients. Fermentation processes can be categorized on the basis of feeding bioreactors into batch, fed-batch, and continuous; on the basis of characteristic state of matter into submerged or solid-state fermentation, which will be discussed in the further sections below.

8.2.1.1 Submerged Fermentation

The effective creation of high-quality final products and the achievement of maximum productivity and yield are ensured by the submerged culture of microbial cells in bioreactors. Different kinds of microbes are cultivated in industrial bioreactors that can run in batch, fed-batch, or continuous modes to produce a variety of products. The various methods for submerged microbe culture in bioreactors are briefly covered in the sections that follow.

(a) *Batch Cultivation*

Batch culture is a closed system in which the medium, nutrients, and inoculum are added to the bioreactor at the start of cultivation, usually under aseptic conditions. This means that the volume of the culture broth in the bioreactor is theoretically constant throughout the cultivation process; in practice, however, small variations in the volume of the culture are caused by sampling or adding air or gas to the culture, as well as by a low feed rate of acid/base solutions to maintain the pH at a desired level; overall, these changes are typically disregarded because they are negligible in relation to the bioreactor's working volume.

In batch culture, a known quantity of viable cells is usually introduced into a bioreactor that has already been loaded with nutrient-rich, sterilized media. The cell culture grows according to the four primary stages of the classical growth curve that Monod (1949) outlined after inoculation. Although the cells are metabolically active, they are adjusting their enzymatic apparatus to a new environment, the lag phase is an "inefficient" stage of culture, and it is preferable to shorten it as much as possible. No appreciable increases in biomass concentration, substrate consumption, or product synthesis are discernible. The inoculum's cell concentration and physiological condition, the inoculation and cultivation medium's composition (primarily the carbon and energy source, pH, and temperature), and the inoculum's size all have an impact on how long the lag phase lasts. The exponential (logarithmic growth) phase is defined by rapid cell proliferation (the biomass concentration increases exponentially with time), a constant specific growth rate that is equivalent to the maximum specific growth rate of the culture in the absence of growth limitation (the growth rate is unrestricted because all nutrients are present in excess but do not reach growth-inhibiting concentrations), rapid consumption of the carbon and energy source, and a high rate of primary metabolite production. By the end of the exponential phase, the nutrients have been depleted (in aerobically growing cultures, this is indicated by a sharp rise in the concentration of dissolved oxygen), which leads to a gradual decrease in the particular growth rate and a shift to the stationary phase, which is crucial for the production of secondary metabolites and is marked by growth stagnation and the use of endogenous carbon and energy stores.

(b) *Fed-Batch Cultivation*

Fed-batch culture is a semi-open method in which the product is kept inside the bioreactor while one or more nutrients are introduced aseptically and gradually; in other words, the volume of the culture broth in the bioreactor grows during this period. The ability to prolong product synthesis, increase cell densities, and thereby increase the amount of product, which is typically proportional to the biomass concentration; the ability to increase yield or productivity through controlled sequential addition of nutrients; and the feature of prolonged productive cultivation over the "unprofitable periods" when the bioreactor would normally be prepared for a new batch are the main benefits of fed-batch over batch cultures.A substrate can be added to the bioreactor in a variety of ways, such as a medium containing carbon and other nutrients or as a concentrated solution of just carbon and energy. The culture performance can be significantly improved by selecting the right nutrient feeding rate, which affects the rate of product formation, cell physiology, and cellular growth rate. The two most popular feeding techniques are (a) discontinuous feeding, which is accomplished by regular or irregular substrate pulses, and (b) regular continuous feeding of nutrients that are planned based on a predetermined profile or feedback control of online measured variables related to cell growth and metabolism, such as pH, CO_2, biomass concentration, dissolved oxygen concentration, and evolution rate.

(c) *Continuous Cultivation*

Continuous culture is an open system where nutrients are continuously and aseptically added to the bioreactor while the culture broth—which contains cells and metabolites—is simultaneously removed; in other words, the volume of the culture broth remains constant because the feed-in and feed-out rates remain constant. Although other forms of continuous operation, such as turbidistat (a constant concentration of biomass controlled by the dilution rate) or nutristat (a constant parameter related to cell growth controlled by the dilution rate), can be used, continuous culture is often used interchangeably with a chemostat, which is represented by a constant specific growth rate of cells, which is equal to the dilution rate and controlled by the availability of the limiting nutrient.

8.2.1.2 Solid-State Fermentation

The terms "solid substrate fermentation" (SSF) and "solid substrate cultivation" (SSC) refer to systems in which microbes are cultivated on the surface of a concentrated water-insoluble substrate with a low concentration of free water. This substrate typically contains polysaccharides as a source of carbon and energy. This method originated in microbes. Its use is restricted to the manufacturing of industrial enzymes, specific food items, and feed additives in Western nations, where it has not been extensively utilized. It has been utilized for centuries in Eastern nations to produce traditional foods like sake, koji, miso, and soy sauce using various substrates and microbes. The process is characterized by very low water activity (Holker and Lenz 2005).

8.2.2 Optimization of Production Conditions

Before producing metabolites on a large or commercial scale, one of the most studied aspects is medium optimization, which also presents a number of difficulties. Prior to the 1970s, media optimization was accomplished through the use of traditional techniques, which were costly, time-consuming, and involved numerous tests with varying degrees of accuracy. However, media optimization has grown more dynamic, effective, efficient, cost-effective, and reliable in producing results since the development of contemporary mathematical and statistical tools. The best fermentation conditions (pH, temperature, agitation speed, etc.) and the right medium components (carbon, nitrogen, etc.) must be determined and optimized in order to design a production medium. Additionally, maximum product concentration might be attained by refining the previously mentioned parameters.

8.2.2.1 Carbon

The most crucial component of the medium is carbon, which provides the microorganisms with energy and is crucial for their growth and the generation of primary and secondary metabolites. The production of primary or secondary metabolites and/or biomass formation is frequently influenced by the rate at which the carbon source is digested.

8.2.2.2 Nitrogen

Nitrogen is also one of the most important parameters that must be at optimum levels throughout the fermentation process. Both inorganic and organic nitrogen sources are utilized by microbes for the production of metabolites. The use of specific amino acids can enhance the production of secondary metabolites, whereas, in the case of unsuitable amino acids, the production can be affected adversely (Singer and Nelder 2009).

8.2.2.3 Phosphate

Another essential element is phosphate, which is needed for the synthesis of nucleic acids and the phospholipids found in microbial cell membranes. The amount of phosphate that needs to be added to the fermentation medium is determined by the microbes' requirements, the broth's composition, and the type of product that is intended. Often, phosphate is desired in lower levels at the initiation of the fermentation process in order to produce optimum metabolites, as higher levels may hinder the production of secondary metabolites (Singh et al. 2017).

8.2.2.4 Optimization Techniques

For optimization of media, there are basically two techniques that can be followed, i.e., traditional one-factor-at-a-time and advanced (statistical/mathematical). In traditional techniques, one factor at a time can be varied, and the rest of the variables are kept constant. It can be followed by either removal or supplementation or replacement of one variable at a time in the medium. This technique is easy but time-consuming and also requires more skilled people to perform experiments. Therefore, advanced techniques, i.e., statistical approaches, were formulated. Advanced optimization techniques either screen significant components (Taguchi and PBD) or optimize the concentration of components (response surface methodology, artificial neural network, genetic algorithm, Nelder–Mead simplex, and fuzzy logic) (Singh et al. 2017). As statistical approaches have advanced, medium optimization has taken on new dimensions. These techniques contribute to the process's overall economics by increasing process efficiency, lowering labor costs, and reducing process time. Due to their biological character, microbial processes have a comparatively high number of inherent variability. The microbial response networks are complicated, with multiple stimuli influencing distinct network components. The trustworthiness of the data gathered during the tests is better understood by means of logical experimental design and statistical analysis of the findings.

8.2.3 Bioreactor Design and Scale-Up

8.2.3.1 Bioreactor Types Suitable for Pigment Production

Microbial pigment synthesis is made possible by different kinds of bioreactors, each designed to meet the metabolic requirements of distinct microorganisms. Stirred-Tank Bioreactors are one of the most common system available because of their highly efficient mixing oxygen mass transfer mechanisms, hence being well-suited for pigments such as prodigiosin, melanin, and carotenoids. For example, *Serratia marcescens* grew prodigiosin effectively under optimized aeration in STRs (Venil et al. 2014).

Airlift bioreactors are shear-reducing and energy-efficient. They are particularly well suited for photosynthetic microorganisms like *Spirulina platensis*, which showed increased phycocyanin production with controlled aeration and light (Silva et al. 2018). Bubble column reactors also provide low shear and easy operation, which are well suited for pigment-producing yeasts like *Rhodotor ulaglutinis*, which produce carotenoids efficiently under such conditions (Bhosale 2004). Packed-Bed Bioreactors, which may be employed with immobilized fungal cells, offer stable long-term pigment production. *Monascus purpureus*, for instance, exhibited good red pigment yield in such systems under solid-state conditions (Babitha et al. 2007). Fluidized-Bed Bioreactors with excellent mass transfer and mixing are ideal for immobilized or suspended cultures. They have been used to boost *Haematococcus pluvialis* astaxanthin production under stress (Zgheib et al. 2018). Lastly, Tray Bioreactors or Solid-State Fermentation (SSF) systems are designed to be optimized for filamentous fungi and yeasts with low moisture requirements. Such systems are suitable for the production of red and yellow pigments such as monascin and ankaflavin from *Monascus* spp. (Mussagy et al. 2019). Each bioreactor system has specific advantages depending on scale of operation, type of organism, and properties of the pigment, and therefore the selection of reactor is an important consideration in effective bioproduction as shown in Fig. 8.1.

8.2.3.2 Scale- Up Challenges and Strategies

Scale-up of the biocolor production is crucial for the commercial application of biocolors. Figure 8.2 illustrated the scale-up process of biocolor production. Scaling up microbial pigment production poses major engineering and biological hurdles as described below.

- *Oxygen transfer limitation* stands out as a crucial challenge in aerobic pigment-producing microbes such as *Serratia marcescens* and *Monascus* spp. Larger volumes lead to decreased oxygen solubility and distribution often resulting in lower yields. To maintain productivity during scale-up, it is essential to optimize the volumetric oxygen transfer coefficient (kLa) (Singh 2017).
- *Shear sensitivity* creates another problem; mechanical stress can harm some pigment-producing organisms damaging cells or reducing pigment biosynthesis. This issue becomes relevant in stirred-tank reactors at industrial scales. Systems

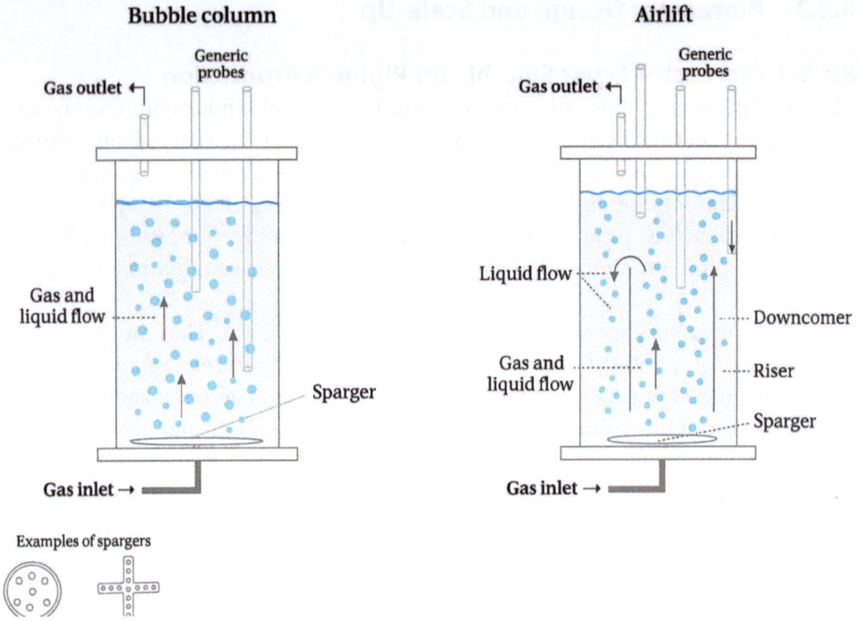

Fig. 8.1 Different types of bioreactors that can be applied in bioprocesses (de Mello et al. 2024)

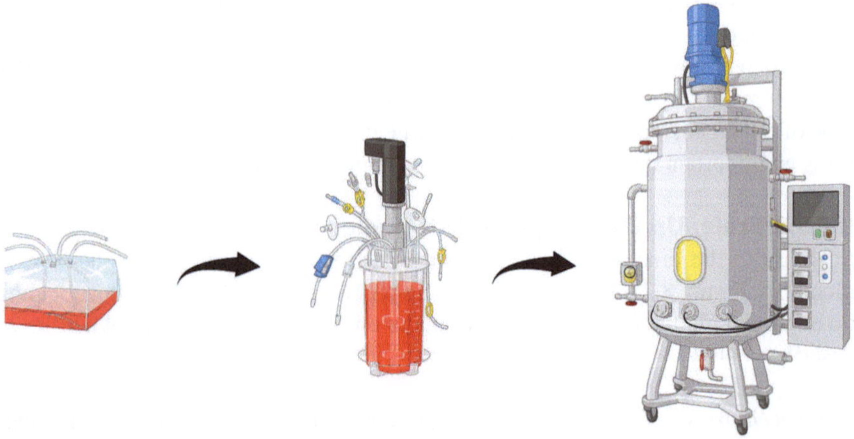

Fig. 8.2 Bioprocessing scale-up

like airlift or bubble column reactors offer gentler hydrodynamics as alternatives (Mussagy et al. 2019).

- *Mixing and hydrodynamics* also create significant scale-up challenges. Uniform distribution of nutrients and gases is more difficult to achieve in large-scale reactors, which may result in substrate gradients and metabolic stress. Computational fluid dynamics (CFD) simulations and optimized agitation systems are essential tactics for overcoming these challenges (Xing et al. 2009).

- *Process control and monitoring* increase with scale, particularly for parameters such as pH, temperature, and dissolved gases. Mishra et al. (2023) highlight the need for Process Analytical Technology (PAT) and automation-based feedback control to maintain consistent production.
- *Sterility and contamination* threats are amplified at scale with longer pipes, intricate equipment geometries, and greater operator participation. Strong clean-in-place (CIP) and steam-in-place (SIP) systems must be implemented in pharmaceutical and food-grade bioreactors (López et al. 2018).

8.2.4 Methods to Address Scale-Up Problems Include

- To maintain nutrient equilibrium while reducing product inhibition, systems operate in fed-batch or continuous modes.
- Structured supports or immobilized cells function within packed-bed and fluidized-bed reactors to maintain pigment stability over extended periods.
- Anand et al., in 2022, explore the use of process analytical technologies (PAT) to achieve real-time control of pH, dissolved oxygen (DO), and substrate levels.
- Laboratory reactors using scale-down models can simulate industrial stress conditions, which helps optimize scale-up parameters early while reducing commercial production risks (Martinez et al., 2022).

8.2.4.1 Process Monitoring and Control Downstream Processing

In large-scale production of microbial pigments (biocolors), closely monitoring and controlling the process is very important because it directly affects how much pigment is made, its quality, and its consistency as shown in Fig. 8.3. In the early stages (upstream), things like pH, oxygen levels, stirring speed, temperature, and nutrients are tracked in real time using sensors in advanced bioreactors. These conditions are managed by systems like PLCs, SCADA, or DCS to keep the microbes healthy and producing pigment efficiently (Lee et al. 2023). The amount of pigment made is checked using tools like spectrophotometers or HPLC, which help spot any problems early. Biomass growth and activity are also tracked using sensors that measure optical density and redox potential (Lv et al. 2017).

In the later stages (downstream), control systems manage pigment extraction using solvents, pH adjustments, filtration, and drying methods like freeze-drying or spray drying. Technologies like PAT and QbD help make the process more reliable and meet safety and quality standards especially important for pigments used in food and medicine (Benner et al. 2022; Brahma et al. 2022). Overall, advanced monitoring and control help improve product quality, reduce waste, and prevent pigment damage during processing. Additionally, incorporating real-time analytics into downstream equipment—online spectroscopic monitoring in chromatography, for example enhances decision-making and minimizes post-process quality testing (Pruvost et al. 2016). End-to-end automation and smart bioprocessing strategies

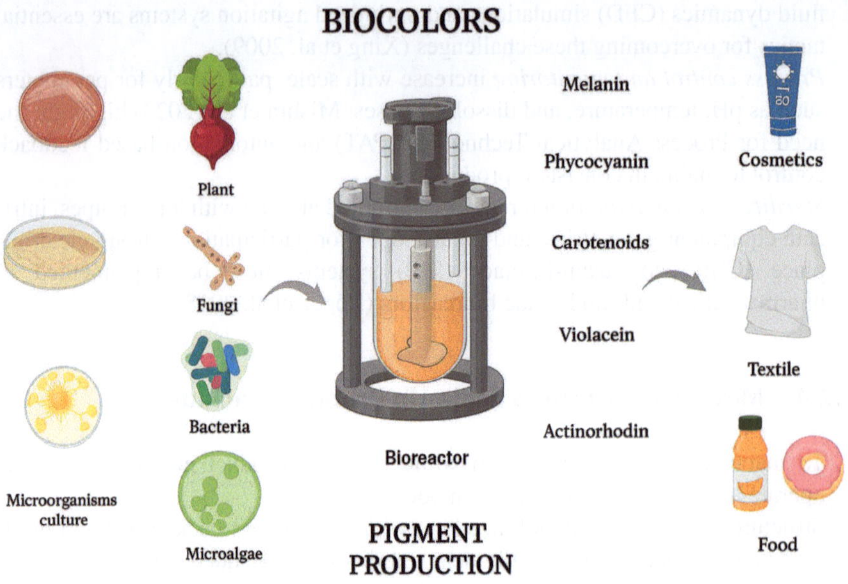

Fig. 8.3 Biocolor production and utilization in industry

have, therefore, become critical for economically sustainable and scalable microbial pigment production.

8.2.4.1.1 Harvesting of Biomass

Harvesting the microbial biomass is a key step that affects how much pigment (biocolor) you get and how pure it is. To extract and purify these pigments, the first thing you need to do is separate the cells from the fermentation liquid. On an industrial scale, this is usually done using either centrifugation or filtration. The best method depends on things like the shape of the cells, whether the pigment is inside or outside the cells, how thick the culture is, and practical factors like equipment costs and energy consumption.

8.2.4.2 Centrifugation

Centrifugation is commonly used for capturing microbial biomass, especially when acquiring intracellular pigments such as carotenoids, prodigiosin, and melanin from bacteria and fungi. Centrifugation also serves as a means of separating cells from liquid growth media by using centrifugal force, with denser particulates (in this case cells) sedimenting quickly relative to other phases. Industry standards for centrifugation include disc-stack and tubular bowl centrifuges used for large operating volumes characterized by continuous flow and high throughput. Emerging results suggest that centrifugation at a given g-force or residence time can achieve high process yields in biomass recovery while achieving reasonable pigment stability. For example, Venkatachalam et al. (2023) performed optimization of basic parameters for centrifugation of *Talaromyces albobiverticillius* and saw improved pigment yield when using high-speed centrifugation methods. While the method is simple, many high-speed centrifugal systems subject pigments and biomass cells to high thermal forces, which may negatively impact pigment stability, especially with heat-sensitive pigments like violacein and some anthocyanins (Brahma et al. 2022). In terms of design, some companies produce low-temperature centrifuges with the aim to increase pigment stability during the recovery stages (4–10 °C). Furthermore, with flocculated sediments, centrifugation and the use of sedimentation tanks significantly improve downstream separation in terms of reducing costs and energy use. With these effluent designs, it may also be possible to link centrifugal separation with continuous fermentation systems and, as a continuous process, reduce downtime for pigment recovery during the microbial production phases (Benner et al. 2022).

8.2.4.3 Filtration

Filtration is often used to collect microbial biomass, particularly in processes where the pigment is secreted or where shear-sensitive organisms are used. The microbes are retained on porous membranes while the liquids are removed by vacuum or pressure. Microfiltration, ultrafiltration, and tangential flow filtration are the most common for large-scale pigment operations. Compared to centrifugation, the most important advantages of filtration are its lower energy consumption and gentler processing conditions, which help in maintaining the stability of pigments. Venkatachalam et al. (2023) showed that effective separation of biomass was achieved through cross-flow microfiltration for cultures of red-pigment-producing fungi, where intact hyphal biomass and clarified broth could be recovered for subsequent extraction. Moreover, new membrane materials with anti-fouling features and clean-in-place (CIP) technology increase the lifespan and operational efficiency, reducing frequent stilling, known as fouling, in high cell density broths (Benner et al. 2022).

Pre-treatment methods such as bio-flocculation or pH correction are also employed to improve filterability and minimize membrane clogging. Filtration is especially suitable for pigment-producing yeasts and microalgae, in which the extracellular pigment is trapped in the filtrate and cell mass is recovered separately for biomass valorization (Shaikh et al. 2016). Additionally, filtration systems are increasingly being coupled with bioreactors as part of continuous bioprocessing trains, allowing for real-time separation and minimizing overall process duration.

8.2.5 Extraction of Pigments

Solvent extraction is an established method for selective recovery of soluble natural biomolecules using a range of organic solvents. Various compounds from a specific composition are separated using this purification technique in accordance with their relative solubility in two distinct immiscible fluids. Solubilization occurs within the molecular structure as a result of the solvents' easy diffusion into the discarded biomass. After that, the solvent flows out of the solid food waste to separate and differentiate the products from the undesirable particulates (Ng et al. 2020). Polar pigments like betalains and anthocyanins are mostly extracted using polar solvents like ethanol, water, methanol, or any of these in combination. While chlorophylls and carotenoids are examples of non-polar pigments that are processed with non-polar solvents such as acetone, petroleum ether, or hexane. However, queries have been raised concerning the safety of the end products produced using organic solvents being utilized for extraction for the harmful leftovers or residues generated from it and their impact on human health (Panda and Manickam 2019).

Therefore, a novel approach of using green extraction methodologies is being explored for the production of biocolors. Green extraction methodologies for pigment extraction emphasize environmentally and ecologically sustainable process protocols that minimize usage of energy, avoid using hazardous and toxic solvents, and sustain the biological activity of the pigment. These technologies enable effective pigment extraction from its sources (plant or microbes) by combining thermal and non-thermal processes in combination with eco-friendly, non-toxic solvents. These technologies can overcome the limitations of traditional extraction methods by reducing extraction time and solvent use while also increasing the extraction yield of natural pigments, including anthocyanins and chlorophyll (Guo et al. 2020). One of the most popular and widely accepted methodologies in this category is supercritical fluid extraction (SFE). It is a solvent extraction technique majorly employed to extract compounds from solid matrices that uses supercritical fluids existing at temperatures and pressures above their critical point and exhibiting properties of both a liquid and a gas. This uniquely favorable setting makes it easier to extract bioactive materials like pigments by increasing solvent strength and a greater dissolution of organic compounds (Wrona et al. 2017).

Additionally, supercritical fluids minimize the contamination risk of the extracts due to solvent and offer superior efficacy, rapid recovery in less time, and reduced solvent consumption and are considered safe. But it is noted to be particularly effective for lipid-soluble pigments such as carotenoids and chlorophylls and has high operational costs and capital expenses, limiting its utilization (Ramzan et al. 2025). Another important category of such a sustainable approach is green solvents, which have minimal detrimental impact on the environment. A solvent must not have any negative impact on health and should be regarded as safe and not cause any harm from its manufacturing, usage, and disposal, such as the depletion of non-renewable resources, in order to be deemed ecologically benign (Sen et al. 2019). Since a proportion of solvents can be considered environmentally friendly, simple to create, and have adjustable properties that enable them to serve as task-specific compounds, ionic liquids (IL) and eutectic solvents have been categorized as alternative solvents that can replace or limit the use of hazardous, toxic, and organic solvents in many fields.

They are distinguished by their distinct physicochemical traits and significant thermal and chemical stability, non-flammability, and negligible vapor pressure (Anastas and Eghbali 2010). Solvents having low transition temperature combinations lower when compared to ideal liquid mixtures with a eutectic temperature are known to be eutectic solvents. Conversely, the melting point of ILs, which are non-molecular substances with large organic cation groups and a small proportion of inorganic anions, is lower than 100 °C. Their ionic forms and other dissolved molecules interact with each other and offer a high degree of solubility for a broad range of chemicals. ILs are being used in a wider approach as analytical solvents since they are regarded as green substitutes for volatile organic solvents. Additionally, ILs readily break down the cellulose wall, forcing the extraction procedure. Several techniques, including the back extraction process and anti-solvent precipitation, have been developed to recover the carotenoids from ILs. Other approaches also include distillable ILs, ion-exchange resins, thermo-responsive polymeric ILs, and microporous resins. Recent research by Murador et al. (2019) found that chlorophylls and carotenoids extracted from orange peels using an IL-mediated process protocol had comparable bioefficacy to those extracted using organic solvents.

A eutectic combination of lactic acid and DL-menthol via UAE was used by Silva et al. to extract lycopene from tomato waste. The results suggested that lycopene may be extracted in a high capacity range (145 mg/100 g) under ideal conditions. When compared to ILs, deep eutectic solvents (DESs) are less expensive to create since their raw components used for creating them are less expensive, less hazardous, and many times biodegradable, for which they are being used progressively (Grillo et al. 2021). Such green solvents may also be recycled, and purification of the bioactive compounds that are extracted using these solvents is simple. Furthermore, analytical practice has recently been exposed to new extraction solvent types known as natural deep eutectic solvents (NADESs). For extraction and storage of anthocyanin from plant sources, NADESs offers a productive and environmentally friendly approach, substituting organic solvents. Compared to an organic solvent derived from mulberries, the extraction level attained with NADESs (6 mg/g FW) was 1.24 times greater for anthocyanins (Guo et al. 2020). Also, in order to obtain anthocyanins from mulberry residues, Grillo et al. (2021) investigated two green extraction techniques (MAE and UAE) employing five different NADES. The performance of both technologies was better than that of traditional extraction, estimating 26 and 21 mg/g of anthocyanins from MAE and UAE, respectively. The reported studies suggested that extraction by NADESs leads to more stable anthocyanin extracts, which can be directly applied as food colorant. This suggests that establishing alternative methods that are sustainable and eco-friendly to extract pigments is crucial and needed, especially given the potential applications of these natural pigments in the food industry.

8.2.6 Stabilization and Formulation

Biocolors have a number of disadvantages, such as stability, liability, and susceptibility to extraction and extraction techniques, as well as environmental factors. The primary determinants of biopigment results include plant enzymes such as PPO and peroxidase, as well as media variables such as pH, light, temperature, oxygen concentration, and water activity. Due to their sensitivity to a range of environmental conditions, such as oxygen, light, pH, and temperature, natural pigments, such as carotenoids, chlorophylls, and anthocyanins, frequently show poor stability. Consequently, these pigments are prone to degradation during the extraction and storage processes. Anthocyanin color and stability are influenced by the quantity and position of -OH groups, the extent of methylation of -OH groups, glycosylation, and pH level. The chemical structure of betalains, another prevalent food biocolor, determines its stability. The intensity and saturation of the red hue of betalain sources are determined by the ratio of betacyanins to betaxanthins. Because of their structural groups, betacyanins are more resistant to oxidation than betaxanthins at high temperatures and acidic pH, while betaxanthins are more resilient against hydrolytic enzymes and at pH 7. However, betacyanins are more stable than betaxanthins because they have a lower oxidation-reduction potential. For low-acid diets, betalains are advantageous due to their enhanced stability (Bashir et al. 2024). The food industry uses a variety of encapsulation techniques to prevent this degradation, including encasing the bioactive ingredient (core) in a protective substance (encapsulant).

The method of microencapsulation involves encasing sensitive particles in a resilient coating substance that should provide an unbroken layer to shield the target ingredient. In food science, this procedure has been extensively researched to improve stability and regulate the release of microbes, flavorings, colors, antioxidants, nutrients, enzymes, and preservatives. The purpose of dye encapsulation is to shield these substances from oxidation, heat, light, and moisture, extending their shelf life and enhancing their stability throughout processing (Gaonkar et al. 2014). Microparticles have been created using a variety of methods. These technologies fall into one of the following categories: physical (extrusion, multi-hole centrifugal extrusion, cocrystallization, fluidized bed coating, spray drying, spray chilling, and freeze-drying), chemical techniques (interfacial polymerization and complexation for molecular inclusion), and physicochemical techniques (emulsion system, liposome production, coacervation, and organic phase separation) (Krithika et al. 2014) (Table 8.1).

8.3 Conclusion

Large-scale production and extensive biprocessing for biocolor production are essential and much needed initiative towards creating mindful and considerate industrial colorant alternatives for synthetic colors. It addresses the consumer demands for scalability and expansion, cost-effectiveness, and consistency while providing synthetic dye substitutes. In the end, embracing large-scale bioproduction promotes a regenerative economy and limits dependency on fossil fuels by supporting ecological sustainability and creating new prospects for a variety of uses in the food, textile, pharmaceutical, and cosmetic industries. Therefore, to fully exploit the commercial and ecological potential of microbial and natural biocolors, further investment and upscaling in large-scale bioprocessing is needed.

Table 8.1 Research studies for biocolor production and stabilization

Extraction/processing technique	Source	Processing conditions	Results/Yield (/100 g d.w.)	References
Solvent extraction method	Apple peel	80% acetone/ethanol	169.80 g Cyanidin-3-glucoside	Wolfe and Liu (2003)
	Red pitaya peel	80% acetone	13.80 g betanin	Rodriguez et al. (2016)
	Prickly pear pericarp	80% methanol +0/1% HCl	4.55 mg betanin	Osorio-Esquivel et al. (2011)
Supercritical CO_2 extraction	Grape extract	25-50 ml/min CO_2 flow rate	80–85% cyaniding-3-glucoside recovery	Bleve et al. (2008)
	Grape extract	45-46 °C, 160-165 kg/cm^2, 6% ethanol	1.18 mg/ml	Ghafoor et al. (2010)
Ultrasound-assisted extraction	Jaboticaba pomace	Water, 531 W CM-2, 20 °C 15 min	510.35 mg of cyaniding 3-glucoside	Souza et al. (2017)
	Dragon fruit	80% acetone, 15 min	101.04 mg/100 g	Mello et al. (2014)
	Beetroot	50% aqueous ethanol/0.5% acetic acid, 50–60 Hz, 125 W	37.22 mg/100 g	Vulic et al. (2014)
Submerged fermentation	Cultures on bread and rice	Carbon/nitrogen ratio 50 g/g good for growth and 7-9 g/g good for pigmentation	*Monascus*, also known as Anka a red-colored pigment, is obtained	Sutthiwong et al. (2013)
Solid-state fermentation	Inoculation of *Monascus* culture on steamed rice grains	Fermentation for 20 days in aerated and temperature-controlled room	Production of red pigment	Sutthiwong et al. (2013)
Fermentation	*Penicillium oxalicum*	pH 5.6–6.2; temperature: 27-29 °C	1.5–2.0 g/L of Arpink red pigment obtained after 3–4 days of incubation	Sardaryan et al. (2004)
	Porphyridium aerugineum	pH 4–5, pigment then produced is stable at 60 °C for 40 min	Obtained marine blue pigment, Phycocyanin	Sutthiwong et al. (2013)
	Blakeslea trispora	Submerged batch fermentation	Production of β-carotene	Papaioannou and Liakopoulou-Kyriakides (2010)

References

Anastas P, Eghbali N (2010) Green chemistry: principles and practice. Chem Soc Rev 39(1):301–312

Babitha S, Soccol CR, Pandey A (2007) Solid-state fermentation for the production of *Monascus* pigments from jackfruit seed. Bioresour Technol 98(8):1554–1560

Barnett JR, Miller S, Pearce E (2006) Colour and art: a brief history of pigments. Opt Laser Technol 38(4–6):445–453

Bashir I, Pandey VK, Dar AH, Dash KK, Shams R, Mir SA, Fayaz U, Khan SA, Singh R, Zahoor I (2024) Exploring sources, extraction techniques and food applications: a review on biocolors as next-generation colorants. Phytochem Rev 23(4):1–26

Benner P, Meier L, Pfeffer A, Krüger K, Oropeza Vargas JE, Weuster-Botz D (2022) Lab-scale photobioreactor systems: principles, applications, and scalability. Bioprocess Biosyst Eng 45(5):791–813

Bhosale P (2004) Environmental and cultural stimulants in the production of carotenoids from microorganisms. Appl Microbiol Biotechnol 63:351–361

Bleve M, Ciurlia L, Erroi E, Lionetto G, Longo L, Rescio L et al (2008) An innovative method for the purification of anthocyaninsfromgrape skin extracts by using liquid and sub-critical carbon dioxide. Sep Purif Technol 64:192–197

Brahma D, Dutta D, Mukherjee S (2022) Mass production of valuable pro-vitamin a pigment from a microbe, cost analysis and targeting it for health benefiting purpose. In: Industrial microbiology based entrepreneurship: making money from microbes. Springer, Singapore, pp 147–178

de Mello AFM, de Souza Vandenberghe LP, Herrmann LW, Letti LAJ, Burgos WJM, Scapini T et al (2024) Strategies and engineering aspects on the scale-up of bioreactors for different bioprocesses. Syst Microbiol Biomanufact 4(2):365–385

Gaonkar A, Vasisht N, Khare A, Sobel R (2014) Microencapsulation in food industry: a praactical implication guide

Ghafoor K, Park J, Choi Y (2010) Optimization of supercritical fluid extraction of bioactive compoundsfrom grape (Vitislabrusca B.) peel using response surface methodology. Innovative Food Science and Emerging Technology, pp 485–490

Grillo G, Calcio Gaudino E, Rosa R, Leonelli C, Timonina A, Grygiškis S, Cravotto G (2021) Green deep eutectic solvents for microwave-assisted biomass delignification and valorisation. Molecules 26(4):798. https://doi.org/10.3390/molecules26040798

Guo L, Kong D, Yao K, Li J, Li H, Lan N, Hua Y (2020) Optimization and characterization of pigment production from boletus edulis bull.: Fr. By ultrasonic-assisted extraction. J Food Process Preserv 44(7):e14534. https://doi.org/10.1111/jfpp.14534

Holker U, Lenz J (2005) Solid state fermentation—are there any biotechnological advantages? Curr Opin Microbiol 8:301–306

Krithika V, Radhai S, Ravindra N, Thirupathi V (2014) Microencapsulation of paprika (Capsicum annum L) oleoresin by spray drying. International Journal of Scientific Engineering and Research:971–980

Lee WK, Ryu YK, Kim T, Park A, Lee YJ, Sunwoo IY et al (2023) Enhanced photosynthetic pigment production using a scaled-up continuously circulated bioreactor. Mar Drugs 21(11):576

Li LL, Yuan H, Liao F, He B, Gao SQ, Wen GB et al (2017) Rational design of artificial dye-decolorizing peroxidases using myoglobin by engineering Tyr/Trp in the heme center. Dalton Trans 46(34):11230–11238

Lopez-Ramirez N, Volke-Sepulveda T, Gaime-Perraud I, Saucedo-Castañeda G, Favela-Torres E (2018) Effect of stirring on growth and cellulolytic enzymes production by Trichodermaharzianum in a novel bench-scale solid-state fermentation bioreactor. Bioresour Technol 265:291–298

Lv J, Zhang BB, Liu XD, Zhang C, Chen L, Xu GR, Cheung PCK (2017) Enhanced production of natural yellow pigments from *Monascus purpureus* by liquid culture: the relationship between fermentation conditions and mycelial morphology. J Biosci Bioeng 124(4):452–458

Lyu X, Lyu Y, Yu H, Chen W, Ye L, Yang R (2022) Biotechnological advances for improving natural pigment production: a state-of-the-art review. Bioresour Bioprocess 9(1):8
Mareque C, da Silva TF, Vollú RE, Beracochea M, Seldin L, Battistoni F (2018) The endophytic bacterial microbiota associated with sweet sorghum (Sorghum bicolor) is modulated by the application of chemical N fertilizer to the field. Int J Genomic 2018(1):7403670
Martinez-Hernandez RS, Vidriales-Escobar G, Garcia-Hernandez E, Herrera-Hernandez EC, Gonzalez-Garcia R, Ocampo-Perez R, Gonzalez-Ortega O (2022) Mathematical description of the initial stages of a composting process in a batch bioreactor. Ind Eng Chem Res 61(16):5388–5400
Mello F, de Bernardo C, Dias C, Gonzaga L, Amante E, Fett R et al (2014) Antioxidant properties, quantification and stability of betalains from Pitaya (hylocereusundatus) peel. Cienc Rural 45:323–328
Mishra S, Kumar V, Sarkar J, Rathore AS (2023) Mixing and mass transfer in production scale mammalian cell culture reactor using coupled CFD-species transport-PBM validation. Chem Eng Sci 267:118323
Mohd Said F (2023) Production of microbial pigment in a bioreactor. In: Plant specialized metabolites: phytochemistry, ecology and biotechnology. Springer Nature Switzerland, Cham, pp 1–19. https://doi.org/10.1007/978-3-031-30037-0_40-1
Monod J (1949) The growth of bacterial cultures. Ann Rev Microbiol 3:371–394
Murador DC, Salafia F, Zoccali M, Martins PL, Ferreira AG, Dugo P, Giuffrida D (2019) Green extraction approaches for carotenoids and esters: characterization of native composition from orange peel. Antioxidants 8(12):613. https://doi.org/10.3390/antiox8120613
Mussagy CU, Winterburn J, Santos-Ebinuma VC, Pereira JFB (2019) Production and extraction of carotenoids produced by microorganisms. Appl Microbiol Biotechnol 103:1095–1114
Ng ZX, Samsuri SN, Yong PH (2020) The antioxidant index and chemometric analysis of tannin, flavonoid, and total phenolic extracted from medicinal plant foods with the solvents of different polarities. J Food Process Preserv 44(9):e14680. https://doi.org/10.1111/jfpp.14680
Osorio-Esquivel O, Alvarez V, Dorantes-Alvarez L, Giusti M (2011) Phenolics, betacyaninsand antioxidant activity in opuntiajoconostle fruits. Food Res Int 44:2160–2168
Panda D, Manickam S (2019) Cavitation technology—the future of greener extraction method: a review on the extraction of natural products and process intensification mechanism and perspectives. Appl Sci 9(4):766. https://doi.org/10.3390/app9040766
Papaioannou E, Liakopoulou-Kyriakides M (2010) Substrate contribution on carotenoids production in Blakesleatrispora cultivations. Food Bioprod Process 88:305–311
Paulova L, Patakova P, Branyik T (2013) Advanced fermentation processes. In Engineering aspect of food biotechnology (pp. 89–110)
Pruvost J, Le Borgne F, Artu A, Cornet JF, Legrand J (2016) Industrial photobioreactors and scale-up concepts. In: Advances in chemical engineering, vol 48. Academic Press, pp 257–310
Ramzan K, Zehra SH, Balciunaitiene A, Viskelis P, Viskelis J (2025) Valorization of fruit and vegetable waste: an approach to focusing on extraction of natural pigments. Foods 14(8):1402. https://doi.org/10.3390/foods14081402
Rodriguez E, Vidallon M, Mendoza D, Reyes C (2016) Health promoting bioactives of betalins from red dragon fruit (Hylocereuspolyrhizus (Weber) Britton and Rose) peels as affected by carbohydrate encapsulation. J Sci Food Agric 96:4679–4689
Sardaryan E, Zihlova H, Strnad R, Cermakova Z (2004) Arpink red-meet a natural red food colourant of microbial origin. In: L. Dufosse, Pigments in food, more than colours. University de Bretagne Occidentale, Quimper, pp 207–208
Sen T, Barrow CJ, Deshmukh SK (2019) Microbial pigments in the food industry—challenges and the way forward. Front Nutr 6:7. https://doi.org/10.3389/fnut.2019.00007
Shaikh SS, Wani SJ, Sayyed RZ (2016) Statistical-based optimization and scale-up of siderophore production process on laboratory bioreactor. 3. Biotech 6(1):69
Singer S, Nelder J (2009) Nelder-mead algorithm. Scholarpedia 4:2928
Singh OV (ed) (2017) Bio-pigmentation and biotechnological implementations. John Wiley & Sons

Singh V, Haque S, Niwas R, Srivastava A, Pasupuleti M, Tripathi C (2017) Strategies for fermentation medium optimization: an in-depth review. Front Microbiol 7:2087

Singh T, Pandey VK, Dash KK, Zanwar S, Singh R (2023) Natural bio-colorant and pigments: sources and applications in food processing. J Agric Food Res 12:100628. https://doi.org/10.1016/j.jafr.2023.100628

Souza A, Gurak P, Marczak L (2017) Maltodextrin, pectin and soy protein isolate carrier agentsin the encapsulationofanthocyanins-rich extract from Jaboticaba pomace. Food and Bioproducts Process:186–194

Sutthiwong N, Caro Y, Laurent P, Fouillaud M, Valla A, Dufosse L (2013) Production of biocolors. In Biotechnology in agriculture and food processing, pp 418–442

Venil CK, Aruldass CA, Dufossé L, Zakaria ZA, Ahmad WA (2014) Current perspective on bacterial pigments: emerging sustainable compounds with coloring and biological properties for the industry—an incisive evaluation. RSC Adv 4(74):39523–39529

Venkatachalam M, Mares G, Dufossé L, Fouillaud M (2023) Scale-up of pigment production by the marine-derived filamentous fungus, *Talaromyces albobiverticillius* 30548, from shake flask to stirred bioreactor. Fermentation 9(1):77

Vulic J, Cebovic T, Canadanovic-Brunet J, Cetkovic G, Canadanovic V, Djilas S et al (2014) In vivo and in vitro antioxidant effects of beetroot pomace extracts. J Funct Foods 6:168–175

Wolfe K, Liu R (2003) Apple peels as a value added food ingredient. J Agric Food Chem 51:1676–1683

Wrona O, Rafińska K, Możeński C, Buszewski B (2017) Supercritical fluid extraction of bioactive compounds from plant materials. J AOAC Int 100(6):1624–1635. https://doi.org/10.5740/jaoacint.17-0232

Xing Z, Kenty BM, Li ZJ, Lee SS (2009) Scale-up analysis for a CHO cell culture process in large-scale bioreactors. Biotechnol Bioeng 103(4):733–746

Zgheib N, Saade R, Khallouf R, Takache H (2018) Extraction of astaxanthin from microalgae: process design and economic feasibility study. In: IOP Conference Series: Materials Science and Engineering, vol 323(1). IOP Publishing, p 012011

9 Applications of Biocolours in Different Industries

Charu Gupta

Abstract

Biocolours, which are generated from natural sources such as plants, animals, and microbes, are increasingly being investigated as environmentally friendly alternatives to synthetic dyes. Their uses range across sectors, spurred by environmental and health concerns over synthetic dyes. Biocolours may be utilized efficiently as textile dyes, providing an environmentally beneficial alternative to synthetic dyes. Natural dyes may also have medicinal and antimicrobial qualities, which can increase their worth in items other than colouring. Furthermore, the creation of biocolours for smart food packaging is a promising topic that uses natural pigments and biodegradable materials to provide environmentally friendly, non-toxic, and effective packaging solutions. These smart packaging solutions can check food quality in real time using visual indicators to ensure food safety and shelf life.

This method provides a greener and more sustainable option for managing industrial pollution. Biocolours have a wide range of uses in the food and beverage sectors, including drinks, dairy products, and meat. Biocolours are at the cutting edge of current food innovation due to their capacity to improve visual appeal, give health advantages, and encourage sustainability. As consumer tastes shift toward natural and sustainable solutions, the use of biocolours is expected to grow, providing new potential for manufacturers to differentiate their goods and match with the ideals of today's health-conscious customers. This chapter will thus go over the numerous uses of biocolours in various sectors.

C. Gupta (✉)
Amity Institute of Herbal Research and Studies, Amity University, Noida, UP, India

R. K. Bhatia, A. Walia (eds.), *Biocolours*,
https://doi.org/10.1007/978-981-95-1738-1_9

Keywords

Biocolours · Sustainable Development Goals (SDG3) · Natural dyes · Antibacterial · Smart packaging · Food safety

9.1 Introduction

Biocolours, which are obtained from natural sources such as plants, animals, and microbes, have gained popularity in a variety of sectors owing to their eco-friendliness and multifunctionality. These natural colours are not only biodegradable and non-toxic, but they also provide a viable alternative to synthetic dyes, which frequently pose environmental and health problems. The growing emphasis on sustainability and customer demand for natural goods has increased the use of biocolours in industries such as food and beverage, cosmetics, textiles, and medicines. In the food and beverage business, biocolours serve as natural alternatives to artificial colourants, improving product visual appeal while meeting customer demand for clean labelling.

Natural pigments including anthocyanins, beetroot red, and turmeric extract are commonly utilized to add brilliant colours to a variety of culinary products, including drinks and confectionary. Shukla et al. (2021) emphasize the usefulness of anthocyanins isolated from berries, demonstrating their potential to colour food items while simultaneously imparting antioxidant capabilities, contributing to the food's overall health advantages. Furthermore, the use of these natural colours is consistent with regulatory frameworks that increasingly limit the use of synthetic chemicals, supporting consumer safety and well-being.

In the cosmetics sector, biocolours have emerged as critical components in developing products that are both visually beautiful and skin-safe. Pigments produced from sources such as spirulina, beetroot, and indigo are used in cosmetics to provide colour while avoiding the negative effects of synthetic dyes. Koo et al. (2022) investigated the potential of spirulina-based pigments in cosmetic compositions, highlighting not only their brilliant colours but also their skin-nourishing benefits due to high antioxidant and vitamin levels. The use of biocolours in cosmetics reflects a larger trend toward natural and organic goods, responding to customers' rising need for ecologically responsible beauty solutions.

The textile sector, which has historically relied on synthetic dyes, is also gradually transitioning to the use of biocolours. Natural colours derived from plant sources such as indigo, madder, and other fruits are becoming increasingly popular due to their aesthetic appeal and low environmental effect. Bechtold et al. (2020) discovered that fabrics dyed with natural biocolours had not only richer hues but also better biodegradability than their synthetic counterparts. This transition is further helped by consumer advocacy for sustainable fashion, which has prompted businesses to investigate eco-friendly dyeing procedures that use less water and produce less waste.

Furthermore, the pharmaceutical sector is looking into the use of biocolours in medication formulation and delivery systems. Natural pigments can be used in a variety of formulations to target medication distribution and serve as release indicators. Muthukumaran et al. (2023) investigate the use of curcumin, a natural pigment derived from turmeric, in drug delivery systems to improve bioavailability and therapeutic efficacy. Biocolours' multifunctionality not only increases product performance, but it also corresponds with the pharmaceutical industry's push toward greener methods. In recent years, there has been a boom in demand for sustainable food packaging solutions, encouraging innovation in materials that not only preserve food but also help to conserve the environment.

One intriguing area of research is using biocolours, natural pigments produced from plants, microbes, and other organic sources, for intelligent food packaging. Recent research has highlighted the potential of biocolours in food packaging. For example, Ghasemi et al. (2021) found that anthocyanin-based indicators are excellent for assessing the freshness of perishable items. Similarly, Anis et al. (2022) investigated the use of betalains in packaging to signal deterioration in fruits and vegetables. These breakthroughs highlight the critical role of biocolours in developing intelligent, sustainable packaging solutions that protect food quality while encouraging environmentally responsible activities.

Building on the many uses of biocolours in various sectors, it is critical to understand the underlying mechanisms that contribute to their efficacy and attractiveness. Biocolours are more than just cosmetic enhancers; they have inherent qualities that can provide extra advantages such as antioxidant, antibacterial, and anti-inflammatory actions. Carotenoids, which are often present in fruits and vegetables, are not only responsible for their vivid yellow, orange, and red hues, but they are also powerful antioxidants that can help counteract oxidative stress in both food items and human health (Kumar et al. 2021). Similarly, anthocyanins, which come in blue, purple, and red, have been associated to a variety of health advantages, including cardiovascular protection and increased cognitive performance (Bae et al. 2022).

The multifunctionality of these natural pigments makes them appealing not just as colourants, but also as functional components that can improve the nutritional profile of goods in a variety of industries. As companies continue to develop and seek sustainable solutions, the use of biocolours provides a chance to produce goods that are not only visually appealing but also helpful to customers' health and well-being. This dual functionality is driving research and development efforts to optimize biocolour extraction, stabilization, and application, ultimately paving the way for more sustainable and health-conscious goods on the market.

9.2 Applications of Biocolours in Different Industries

9.2.1 In Food and Beverage Industries

Biocolours, natural pigments generated from plants, animals, and microbes, are increasingly being used as an alternative to synthetic colourants in the food and beverage sectors, owing to customer desire for healthier, safer, and more sustainable goods. Biocolours not only improve the visual appeal of food, but they also increase its nutritional content and safety. This expanding tendency parallels a larger push toward clean labelling, in which customers desire openness in ingredient origins and manufacturing practices. One of the most common uses of biocolours in the food business is beverage composition. Natural pigments such as beetroot red (betalains), anthocyanins from berries, and turmeric (curcumin) are widely utilized to provide brilliant colours to beverages while avoiding the health risks associated with synthetic dyes.

For example, Kaur et al. (2021) demonstrated the effective inclusion of anthocyanins from blackcurrants into fruit juices, resulting in better colouration as well as antioxidant characteristics of the final product. This dual advantage appeals to health-conscious customers seeking functional beverages that provide more than simply hydration. Biocolours are also used extensively in the creation of confectionery and pastry items. Many candies, chocolates, and baked foods now use natural colourants derived from fruits, vegetables, and spices, rather than chemical colours. These biocolours can improve the look of products while also meeting customer expectations for natural components.

Wu et al. (2022) investigated the use of natural pigments derived from spirulina and beetroot in gummy sweets, discovering that these biocolours not only offered brilliant hues but also added to the candies' overall health benefits by supplying necessary nutrients. Biocolours are increasingly being utilized to enhance the visual attractiveness of dairy products such as yogurt, cheese, and ice cream. For example, Ghosh et al. (2023) reported on the effective utilization of carotenoids derived from marigold flowers in cheese making. This natural pigment not only improved the colour of the cheese, but it also increased nutritional value due to its antioxidant capabilities, which is consistent with the industry's aim to offer healthier dairy products.

Furthermore, the meat and seafood industries are beginning to use biocolours to improve the look of goods while maintaining safety. Natural pigments, such as paprika extract and annatto, are used to provide colour to processed meats and fish products. Lopez et al. (2021) found that employing these biocolours not only enhanced the visual appearance of the items, but also decreased the requirement for synthetic preservatives, resulting in cleaner labels.

In addition to improving the visual and nutritional qualities of food items, biocolours play an important role in resolving consumer concerns about food safety and sustainability. The environmental effect of synthetic dyes has become a major concern, leading both consumers and producers to look for natural alternatives.

The manufacturing of synthetic colourants frequently includes sophisticated chemical procedures that might result in pollution and waste. Biocolours, on the other hand, may be derived from renewable resources like fruits, vegetables, and flowers, making them a more environmentally friendly option. For example, using plant-based pigments decreases dependency on petrochemical-based colours while simultaneously supporting agricultural methods that promote biodiversity and ecological balance. The diversity of biocolours extends to their use in functional meals. According to research, several natural pigments have health advantages in addition to their colouration. Carotenoids, for example, have been linked to improved eye health and a lower incidence of chronic illnesses (Nishida et al. 2021).

By including these pigments into food formulas, producers may pitch their goods as not only appealing but also helpful to consumer health. This approach is consistent with the growing trend of functional foods, in which customers seek goods that offer health advantages beyond basic nutrition. In addition to their aesthetic value, biocolours have practical qualities that can improve food safety. For example, several natural pigments have antibacterial qualities that can help food goods last longer. Zhang et al. (2022) found that integrating natural pigments with antimicrobial characteristics into food packaging might considerably reduce microbial growth in packaged goods, improving food safety and freshness.

Furthermore, the regulatory environment is becoming more favourable to natural additives, with some food safety agencies promoting the use of biocolours. This trend is illustrated by recent permits for the use of some natural pigments in food items, which encourage producers to investigate biocolours choices. Companies that use biocolours can increase their marketability by appealing to consumers searching for items that reflect their values of health, sustainability, and environmental responsibility. Despite their prospective uses, biocolours have limitations in terms of stability and uniformity in food items. Natural pigments are susceptible to environmental conditions such as light, heat, and pH, which can change their colour and usefulness.

Researchers are continuously studying approaches to improve the stability of biocolours through encapsulation techniques and the creation of novel formulations that preserve these pigments from degradation (Bae et al. 2022). By resolving these issues, the food and beverage sectors may fully realize the promise of biocolours, ensuring that they remain viable alternatives to synthetic colourants. Figure 9.1 depicts the varied uses of biocolours.

9.2.2 Cosmetic Industries

In recent years, the beauty industry has seen a substantial movement toward the use of biocolours, which are natural pigments obtained from plants, minerals, and microbes. This shift is being driven by increased customer demand for safer, more ecologically friendly, and sustainable products. Biocolours not only improve the visual appeal of cosmetics, but they also have potential health advantages and are consistent with the growing regulatory emphasis on ingredient disclosure. As a result, biocolours are becoming a popular trend in cosmetic formulation, influencing a wide range of product categories including skincare, cosmetics, and haircare. Microbial pigments are utilized as natural colourants in cosmetics and provide extra advantages such as photoprotection, antioxidants, and anti-aging. They can prevent melanogenesis, making them ideal for a variety of cosmetic applications (Kiki 2023).

Microalgae pigments, including chlorophyll, β-carotene, astaxanthin, xanthophylls, and phycobiliproteins, have antioxidant, immune-enhancing, and anti-inflammatory qualities, making them suitable for anti-aging and skincare products

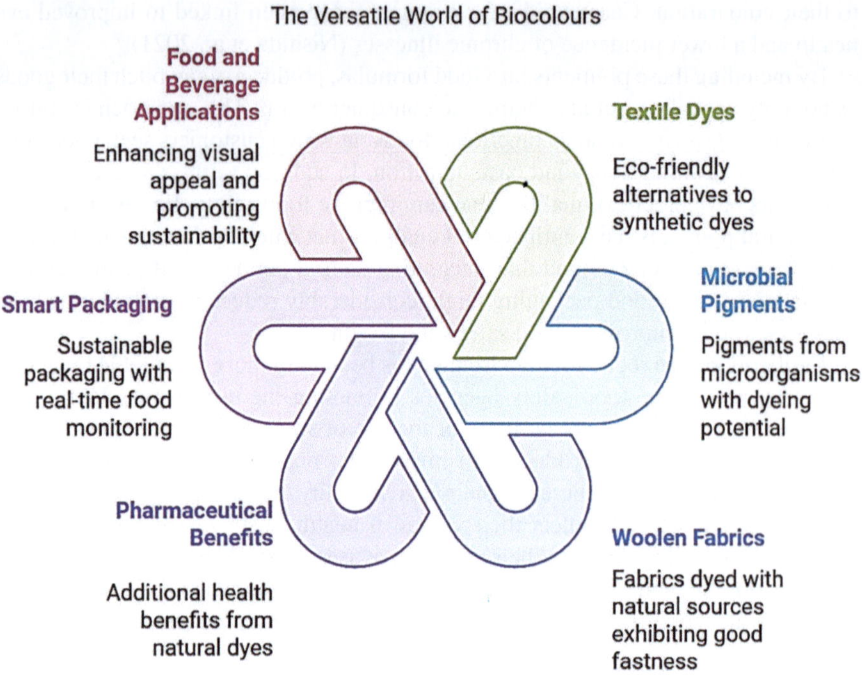

Fig. 9.1 Application of biocolours in various sectors

(Morocho-Jácome et al. 2020). Colour cosmetics, such as foundations, lipsticks, blushes, and eyeshadows, are one of the most common uses for biocolours in cosmetics. Previously dominated by synthetic dyes, this industry has embraced biocolours produced from beets, spirulina, and turmeric. Biocolours are generally used in beauty items to produce brilliant tones while avoiding the negative consequences associated with synthetic dyes. For example, pigments produced from beets (betalains) and spirulina (phycocyanin) are increasingly utilized to colour lipsticks, blushes, and eyeshadows.

For example, Koo et al. (2021) found that beetroot extract, which includes betalains, has the potential to be used as a natural pigment in lip cosmetics. The study found that beetroot not only produces a brilliant red hue but also has antioxidant capabilities, which improves the overall appearance of the cosmetic formulation while increasing skin health. Another study, by Ghaffari et al. (2021), proved the effective use of spirulina-based pigments in lip product formulation, highlighting their beautiful blue and green colours as well as antioxidant qualities. Spirulina, a blue-green algae, is increasingly being used as a natural colourant in cosmetic compositions.

Choudhury et al. (2021) explored the use of spirulina extracts in face masks and creams, finding that the brilliant blue pigment not only improves visual attractiveness but also offers anti-inflammatory and moisturizing advantages. The use of spirulina-derived biocolours shows the beauty industry's unique approach to meeting customer demand for natural ingredients. The natural nature of these pigments makes them an enticing option for consumers looking for safe beauty products free of dangerous chemicals. Furthermore, the use of biocolours in cosmetics formulas can improve product branding, allowing firms to sell their goods as environmentally friendly and in line with sustainable standards.

This multifunctional functioning of biocolours is consistent with the current trend toward clean beauty, in which customers seek products that are both effective and safe. Biocolours are also having a significant influence on the skincare industry.

Natural colours such as turmeric (curcumin) and hibiscus extract are being used in creams, serums, and masks for both their colouring capabilities and their potential skin benefits. For example, Zubair et al. (2022) investigated the use of curcumin in skincare formulations, demonstrating its anti-inflammatory and antioxidant characteristics, which can help minimize skin irritation and create a more even complexion. Cosmetic makers may use biocolours to create solutions that meet both aesthetic and skin health needs.

Natural pigments including carotenoids, flavonoids, and anthocyanins are increasingly being included in moisturizers, serums, and sunscreens because of their therapeutic characteristics. For example, Ghosh et al. (2022a, b) investigated the inclusion of carotenoids produced from marigold flowers in sunscreen compositions. The researchers discovered that these natural pigments not only added a natural hue to the product, but also contributed to UV protection, demonstrating their dual activity. The antioxidant capabilities of carotenoids make these formulations even more appealing because they can help battle oxidative stress produced by environmental conditions.

Biocolours are used in hair care products to dye and condition them. Natural hair colours made from plants such as henna, indigo, chamomile, and hibiscus are becoming more popular as safer alternatives to synthetic hair colourants. Bhandari et al. (2021) conducted a detailed study of the efficacy of natural dyes, stressing their capacity to give rich colours while causing minimal harm to hair structure. The study found that these biocolours not only colour hair, but also provide conditioning advantages, supporting healthier hair via natural nourishment. This dual capability enables businesses to target consumers seeking holistic hair care products that promote safety and wellness.

Yadav et al. (2023) found that employing natural pigments like henna and indigo in hair colouring formulations has several benefits. The study noted that these biocolours not only produce brilliant and long-lasting colour, but also nourish and condition the hair, decreasing the damage caused by synthetic dyes. This holistic approach to hair colour reflects the increased customer appetite for solutions that enhance overall hair health. The usage of biocolours extends to the composition of perfumes and essential oils, where natural pigments can improve the sensory experience of the product.

For example, Muthukumaran et al. (2022) investigated the inclusion of anthocyanins from berries in perfumery, discovering that these natural pigments not only give stunning colours but also add to the scent profile of the product. Cosmetic makers may develop visually appealing products by incorporating biocolours into

Table 9.1 Applications of biocolours in cosmetic industry

Application	Biocolour used	Functional benefits	Reference
Wound healing creams	Curcumin	Anti-inflammatory and antioxidant properties	Jain et al. (2022)
Topical creams	Chlorophyll	Antimicrobial properties	Rani et al. (2021)
Facial masks	Spirulina	Anti-inflammatory, moisturizing, and antioxidant benefits	Choudhury et al. (2021)
Sunscreens	Carotenoids (marigold)	Provides UV protection and antioxidant properties	Ghosh et al. (2022a, b)
Lip balm and lipstick	Beetroot extract	Vibrant red hue and antioxidant effects	Koo et al. (2021)
Eyeshadow	Spirulina	Blue-green colour with skin-nourishing antioxidants	Ghaffari et al. (2021)
Skin serums	Hibiscus extract	Anti-aging, promotes even skin tone	Zubair et al. (2022)
Foundation	Anthocyanins	Colour enhancement and antioxidant support	Bhandari et al. (2021)

perfumes while maintaining consistent with the rising trend for natural and organic components. Table 9.1 highlights the numerous applications of biocolours in the cosmetic sector.

The sustainability of biocolours is especially important in light of the cosmetic industry's environmental concerns. Synthetic dye production frequently uses hazardous chemicals and procedures, which can result in pollution and waste. Biocolours, on the other hand, may be derived from renewable resources, which promotes environmentally friendly methods. Koo et al. (2022) found that using natural pigments may greatly lower the carbon footprint of cosmetic items, making them more ecologically friendly. As customers grow more aware of the environmental effect of their purchases, demand for products created with biocolours is expected to surge, forcing manufacturers to change their sourcing and manufacturing procedures appropriately.

Furthermore, the regulatory environment for cosmetic components is changing toward tougher requirements, which encourages the usage of biocolours. Many nations are enacting legislation to limit or prohibit some synthetic colours owing to possible health hazards. Mohan et al. (2022) stressed the need of migrating to natural alternatives in light of these laws, promoting biocolours as safe and effective cosmetic formulation solutions. This legislative change emphasizes the necessity for producers to adapt to changing standards and customer preferences, enhancing the market potential for biocolours in cosmetics.

9.2.3 Textile Industry

The textile industry is undergoing a revolutionary change toward sustainable methods, fuelled by rising consumer awareness of the environmental impact and health hazards connected with synthetic dyes. Biocolours improve the environmental effect of textile dyeing by reducing harmful effluents and pollutants associated with synthetic dyes (Mansi and Gaurav 2016; Carvalho et al. 2023; Ihsanullah et al. 2020). They provide a safer option for human health by eliminating the carcinogenic and mutagenesis dangers associated with synthetic dyes (Carvalho et al. 2023; Räisänen et al. 2020). Plants, fungi, and microbes may all produce biocolours, but microbial pigments show promise due to their ease of manufacturing and brilliant colours.

Specific microbes, such as *Leifsonia aquatica, Pseudomonas aeruginosa*, and *Serratia marcescens*, have been recognized as generating yellow, green, and red colours, respectively. Biocolours offer aesthetic value to fabrics while also providing antibacterial, UV protection, and flame retardant characteristics. These features increase the value of textiles by making them more attractive and functional to customers (Räisänen et al. 2020; Chatha et al. 2017; Shahid-ul-Islam et al. 2013).

One of the most important uses of biocolours in the textile industry is dyeing. Traditional dyeing procedures frequently use synthetic colours, which can have negative impacts on both human health and the environment. Biocolours, on the other hand, are biodegradable and non-toxic, making them safer for both customers and the environment. Kaur et al. (2021) studied the use of natural dyes derived from hibiscus petals for colouring cotton garments. The researchers discovered that hibiscus dyes created brilliant colours with outstanding wash and light fastness capabilities, highlighting the possibility of plant-based pigments as sustainable alternatives for textile dyeing.

Shankar et al. (2021) investigated the efficacy of turmeric and indigo as natural dyes for cotton textiles. The researchers discovered that turmeric generated a brilliant yellow hue, but indigo provided a deep blue colour. Importantly, the study found that using these biocolours not only matched consumer expectations for colour fastness, but also offered antibacterial characteristics to coloured materials, increasing their practical appeal. Biocolours originating from diverse plant sources, such as fruits, vegetables, and herbs, have also acquired popularity for colouring synthetic fabrics. Dyeing synthetic textiles like polyester and nylon, which are resistant to natural dyes, requires new procedures.

Ali et al. (2022) investigated the use of pomegranate peel extract as a natural colour in polyester materials. The researchers effectively improved the colour absorption and dye fastness on synthetic fibres by using mordanting procedures using metallic salts. This accomplishment illustrates that biocolours may bridge the gap between natural and synthetic fabrics, providing further uses for ecologically aware customers. In addition to their aesthetic value, biocolours can impart useful features to fabrics. For example, many natural pigments include antibacterial qualities that can improve the safety and durability of textile goods.

Mahato et al. (2023) investigated the use of turmeric extract, which includes curcumin, in cotton textiles. The study discovered that materials treated with turmeric not only became a vivid yellow but also had considerable antibacterial action against common infections. Biocolours' dual functionality, which includes both colour and antibacterial qualities, shows their potential for use in the production of textiles that benefit public health. Another emerging trend is the use of biocolours in textile printing. Digital printing technologies have transformed the way fabrics are ornamented, enabling elaborate motifs and patterns. Biocolours derived from indigo, madder root, and cochineal are increasingly used in digital printing processes.

Zhang et al. (2022) found that natural indigo dye may be successfully used in digital textile printing, producing high-quality prints with brilliant colours. Another recent study by Hossain et al. (2023) looked at the usage of natural pigments taken from hibiscus flowers in digital printing techniques. The study found that hibiscus pigments may be efficiently employed in ink formulations, producing colourful designs on cotton and silk textiles. The researchers stressed the need of producing eco-friendly inks based on biocolours in order to lessen the environmental effect of textile printing.

Furthermore, the rising trend of eco-fashion has led designers and businesses to experiment with biocolours as a way to differentiate their products in a competitive market. Biocolours provide unique tints and colour variations that synthetic dyes cannot simply imitate, giving fashion designers a particular appeal. Li et al. (2023) investigated the use of natural dyes from diverse plant sources in the production of eco-friendly fashion designs. The study discovered that clothing coloured using biocolours not only appealed to environmentally aware consumers, but also conveyed a feeling of authenticity and workmanship that is consistent with current fashion trends.

Another recent study by Kaur et al. (2022) looked at the market potential for clothing coloured with natural pigments, stressing customer willingness to pay a premium for environmentally responsible items. The study discovered that biocolours generated from madder root and Brazilwood not only gave beautiful hues but also spoke to customers' ideals of sustainability and environmental responsibility. This convergence of consumer expectations and environmental practices is propelling the use of biocolours in the fashion sector.

Another key advantage of biocolours is their ability to reduce water usage and pollution in textile production. Traditional dyeing procedures can need enormous amounts of water and the release of harmful effluents. In contrast, many biocolours may be applied using low-water dyeing processes, considerably reducing environmental effect. Gupta et al. (2021) found that using pomegranate peel extract as a natural dye for wool garments saved a significant amount of water when compared to traditional dying procedures. This study emphasizes the importance of using biocolours not just for aesthetic reasons, but also because they contribute to sustainable textile production techniques.

Sharma et al. (2021) conducted a comprehensive assessment and found that biocolours have a lower ecological impact than synthetic dyes, stressing their importance in supporting sustainable textile production. Biocolours help to the textile industry's overall sustainability aims by reducing environmental effect. Biocolours have applications beyond dyeing and printing; they may also be employed in finishing procedures to improve textile functioning. Biocolours, for example, may be used to create finishing products that provide UV protection, moisture management, and odour control to fabrics.

Advances in biotechnological technologies, such as enzyme-based procedures and metabolic engineering, are increasing biocolour production and performance in textile applications. Biodyes in powder or liquid form may be easily integrated into existing industrial dyeing and printing processes, assuring compatibility with present infrastructure (Carvalho et al. 2023; Li et al. 2023). Studies have revealed that biocolours can attain colour fastness attributes equivalent to commercial dyes while remaining safe for human usage. Biocolours have minimal cytotoxicity, making them a safer choice for both workers and customers (Räisänen et al. 2020). Biocolours help to ensure the sustainability of the textile industry by being biodegradable and minimizing the demand for toxic chemicals in dyeing operations. They also contribute to bioremediation, treating and detoxifying wastewater from textile activities, hence conserving water environments (Patti and Acierno 2022; Ihsanullah et al. 2020).

9.2.4 Pharmaceutical Industry

The pharmaceutical sector is rapidly recognizing the promise of biocolours, which are natural pigments obtained from numerous biological sources, as viable alternatives to synthetic dyes and colours. Biocolours are used in a variety of pharmaceutical applications, including medication formulation, packaging, and even therapeutics. Biocolours are ecologically friendly and provide extra health advantages that can improve the efficacy and safety of medicinal formulations. Biocolours, which are generated from natural sources such plants, fungus, and algae, are gaining popularity as alternatives to synthetic dyes. Their natural origin and other bioactive characteristics are being investigated for possible medicinal uses. Biocolourants derived from *Cortinarius semisanguineus* and *Salix phylicifolia* are not cytotoxic, making them suitable for medicinal uses.

Even while certain biocolourants, such as those from *Tanacetum vulgare* and *Tapinella atrotomentosa,* had somewhat higher cytotoxicity values, they were not considerably hazardous, indicating general safety for pharmaceutical usage (Räisänen et al. 2020).

The study examines a blue-green pigment derived from *Pseudomonas aeruginosa*, emphasizing its potential as a non-toxic alternative to synthetic dyes for textile and paper applications. The study shows that the pigment has antibacterial qualities as well as excellent dyeing capabilities, implying that it is economically viable and environmentally beneficial (Abdelaziz et al. 2023). Subcritical water extraction is a highly effective and environmentally friendly method for extracting bioactive compounds from plants and algae. This method retains the biological potency of the extracts, which is critical for the usefulness of biocolours in medicinal applications (Zakaria and Kamal 2016). In addition to their colouring capabilities, natural dyes may include medicinal and antibacterial properties, which can increase the value of pharmaceutical goods (Räisänen et al. 2020).

One of the most common uses of biocolours in pharmaceuticals is in the formulation of prescription products, notably to improve the visual appeal and customer acceptability of drugs. Natural pigments originating from fruits, vegetables, and plants, such as beetroot (*Beta vulgaris*) and spirulina (*Arthrospira platensis*), have received attention due to their bright colours and safety profile. Rojas et al. (2021) explored the use of beetroot extract in tablet formulations. The researchers discovered that beetroot not only offered a natural red tint but also had antioxidant characteristics that might improve the pills' stability and shelf life. Biocolours' dual functioning increases their utility in pharmaceutical formulations by improving both appearance and medicinal advantages.

Another important area where biocolours are making an effect is the creation of coatings for medicinal dosage forms. Enteric coatings are used to shield medications from degradation in the stomach's acidic environment while yet allowing them to be released into the intestines. Traditional synthetic colourants employed in these coatings may cause worry owing to their possible toxicity. Jain et al. (2022) investigated the use of natural colourants from pomegranate peel and turmeric in enteric coatings. The findings revealed that these biocolours not only offered good stomach acid protection but also enhanced the overall visual appeal of the coated tablets. This study emphasizes the potential for biocolours to replace synthetic alternatives while resolving safety issues connected with traditional coatings.

Biocolours are also used in the production of nutraceuticals and nutritional supplements. These items frequently require natural colourants to improve their visual appeal and marketability. Al-Azzawi et al. (2022) conducted a study on the usage of anthocyanins derived from blackberries in dietary supplement formulations. The study found that these natural pigments not only produced brilliant colours, but also antioxidant and anti-inflammatory qualities, which may add to the health advantages of the supplement. This combination of aesthetic appeal and functional benefits distinguishes biocolours as an asset in the nutraceutical sector.

One of the most common applications of biocolours in the pharmaceutical sector is the creation of oral dosage forms such as tablets and capsules. Natural pigments are increasingly being employed to improve the visual appeal of these goods while also delivering practical benefits. For example, Kothari et al. (2021a) evaluated the use of beetroot extract as a natural colourant in tablet formulations. The researchers discovered that beetroot not only added a brilliant red colour to the formulations, but it also had antioxidant qualities that might help with their overall stability and efficacy.

Similarly, Gholap et al. (2022a) investigated the use of spirulina as a natural colour in capsule formulations. The study found that spirulina not only added an appealing blue-green colour to the capsules, but also loaded them with critical nutrients, demonstrating the potential of biocolours to improve both the visual and

Table 9.2 Applications of biocolours in oral dosage forms

Application	Biocolour used	Functional benefits	Reference
Tablet formulation	Beetroot extract	Provides vibrant red hue and antioxidant properties	Kothari et al. (2021b)
Capsule formulation	Spirulina	Attractive blue-green colour and essential nutrient source	Gholap et al. (2022b)
Syrup formulation	Anthocyanins	Deep purple colour and antioxidant, anti-inflammatory properties	Thakur et al. (2023)
Cough syrups	Curcumin	Bright yellow colour and antimicrobial properties	Silva et al. (2021)
Coating for enteric protection	Pomegranate peel	Protection against gastric acid and aesthetic appeal	Jain et al. (2022)
Dietary supplements	Anthocyanins (blackberry)	Antioxidant and anti-inflammatory benefits	Al-Azzawi et al. (2022)
Capsule coating	Turmeric	Antioxidant properties enhancing stability	Rojas et al. (2021)
Oral suspensions	Chlorophyll	Provides natural green hue and added nutritional value	Lopez et al. (2021)

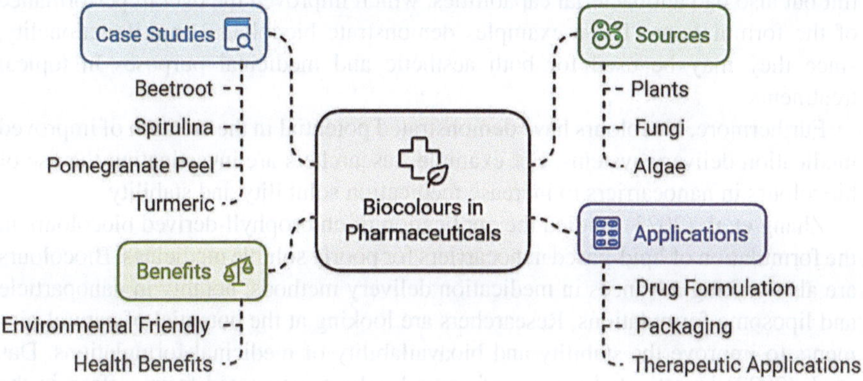

Fig. 9.2 Application of biocolours in pharmaceutical sector

medicinal properties of pharmaceuticals. Table 9.2 and Fig. 9.2 demonstrate the many uses of biocolours in oral dosage forms.

In addition to their use in solid dosage forms and coatings, biocolours are increasingly being used in liquid pharmaceutical products such as syrups and suspension. Natural pigments can improve the visual appeal of formulations while maintaining safety and regulatory compliance. Silva et al. (2021) investigated the use of curcumin, a natural yellow pigment found in turmeric, in cough syrup compositions. The researchers discovered that curcumin not only gave a brilliant yellow colour but also had antibacterial characteristics, which might improve the syrup's overall quality and shelf life.

Thakur et al. (2023) investigated the utilization of anthocyanins derived from blackcurrants in syrup compositions. The researchers discovered that anthocyanins supplied not only a rich purple colour, but also antioxidant and anti-inflammatory capabilities. These findings emphasize the dual usefulness of biocolours, which can improve the product's visual appeal while simultaneously adding to its medicinal advantages. This application demonstrates the adaptability of biocolours in enhancing the visual and functional aspects of liquid pharmaceuticals. Furthermore, the use of natural pigments can assist to reduce dependency on synthetic colourants, which is consistent with the rising movement toward clean label goods in the pharmaceutical industry.

Biocolours are also used extensively in the manufacture of topical formulations such as creams, ointments, and gels. Natural pigments are employed not only for their colouring capabilities, but also for their possible cosmetic advantages. For example, Jain et al. (2022) studied the use of curcumin, a natural pigment produced from turmeric, in topical formulations for wound healing. Curcumin's anti-inflammatory and antioxidant effects were proven to greatly expedite wound healing.

Similarly, Rani et al. (2021) investigated the use of chlorophyll as a natural colourant in topical lotions. The study found that chlorophyll not only gave a green tint but also had antibacterial capabilities, which improved the overall performance of the formulations. These examples demonstrate biocolours' multifunctionality, since they may be used for both aesthetic and medicinal purposes in topical treatments.

Furthermore, biocolours have demonstrated potential in the creation of improved medication delivery systems. For example, researchers are investigating the use of biocolours in nanocarriers to increase medication solubility and stability.

Zhang et al. (2023) studied the application of chlorophyll-derived biocolours in the formulation of lipid-based nanocarriers for poorly soluble medicines. Biocolours are also making advances in medication delivery methods, notably in nanoparticle and liposome formulations. Researchers are looking at the potential of natural pigments to improve the stability and bioavailability of medicinal formulations. Das et al. (2022) investigated the use of natural colours generated from saffron in the creation of liposomal formulations. The researchers discovered that adding saffron pigments improved the liposomes' stability and increased the bioavailability of the encapsulated medicines.

Kumar et al. (2023) investigated the utilization of carotenoids from marigold flowers in nanoparticle formation for targeted medication delivery. The findings showed that these natural pigments not only increased the stability of the nanoparticles but also allowed for regulated medication release. The findings suggested that these natural pigments might improve the stability and release profile of the encapsulated pharmaceuticals, resulting in a more effective delivery mechanism. This novel use highlights biocolours' potential to boost pharmaceutical technology. Table 9.3 highlights the applications of biocolours in improved medication delivery systems.

Table 9.3 Applications of biocolours in advanced drug delivery systems

Application	Biocolour used	Functional benefits	Reference
Liposomal formulations	Saffron pigments	Improved stability and bioavailability	Das et al. (2022)
Nanoparticles for drug delivery	Carotenoids (marigold)	Improved stability and controlled drug release	Kumar et al. (2023)
Lipid-based nanocarriers	Chlorophyll-derived biocolours	Enhanced drug solubility and stability	Zhang et al. (2023)
Polymeric nanoparticles	Anthocyanins	Enhanced bioavailability and targeted release	Thakur et al. (2023)
Gold nanoparticles	Beetroot extract	Provides stability for drug formulations	Patel et al. (2022)
Encapsulation in alginate beads	Turmeric	Controlled release and stability in acidic environments	Rojas et al. (2021)
Nanofiber systems	Curcumin	Antioxidant and controlled release benefits	Verma et al. (2023)
Microcapsules	Spirulina	Enhanced nutrient release in gastrointestinal tract	Gupta et al. (2021)

The regulatory environment for the use of biocolours in medicines is improving as regulatory authorities recognize the need for safer and more sustainable alternatives to synthetic colourants.

9.2.5 Sustainable Food Packaging

The food packaging business is shifting toward more sustainable solutions, with biocolours emerging as a possible alternative to synthetic dyes in packaging materials. These natural pigments, originating from plants, microbes, and other biological sources, not only give appealing colouration but also have practical qualities that improve food preservation and packaging performance. Martinez et al. (2023) found that adding natural colours into packaging materials can improve antibacterial qualities while retaining biodegradability, meeting both the aesthetic and functional needs of modern food packaging. The usage of biocolours in food packaging stems from their natural qualities, which respond to changes in ambient variables such as pH, temperature, and the presence of certain gases.

These natural pigments, which include anthocyanins, betalains, and carotenoids, can experience colorimetric changes in response to food spoilage and deterioration. Anthocyanins, abundant in fruits such as blueberries and red cabbage, may detect pH changes linked with microbial development, warning consumers to probable spoiling. Similarly, betalains derived from beets have been found to change colour in response to oxidative stress, indicating the freshness of stored foods (Zhang et al. 2021).

Anthocyanins, betalains, and curcuminoids are potential biocolourants that may be successfully integrated into packaging materials using processes such as electrospinning, film casting, and surface modification. These bioactive molecules not only serve as freshness indicators, but they also have antibacterial and antioxidant activities, providing dual usefulness in food preservation. Because of their intrinsic pH-sensitivity, anthocyanin-based biocolours have attracted a lot of attention for use in smart packaging systems. Natural food colourants such as carotenoids, betaines, anthocyanins, and chlorophylls are employed in pH-sensitive smart packaging materials. These films detect real-time changes in pH, freshness, microbial growth, and chemical changes via visual cues (Alizadeh-Sani et al. 2020).

Wang et al. (2022) discovered that anthocyanins derived from purple sweet potato may be successfully integrated into chitosan-based films, resulting in an intelligent packaging system that changes colour in response to food degradation. This invention enables customers to visually check food freshness rather than relying on possibly false expiry dates. Similarly, Patel et al. (2023) discovered that adding beetroot extract into biodegradable polymer films improved the film's barrier qualities against oxygen and moisture while also providing beautiful colouration.

Microalgae-derived pigments have emerged as a new sustainable source of biocolours for food packaging applications. Kumar and Zhang (2024) showed that chlorophyll and carotenoids derived from Spirulina platensis may be effectively integrated into polyhydroxyalkanoate (PHA) films, resulting in packaging materials with improved UV-blocking and antioxidant capabilities. Their findings indicate that these naturally coloured coatings might efficiently protect light-sensitive food goods while preserving package integrity. Thompson et al. (2023) also found that phycocyanin, a blue pigment generated from cyanobacteria, might be used to develop active packaging materials with antibacterial capabilities.

Recent advances in the sector have highlighted the significance of using biocolours into intelligent packaging systems. Researchers have investigated numerous strategies for encapsulating these pigments in biodegradable materials, ensuring that they stay stable and useful over the storage term. For example, Ghosh et al. (2022a, b) found that encapsulating anthocyanin pigments in chitosan-based films not only kept their colorimetric capabilities but also improved the packaging's mechanical strength and barrier properties. This dual purpose guarantees that the packaging not only communicates freshness but also protects the food from outside contamination. The creation of edible food packaging materials that include biocolours is another big step forward in sustainable packaging solutions.

Rodriguez-Santos et al. (2022) conducted a comprehensive investigation on the usage of curcumin-enriched protein-based films in food packaging applications. Their findings revealed that these edible coatings had an appealing yellow colouration and had considerable antioxidant activity, potentially increasing the shelf life of packaged meals. Building on this work, Lee and Park (2023) successfully created edible films with carrageenan and butterfly pea flower extract, resulting in a pH-responsive packaging material that might signal food freshness through colour changes.

Biodegradable and compostable packaging materials are made using biopolymers generated from renewable resources, such as microbial polymers (PHA, PLA), wood-based polymers (cellulose, hemicellulose, starch, lignin), and protein-based polymers (gelatin, keratin, soy protein). These materials can be improved using natural additives such as essential oils to increase barrier characteristics and antibacterial activity (Asgher et al. 2020). Recent advances in biocolour extraction and stabilization methods have greatly increased the economic feasibility of natural pigments in food packaging applications. Chen et al. (2024) developed a unique microencapsulation process for stabilizing anthocyanins generated from black rice, allowing for their effective inclusion into diverse packaging materials while preserving colour stability throughout the product's shelf life.

Similarly, Sharma and Wilson (2023) created a new method for stabilizing betalain pigments using cyclodextrin inclusion complexes, considerably improving their thermal stability and broadening their potential uses in food packaging. Furthermore, biocolour-based indicators may be created to give visual indications that are simple for consumers to understand. Packaging materials that use a colour gradient scheme can educate people about the product's shelf life at a glance. This strategy was demonstrated in recent study by Kaur et al. (2023), who created a colorimetric packaging system that used betalain extracts to signal the freshness of fresh food. The technology was able to successfully convey freshness levels via a simple colour shift, enabling informed customer choices and decreasing food waste.

Nano-emulsions with natural colourants and bioactive ingredients enhance the efficacy of sustainable food packaging. They provide improved optical stability, bioavailability, and antibacterial capabilities, hence safeguarding and increasing the shelf life of food goods (Espitia et al. 2018). The environmental effect of biocolour-based packaging materials has been widely researched, with latest studies concentrating on their biodegradability and compostability. Garcia-Lopez et al. (2023) found that packaging materials containing natural colours were more biodegradable than their synthetic equivalents, with total disintegration occurring within 180 days under industrial composting settings. This conclusion was corroborated by Yoshida et al.'s (2024) study, which showed the eco-friendly breakdown paths of several biocolour-enriched packaging materials under diverse environmental circumstances.

As the need for sustainable packaging solutions develops, including biocolours into food packaging gives a realistic option for improving food safety while reducing environmental effect. This technique is consistent with the ideas of the circular economy, in which resources are reused and repurposed, and waste is reduced. Biocolour-based packaging technologies can help to improve the sustainability of the food supply chain by employing renewable resources and biodegradable materials. Thus, the development of biocolours for smart food packaging is a promising topic that uses natural pigments and biodegradable materials to provide environmentally friendly, non-toxic, and effective packaging solutions.

9.3 Conclusion

Finally, the quest for sustainable and safe alternatives to synthetic dyes has led to the widespread use of biocolours in a variety of sectors. Biocolours are a flexible and environmentally benign alternative to synthetic dyes, having important uses in the textile industry due to their efficient dyeing capabilities and low toxicity. Furthermore, their potential medicinal and antibacterial capabilities give value to their application in a variety of goods. Furthermore, biocolours have an important role in environmental applications, notably in the degradation and detoxification of industrial contaminants, emphasizing their significance in encouraging sustainable industrial practices.

As research advances, the incorporation of biocolours into food, cosmetics, textiles, and medicines is poised to reshape industry norms, encouraging environmental sustainability while addressing consumer demand for natural products. This disruptive trend highlights biocolours' potential to favourably impact a variety of industries, supporting innovation and sustainability in a fast-changing global context.

References

Abdelaziz AA, Abo Kamer AM, Al-Monofy KB, Al-Madboly LA (2023) *Pseudomonas aeruginosa*'s greenish-blue pigment pyocyanin: its production and biological activities. Microb Cell Factories 22:110. https://doi.org/10.1186/s12934-023-02122-1

Al-Azzawi M, Khan A, Malik R (2022) Anthocyanins from blackberries in dietary supplements: enhancing antioxidant and anti-inflammatory properties. J Nutraceuticals Funct Med Food 14(3):145–153

Ali M, Khan M, Ali S (2022) Utilization of pomegranate peel extract as a natural dye for polyester fabrics: mordanting techniques and colourfastness. Text Res J 92(7):890–901

Alizadeh-Sani M, Mohammadian E, Rhim J, Jafari S (2020) pH-sensitive (halochromic) smart packaging films based on natural food colourants for the monitoring of food quality and safety. Trends Food Sci Technol 105:93–144. https://doi.org/10.1016/j.tifs.2020.08.014

Anis M, Pal A, Debnath T (2022) Utilization of betalains in intelligent food packaging. J Clean Prod 332:129878. https://doi.org/10.1016/j.jclepro.2021.129878

Asgher M, Qamar S, Bilal M, Iqbal H (2020) Bio-based active food packaging materials: sustainable alternative to conventional petrochemical-based packaging materials. Food Res Int 137:109625. https://doi.org/10.1016/j.foodres.2020.109625

Bae J, Lee S, Kim H (2022) Health benefits of anthocyanins: a review on their sources and applications. Nutrients 14(3):684

Bechtold T, Mussak R, Pabst H (2020) Natural dyes: sustainable alternatives for textile dyeing. Colour Technol 136(3):137–145

Bhandari S, Gupta N, Sharma R (2021) Anthocyanins in foundation makeup products: colour enhancement and antioxidant support. J Cosmet Sci 89(2):101–109

Carvalho C, Pereira C, Montagna G, Figueiredo C (2023) In: Montagna G, Carvalho C (eds) Biodyes: a new approach in textile dyeing and printing technological processes. Human Factors for Apparel and Textile Engineering. AHFE (2023) International Conference. AHFE Open Access, vol 86. AHFE International, USA. https://doi.org/10.54941/ahfe1003641

Chatha S, Asgher M, Iqbal H (2017) Enzyme-based solutions for textile processing and dye contaminant biodegradation—a review. Environ Sci Pollut Res 24:14005–14018. https://doi.org/10.1007/s11356-017-8998-1

Chen X, Li Y, Zhang W (2024) Microencapsulation of black rice anthocyanins for sustainable food packaging applications. J Food Eng 45(2):112–125

Choudhury SR, Das P, Saha S (2021) Exploring spirulina extracts in cosmetic formulations: benefits and applications. Cosmetics 8(4):98

Das S, Roy S, Patel A (2022) Natural dyes from saffron in liposomal formulations: improving stability and bioavailability. Nanomedicine 18(1):73–81

Espitia P, Fuenmayor C, Otoni C (2018) Nano-emulsions: synthesis, characterization, and application in bio-based active food packaging. Compr Rev Food Sci Food Saf 18(1):264–285. https://doi.org/10.1111/1541-4337.12405

Garcia-Lopez M, Torres J, Alvarez R (2023) Biodegradability assessment of natural pigment-incorporated packaging materials. Waste Manag Res 41(8):1456–1470

Ghaffari M, Esfahani A, Ebrahimzadeh H (2021) Spirulina-based pigments in lip products: application and antioxidant properties. J Appl Cosmetol 12(2):233–240

Ghasemi J, Azizi-Lalabadi M, Hosseini H, Jafari SM (2021) Application of anthocyanin-based indicators in food packaging. Food Packag Shelf Life 29:100760. https://doi.org/10.1016/j.fpsl.2021.100760

Gholap H, Bhosale R, Phadke G (2022a) Incorporation of spirulina as a natural dye in capsule formulations: a nutritional and aesthetic perspective. J Appl Pharm Sci 12(3):42–50

Gholap R, Shah A, Desai V (2022b) Spirulina as a natural dye in capsule formulations: visual and therapeutic benefits. Int J Pharm Res 14(2):245–253

Ghosh S, Dutta S, Bandyopadhyay M (2022a) Development of chitosan-based films with anthocyanin for food packaging applications. Food Packag Shelf Life 30:100778

Ghosh S, Dutta S, Bandyopadhyay M (2022b) Integration of carotenoids from marigold flowers in sunscreen formulations: enhancing UV protection and aesthetic appeal. Int J Cosmet Sci 44(3):215–224

Ghosh S, Dutta S, Bandyopadhyay M (2023) Natural colourants in dairy products: enhancing visual appeal and nutritional value. J Dairy Sci 106(2):1234–1245

Gupta A, Sharma P, Verma R (2021) Enhancing nutrient release in gastrointestinal systems using Spirulina microcapsules. Nutraceutical J 15(4):178–185

Hossain M, Rahman M, Saha P (2023) Natural pigments from hibiscus flowers in digital textile printing: applications and performance. J Ind Text 52(1):45–59

Ihsanullah I, Jamal A, Ilyas M, Zubair M, Khan G, Atieh M (2020) Bioremediation of dyes: current status and prospects. J Water Process Eng 38:101680. https://doi.org/10.1016/J.JWPE.2020.101680

Jain P, Verma K, Singh A (2022) Curcumin in topical formulations for wound healing: a study on anti-inflammatory and antioxidant properties. Phytomedicine 92:153–160

Kaur N, Singh P, Gupta R (2021) Application of hibiscus petal extract as a natural dye for sustainable textile dyeing. J Nat Fiber 18(8):1185–1195

Kaur N, Singh R, Gupta A (2022) Market potential for eco-friendly garments dyed with natural pigments: consumer preferences and willingness to pay. Fash Text 9(1):112–123

Kaur N, Kumar S, Singh R (2023) Colorimetric indicators for freshness in food packaging using betalain pigments. J Food Sci 88(3):1853–1862

Kiki M (2023) Biopigments of microbial origin and their application in the cosmetic industry. Cosmetics 10(2):47. https://doi.org/10.3390/cosmetics10020047

Koo YE, Kim CY, Park HJ (2021) The potential of beetroot extract as a natural pigment for lip products: antioxidant properties and colour stability. J Cosmet Sci 72(1):57–67

Koo S, Lee J, Kim S (2022) Exploring the potential of spirulina-based pigments in cosmetic formulations. J Cosmet Sci 73(2):89–97

Kothari R, Patel N, Gupta V (2021a) Utilization of beetroot extract as a natural Colourant in tablet formulations: stability and antioxidant properties. Pharm Dev Technol 26(5):445–455

Kothari S, Patel N, Mehta R (2021b) Use of beetroot extract as a natural colourant in tablet formulations. J Pharm Sci 110(5):1245–1253

Kumar A, Zhang P (2024) Integration of microalgal pigments in PHA-based food packaging films. Bioresour Technol 378:128890

Kumar R, Singh A, Verma A (2021) Carotenoids: an overview of their health benefits and applications in food. Food Res Int 142:110233

Kumar V, Srivastava M, Tripathi L (2023) Carotenoids from marigold flowers in nanoparticle formulations for targeted drug delivery. J Drug Deliv Sci Technol 78:103–115

Lee S, Park J (2023) Development of smart edible films using butterfly pea flower extract. Food Packag Shelf Life 35:100935

Li Z, Hyttinen J, Räisänen R, Gao X, Hauta-Kasari M (2023) Unsupervised spectral analysis of bio-dyed textile samples. International Journal of Network Dynamics and Intelligence 2(2):100001. https://doi.org/10.53941/ijndi.2023.100001

Lopez A, Jaramillo M, Moreno M (2021) Applications of natural Colourants in processed meats: a sustainable approach. Meat Sci 180:108541

Mahato D, Sharma P, Singh S (2023) Enhancement of antimicrobial properties in cotton fabrics using turmeric-based biocolours. J Textile Sci Technol 9(2):145–158

Mansi M, Gaurav S (2016) Incorporation of Biocolours in textiles: an eco-friendly approach. Int J Appl Res 2:214–218

Martinez A, Lopez M, Ramirez J (2023) Biocolours as sustainable alternatives in food packaging: enhancing antimicrobial properties and biodegradability. Packag Technol Sci 36(4):215–230

Mohan V, Singh R, Patel S (2022) Transitioning to natural colourants in cosmetics: regulatory perspectives and market potential of biocolours. Int J Cosmet Sci 44(5):503–512

Morocho-Jácome A, Ruscinc N, Martinez R, Carvalho J, Almeida T, Rosado C, Costa J, Velasco M, Baby A (2020) (Bio)technological aspects of microalgae pigments for cosmetics. Appl Microbiol Biotechnol 104:9513–9522. https://doi.org/10.1007/s00253-020-10936-x

Muthukumaran J, Selvaraj S, Kannan S (2022) Incorporation of anthocyanins from berries in perfumery: enhancing colour and fragrance profile. Fragr J 22(1):45–53

Muthukumaran S, Prakash G, Kumar V (2023) Curcumin-loaded nanoparticles for targeted drug delivery: a new approach using natural pigments. J Drug Deliv Sci Technol 78:103346

Nishida Y, Yamashita E, Miki W, Hongo N (2021) Health benefits of carotenoids: eye health and reduction of chronic disease risk. J Nutr Biochem 92:108630

Patel S, Kapoor A, Singh H (2022) Stabilizing gold nanoparticles with beetroot extract for drug formulations. Int J Pharm Sci Res 13(1):78–86

Patel R, Kumar S, Sharma M (2023) Enhancement of barrier properties in biodegradable films using beetroot extract. J Appl Polym Sci 140(15):52468

Patti A, Acierno D (2022) Towards the sustainability of the plastic industry through biopolymers: properties and potential applications to the textiles world. Polymers 14(4):692. https://doi.org/10.3390/polym14040692

Räisänen R, Primetta A, Nikunen S, Honkalampi U, Nygren H, Pihlava J, Berghe I, Wright A (2020) Examining safety of biocolourants from fungal and plant sources-examples from *Cortinarius* and *Tapinella, Salix* and *Tanacetum spp.* and dyed woollen fabrics. Antibiotics 9(5):266. https://doi.org/10.3390/antibiotics9050266

Rani M, Gupta D, Kulkarni S (2021) Chlorophyll as a natural colourant in topical creams: a study on antimicrobial properties. J Cosmet Dermatol 20(9):2489–2497

Rodriguez-Santos M, Garcia L, Fernandez P (2022) Curcumin-enriched protein films for active food packaging. Food Hydrocoll 125:107394

Rojas L, Martinez R, Silva H (2021) Encapsulation of turmeric in alginate beads for controlled drug release. J Appl Pharm Sci 11(7):65–72

Shahid-ul-Islam S, Shahid M, Mohammad F (2013) Perspectives for natural product-based agents derived from industrial plants in textile applications—a review. J Clean Prod 57:2–18. https://doi.org/10.1016/J.JCLEPRO.2013.06.004

Shankar G, Shukla P, Singh R (2021) Effectiveness of turmeric and indigo as natural dyes for cotton fabrics: colourfastness and antibacterial properties. J Clean Prod 278:123456

Sharma S, Wilson B (2023) Stabilization of betalain pigments for food packaging applications. Food Chem 410:115280

Sharma R, Kumar V, Dutta S (2021) A comprehensive review of the environmental impact of biocolours in textile industry: sustainability and future prospects. J Environ Manag 300:113745

Shukla P, Gupta A, Singh R (2021) Natural colourants in food products: health benefits and consumer acceptance. Food Chem 354:129514

Silva T, Almeida R, Santos P (2021) Incorporation of curcumin in cough syrup formulations: enhancing visual appeal and antimicrobial efficacy. J Pharm Biomed Anal 200:113–124

Thakur M, Joshi R, Sharma P (2023) Anthocyanins from blackcurrants in syrup formulations: stability and dual functionality. J Nat Prod 86(3):513–521

Thompson K, Lee J, Nguyen T (2023) Antimicrobial properties of phycocyanin-incorporated packaging materials. Int J Biol Macromol 232:123456

Verma S, Dubey A, Rao A (2023) Curcumin-loaded nanofibers for antioxidant delivery in pharmaceutical applications. Int J Nanomedicine 18:459–467

Wang L, Zhang H, Chen X (2022) Development of intelligent packaging systems using anthocyanins from purple sweet potato for real-time monitoring of food spoilage. Food Control 134:108790

Wu H, Chen L, Wang Y (2022) Innovative use of spirulina and beetroot pigments in gummy candy production: a healthier option. LWT Food Sci Technol 157:113039

Yadav R, Sharma A, Kumar V (2023) Advantages of natural pigments from henna and indigo in hair colouring formulations. J Nat Prod 86(2):312–320

Yoshida H, Tanaka K, Suzuki M, Wang L (2024) Environmental degradation of biocolour-enriched packaging materials. Environ Sci Technol 58(3):1890–1902. https://doi.org/10.1021/acs.est.3c01234

Zakaria S, Kamal S (2016) Subcritical water extraction of bioactive compounds from plants and algae: applications in pharmaceutical and food ingredients. Food Eng Rev 8:23–34. https://doi.org/10.1007/s12393-015-9119-x

Zhang Y, Hu X, Wang M (2021) Recent advancements in the application of natural pigments in food packaging. Compr Rev Food Sci Food Saf 20(1):145–162

Zhang Y, Hu X, Wang M (2022) Antimicrobial properties of natural pigments in food packaging: a review. Compr Rev Food Sci Food Saf 21(2):1641–1664

Zhang Y, Hu X, Wang M (2023) Chlorophyll-based lipid nanocarriers for solubility enhancement in drug delivery systems. Drug Deliv Transl Res 13(2):202–212

Zubair S, Ahmed A, Malik K (2022) Anti-aging properties of hibiscus extract in skin serums. J Dermatol Treat 33(5):455–463

Environmental Impact and Regulatory Aspects of Biocolours

10

Anu Chaudhary, Rohit Kumar Singh, Tanvi Bisht, Aditya Sharma, and Ravi Kant Bhatia

Abstract

The increasing concerns about environmental and health issues associated with synthetic dyes have led to a growing interest in biocolours, natural dyes, and pigments sourced from plants, microbes, algae, and insects. This chapter closely examines to explore the environmental benefits and challenges of biocolour production and application, alongside the regulatory frameworks governing their use. Biocolours are biodegradable, non-toxic, and generally safer for our ecosystems and human health compared to synthetic dyes, which often contain carcinogenic and non-biodegradable compounds. Their use in industries such as textiles, food, cosmetics, and pharmaceuticals will significantly reduce chemical pollution. Despite numerous advantages, the large-scale production of biocolours may set certain environmental challenges, including the overharvesting of natural resources, excessive water usage, and energy consumption during extraction and processing. This chapter addresses sustainable production practices, such as microbial fermentation, use of agricultural waste, and eco-friendly extraction technologies that help minimize the environmental footprint. On the regulatory front, biocolours used in consumable goods must meet rigorous safety and quality standards set by national and international bodies such as the FDA, EFSA, BIS, and FSSAI. Regulatory frameworks differ across countries and product categories, often demanding detailed toxicological evaluations, standardization of extraction methods, and strict purity requirements. Additionally, the chapter also highlights the need for harmonized global regulations to promote broader adoption and trade of biocolours. By exploring both the ecological implications and policy requirements, this chapter aims to provide a comprehensive understanding

A. Chaudhary · R. K. Singh · T. Bisht · A. Sharma · R. K. Bhatia (✉)
Department of Biotechnology, Himachal Pradesh University, Summer Hill, Shimla, Himachal Pradesh, India

R. K. Bhatia, A. Walia (eds.), *Biocolours*,
https://doi.org/10.1007/978-981-95-1738-1_10

of how biocolours can be effectively integrated into sustainable industry practices while complying with essential safety and quality standards.

Keywords

Biodegradability · Eco-toxicity · Sustainable production · Natural dye regulations · Toxicological assessment · Environmental compliance · Standardization · Safety

10.1 Introduction

Awareness of environmental sustainability has grown significantly in recent years, and it has led to significant changes across various sectors, particularly in the adoption of bio-based pigments. These innovative colourants, made from organic materials, serve as a compelling alternative to traditional synthetic dyes, which often carry negative environmental impacts. The benefits of bio-based pigments extend beyond their production and use; they also involve key factors such as durability and resilience in changing conditions. This shift towards biocolours indicates a growing awareness of the harmful effects linked to conventional dyeing practices, including water pollution and biodiversity loss. Additionally, it aligns with global efforts to combat climate change and promote sustainable resource management. By thoroughly examining the life cycle of biocolours from the sourcing of raw materials to their application and eventual disposal researchers can better understand their ecological impact compared to synthetic alternatives. This chapter aims to explore the complex environmental implications of biocolours, looking at the benefits they can bring as well as the challenges we face in making the textile and manufacturing industries more sustainable. Biocolours are derived from natural pigments and microbial biosynthesis, and they have significant applications across various sectors. Synthetic dyes, mainly sourced from petrochemicals, have sparked serious concerns due to their environmental and health risks (Jayavani and Shahin Banu 2025). These dyes are commonly found in textiles, food, and cosmetics, but their widespread use contributes to considerable ecological harm. One of the biggest issues is water pollution, as textile industries often release untreated dye waste into waterways, leading to toxic effects on aquatic life (Kusumlata et al. 2024). Many synthetic dyes do not break down naturally, which means they can remain in ecosystems and disrupt biodiversity. Moreover, the presence of hazardous substances like azo dyes and heavy metals worsens soil and water contamination, making the issue even more pressing.

Natural pigments such as carotenoids, flavonoids, anthocyanins, and betalains. Microbial synthesis sources such as melanin from fungi and bacteria, prodigiosin from *Serratia marcescens*, violacein from *Chromobacterium violaceum*, phycobiliproteins from cyanobacteria and algae, etc., are listed in Table 10.1.

Biocolours are used in the food industry as natural colouring ingredients to replace synthetic dyes (e.g., beetroot extract for red, curcumin for yellow). In textiles as environmentally friendly dyes (like indigo from *Indigofera tinctoria*), in

Table 10.1 Microbial pigments: biological sources and therapeutic potential

Pigment	Source	Key properties & applications
Violacein	*Chromobacterium violaceum*	Purple pigment with antibiotic and anticancer properties; facilitates apoptosis in cancer cells; shows antimicrobial efficacy; strong antioxidant potential against free radical damage (linked to ageing and disease) (De Oliveira Barreto et al. 2023)
Prodigiosin	*Serratia marcescens*	Red pigment with antimicrobial and antioxidant properties; initiates apoptosis in cancer cells (potential anticancer agent); supports antibiotic drug development; combats oxidative stress in human cells (Keerthana and Sai Subhashini 2025)
Indigoidine	*Erwinia chrysanthemi*	Blue pigment with strong antioxidant potential; neutralizes free radicals, protecting cells from oxidative damage (linked to degenerative diseases); possible use in anti-inflammatory and tissue-protective formulations (De Oliveira Barreto et al. 2023)
Melanin	*Klebsiella* sp., *Streptomyces* sp.	Dark pigment that absorbs UV radiation; strong antioxidant activity; potential neuroprotective effects; studied for UV damage protection and oxidative stress-related conditions (Keerthana and Sai Subhashini 2025)
Astaxanthin	*Pontibacter korlensis*	Carotenoid pigment with exceptional antioxidant capabilities; reduces inflammation; enhances immune function; marketed as dietary supplement for skin, eye, cardiovascular health; potential anticancer properties by inhibiting tumour cell proliferation (Kuddus et al. 2024)
Pyocyanin	*Pseudomonas aeruginosa*	Virulence factor with antimicrobial activity; generates reactive oxygen species that disrupt microbial functions; toxic, limiting direct therapeutic use; offers insights for biomimetic drug design (De Oliveira Barreto et al. 2023)
Phycobiliproteins	*Arthrospira platensis,* Cyanobacteria, algae	Water-soluble fluorescent pigments that serve as light-harvesting proteins in photosynthesis; classified into phycoerythrin, phycocyanin, and allophycocyanin based on absorption spectra; used in food colouring, cosmetics, and biomedical imaging due to their non-toxic and antioxidant properties (Keerthana and Sai Subhashini 2025)

cosmetics as non-toxic pigments in a variety of products like lipsticks and skincare (like anthocyanins, carotenoids), and in pharmaceuticals as visually appealing and potentially health-promoting substances (like chlorophyll derivatives). There is several health hazards associated with synthetic colours. Mutagenic substances included in some of these dyes may raise the risk of developing cancer (Parmar and Phutela 2015). Additionally, exposure to these dyes may cause skin sensitivities, such as inflammation and allergy. Some synthetic dyes also intervene with hormonal balance and may cause metabolic issues since they are endocrine disruptors. On the other hand, biocolours provide a more environmentally friendly option. Low toxicity, biodegradability, and negligible ecological impact are their defining characteristics (Keerthana and Sai Subhashini 2025).

A growing desire for sustainable and environmentally friendly products is another factor driving the switch from synthetic dyes to natural biocolours, in addition to lowering health hazards. Using natural pigments from plants and fungi offers an intriguing potential that aligns with existing sustainability goals as companies grow more aware of their environmental impact. Consider bio-dyes as an example. These enzymatic pigments can create bold, long-lasting colours while reducing the pollution associated with conventional dyeing techniques. However, there are still challenges to be addressed, such increasing the cost-effectiveness and scalability of production, despite the promise. To make sure pigment yields satisfy consumer standards, extensive safety testing is required as they can fluctuate. Biocolours continue to exhibit enormous potential across several industries as research advances. Microbial pigments are of particular interest to scientists since they have remarkable advantages like improved colour retention and a smaller environmental impact. Biocolours have the potential to quickly emerge as the go-to option for eco-friendly and vibrant dyes with further development. By eliminating waste and lowering dependency on new resources, the textile and cosmetics industries may support a circular economy. One excellent example is the usage of pigments made from agricultural waste, which helps reduce landfill trash while also giving discarded materials another chance on life. This approach is in line with sustainability objectives, especially when using bio-dyes derived from organic materials like turmeric and beetroot (Joshi and Shinde 2017). These dyes are cost-effective and environmentally friendly and demonstrate how nature can offer substitutes to artificial chemicals. These eco-friendly and reasonably priced dyes show how nature may provide lovely substitutes for artificial chemicals. Additionally, it is anticipated that the market for products that use biocolours would increase as more consumers give priority to eco-friendly solutions. This change will push companies to use smarter, environmentally friendly production techniques that strike a balance between market demands and environmental responsibilities.

10.2 Environmental Impact of Biocolours

10.2.1 Sustainability of Raw Materials

Water is necessary for the transportation of chemicals and cleaning supplies throughout the textile manufacturing process. On the other hand, overuse can contaminate wastewater, causing water scarcity and environmental damage. Every year, the textile industry produces around 200 billion litres of dyed effluent, which leads to large expenditures in wastewater treatment and pollution prevention. European directives govern dangerous substances and pollution prevention in order to address these issues. Important regulations include the Dangerous Substances Directive

(67/548/EEC), REACH restrictions on azo colourants (2002/61/EC), and directives focused on wastewater treatment and pollution control, such as the Urban Wastewater Treatment Directive (91/271/EEC) and the Water Framework Directive (2000/60/EC). By encouraging sustainable practices in the textile industry, these policies seek to safeguard human health and the environment (Parisi et al. 2015). The shift to biocolour production goes beyond merely replacing synthetic dyes; it entails revaluating the whole supply chain, from obtaining raw materials to processing, while avoiding new ethical issues. Natural dyes derived from plants like Indigo and Madder can lessen dependency on artificial dyes but may compete with agricultural output for land, which could lead to issues with food security and land use (Beatley 2014).

10.2.2 Production Processes

Biocolour methods are often seen as more environmentally friendly than conventional synthetic dyes, using less energy and operating in milder environments. Their effectiveness varies, though, based on extraction techniques—microbial fermentation, enzymatic extraction, and solvent-assisted methods each have special benefits and drawbacks regarding pigment purity, yield, and energy consumption. For actual sustainability, these processes must be carefully evaluated and optimized (Jayavani and Shahin Banu 2025). Despite using less energy, biocolour production generates wastewater containing organic compounds and trace minerals, which can be hazardous the environment if untreated. Therefore, efficient waste management is essential to the manufacturing of sustainable dyes.

Microbial fermentation has become a common technique for producing pigments like prodigiosin, violacein, and indigoidine. By using bacteria, yeasts, or fungi in a controlled environment, we can assure constant and reliable yields year-round. These microbes' production pathways can be optimized through genetic engineering, thus increasing process efficiency (Jayavani and Shahin Banu 2025). Another promising technique is enzyme-assisted extraction, where specific enzymes degrade plant or microbial cell structures to release pigments while preserving their purity. By combining these extraction processes with modern processing procedures like membrane filtration and chromatography, we can produce high-purity pigments suitable for a range of commercial applications.

10.2.3 Biodegradability and Environmental Fate

Biocolours: Material sourced from natural sources like plants (indigo, saffron), insects (cochineal beetles, lac scale insects), animals (like some varieties of mollusks or shellfish), and minerals (like pyrophyllite, ochre, and clay) without any chemical treatment (Kadolph 2008). These gained attention for their capability to substitute synthetic dyes in variety of industries. One significant advantage of biocolours is their biodegradability, which allows them to break down naturally in the environment, decreasing the chances of persistent pollution. Studies have demonstrated that biocolours can degrade rapidly in aquatic and soil environments by reducing their environmental implications (Tkaczyk et al. 2020). Biocolours are biodegradable, and once they are discarded, they do not create pollution. They are naturally obtained from renewable sources, securing harmless outcomes to the environment. And their preparation required a much smaller amount of chemicals. For example, these are very safe. When ingested, certain natural colours, like carmine in lipsticks, are not harmful to you or cause any adverse effects (Choudhury 2018).

In aquatic environments, biocolours can be broken down by microorganisms such as bacteria and fungi, which facilitate this breakdown through the production of enzymes like laccases and peroxidases, which degrade complex organic molecules into simpler, non-toxic forms (Chung 2016). This rapid breakdown minimizes the potential for bioaccumulation and toxicity to the aquatic organisms (Robinson et al. 2001). On the other hand, synthetic dyes are non-biodegradable due to their intricate chemical makeup and structure, harming the aquatic and soil habitats. Synthetic dyes tend to remain quite stable to common oxidation and reduction processes as per their structure and so are difficult to remove from textile industry effluents. These textile industries are estimated to be the largest producers of coloured wastewater; approximately 20% of the dye used for dyeing textile fibres is not fixed and is disposed of in the wastewater, resulting in an elevated level of pollution (Kant 2012). Industrial effluents limit the amount of light passing through, raise the biochemical oxygen demand, and influence aquatic flora's photosynthetic activities. This negatively impacts aquatic organisms' food source (Sharma et al. 2021).

In soil, microbes such as bacteria and fungi break down pigments. They utilize enzymes that naturally occur, such as laccases, peroxidases, and polyphenol oxidases (Durán and Esposito 2000). Enzymes convert the complex structures of plant-based pigments into simpler, safer byproducts such as carbon dioxide, water, and biomass. The speed of this breakdown varies depending on environmental factors, including temperature, moisture levels, and pH levels. For instance, anthocyanins decompose more quickly in soils that are slightly acidic to neutral. This happens because microbe enzymes work best in these conditions (Khoo et al. 2017). In contrast, synthetic dyes have a negative impact on soil fertility and plant growth. Textile effluent is a major contributor to the introduction of these dyes into our environment. For example, Ahmad et al. (2012) conducted a study on industrial effluent used for irrigation in the Bhairavgarh area of Ujjain city. The study concluded that the textile industry discharges wastewater with high levels of alkalinity, chlorides, cations, and anions, which do not meet regulatory standards. The research indicates that this factory waste, or contaminated water, adversely affects the physical and chemical properties of the soil.

Natural sources such as plants, algae, and tiny organisms produce biocolours. These natural pigments provide a greener and more sustainable alternative to synthetic dyes because they decompose and do not remain in the environment. When these pigments enter the soil through waste streams, composted plant material, or agricultural inputs, soil microbes break them down. Enzymes, particularly oxidative ones like laccases and peroxidases, transform these natural compounds into harmless by-products such as carbon dioxide, water, and biomass (Durán and Esposito 2000). This concise breakdown minimizes the risk of buildup and toxicity in land ecosystems. Unlike synthetic dyes, which often contain long-lasting aromatic rings and toxic heavy metals, biocolours decompose without leaving harmful residues or disrupting microbial communities (Scalbert 1991). Their breakdown does not clutter with fertility of the soil, nutrient cycling, and water quality. This makes them useful in farming, food, and textiles with negligible impact on the environment. By promoting biocolours, we can support sustainable industry practices and protect our soil and water resources in the long run.

10.2.4 Potential Environmental Concerns

Biocolours are becoming increasingly popular as awareness towards environment grows and consumers demanding more eco-friendly products. These come from biological sources like plants, algae, fungi, bacteria, and insects. As a result, biocolours have emerged as a green products and sustainable alternative, particularly in the food, textile, cosmetic, and pharmaceutical industries. Regardless of its multiple advantages, an uncontrolled harvesting of natural sources for pigment extraction poses major dangers for habitat, including the loss of wild plant species, deforestation, genetic diversity, and influences on microbial sources. Many plant-based biocolours arise from wild or semi-cultivated species; as an example, habitat loss and a decrease in wild seed quantities have been caused by the overharvesting of Bixa orellana (plants that bear annatto pigment) in South America (Rodriguez-Amaya 2015).

Large-scale biocolour production is often linked to significant changes in land use, which can result in habitat loss and fragmentation. This may lead to a decline in species richness and alterations in species composition, ultimately impacting ecosystem function. Like, the shift of natural forests to monocultural rubber plantations has well-documented negative effects on biodiversity, such as decreased species richness and abundance across various taxa (Warren-Thomas et al. 2023). The decline in biodiversity, driven by frequent bio-colour production, can significantly impact ecosystems, leading to reduced productivity and stability. Studies show that even modest reductions in grassland plant diversity can affect productivity comparable to substantial changes in nitrogen, water, CO_2 levels, herbivore populations, drought conditions, or fire events. This highlights the urgent need to protect biodiversity within natural habitats (Tilman et al. 2012). It also has influence on several insects and microbes, like Cochineal dye (carminic acid), obtained from insects, and pigments from bacteria (e.g., prodigiosin), which are resource-intensive and require careful farming. Overharvesting or overuse may harm controlled ecological systems such as algae ponds or insect farms. (Damant 2011). While large-scale bio-colour production poses major threats to biodiversity, there are chances for conservation. It is possible to enhance farmland biodiversity while minimizing the negative impacts of biofuel production by adopting ecological guidelines and sustainable practices,

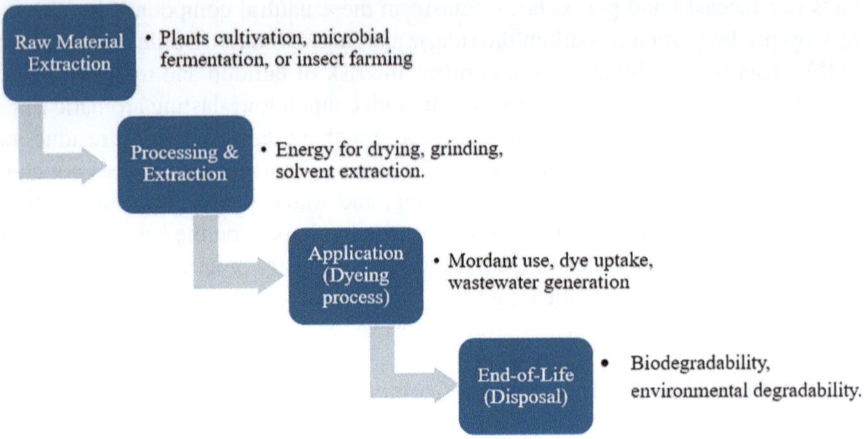

Fig. 10.1 Flowchart of lifecycle analysis (LCA)

such as short-rotation coppice management. Further research is needed to explore these opportunities and develop effective conservation strategies.

10.3 Lifecycle Analysis (LCA)

Lifecycle analysis (LCA) is an essential structure for evaluating ecological effects related with all levels of a product's life, including raw material extraction, manufacturing, usage, and end-of-life disposal. This process is commonly regarded as "cradle-to-grave" analysis (Klüppel 2005). In the textile industry, which is known for its significant contributions to global environmental degradation, LCA is especially helpful for identification and quantifying the sustainability impacts of various production processes as shown in Fig. 10.1.

Dyeing operations utilize considerable resources and generate considerable waste. While synthetic dyes are famous for their bright colours and cost-effectiveness, they pose serious environmental and health risks. In contrast, biocolours, derived from natural or bio-based sources, have emerged as a greener, more sustainable option. This study uses LCA to compare synthetic dyes and biocolours, focusing on key environmental indicators such as carbon footprint, H_2O consumption, energy demand, waste production, and pollution levels. This comparative analysis emphasizes the potential for biocolours to lessen the ecological impact of textile production.

10.3.1 Carbon Footprint

The carbon footprint of biocolours is considerably lower than that of synthetic dyes, essentially because of contrasts in their raw material sources and production processes. Biocolours come from renewable sources like plants, bacteria, and fungi.

These sources usually produce less carbon emissions. For example, natural dyes from plants need less energy for growing, picking, and processing. On the other hand, synthetic dyes come from petrochemicals and greatly increase greenhouse gas emissions throughout their lifecycle (Pizzicato et al. 2023). Synthetic dyes have a particularly high carbon footprint because they rely on fossil fuels and energy-intensive production methods. The manufacturing process involves various stages, including the synthesis of chemical intermediates, the actual dye synthesis, and subsequent finishing treatments, all of which require substantial energy. All these stages add to the cumulative greenhouse gas releases, moreover, increasing the Environmental consequences of man-made dyes use (Li et al. 2023) (Filho et al. 2024). Moreover, emissions are not limited to production alone; post-dyeing processes also play a key role. For example, treating the wastewater produced from synthetic dyeing operations—often through advanced oxidation techniques such as the Fenton process—can release up to 3.73 kg of CO_2-equivalent for every cubic meter of treated effluent, with electricity consumption accounting for about 60% of this amount (Grisales et al. 2018). Overall, these factors illuminate the significantly higher carbon footprint of synthetic dyes and highlight the potential of biocolours as an eco-friendlier, sustainable substitute in textile dyeing.

10.3.2 Water Usage

The textile industry is one of the largest industrial consumers of freshwater, with dyeing processes accounting for a huge portion of this demand (Rathoure et al. 2019). Synthetic dyeing operations are particularly water-intensive, typically requiring between 150 and 150–200 litres of water for each kilogram of textile produced, depending on the type of dye and the application method used (Liu et al. 2024). These processes involve several water-consuming stages, including dye preparation, application, rinsing, and washing. As a result, the industry is responsible for approximately 20% of global industrial water pollution, primarily due to the discharge of untreated or inadequately treated dye effluents into water bodies (Chavan 2011) (Kant 2012). Synthetic dye effluents often contain harmful chemicals, including heavy metals, salts, and fixatives, which pose significant risks to both the environment and human health (C. Li et al. 2024). Treating such wastewater is complex and costly, and in many regions, inadequate regulation, and infrastructure lead to direct environmental contamination. Furthermore, the persistence and toxicity of synthetic dyes in aquatic ecosystems worsen their ecological impact (Poo et al. 2024).

In contrast, biocolours, particularly those derived from natural sources, are generally more environmentally friendly in terms of water pollution. Natural dyes are typically biodegradable and non-toxic, resulting in cleaner wastewater profiles and lower treatment burdens. Some biocolour application methods also permit reduced water usage, especially when traditional or low-water techniques are employed (Mouro et al. 2023) (Pizzicato et al. 2023). However, using biocolours also presents water-related challenges. The extraction of natural dyes from plants or microbial sources can require significant amounts of water and, in some cases, enormous quantities of raw material. Additionally, the use of mordants—chemical agents required to fix dyes to fabric—can introduce further pollutants into wastewater streams if not effectively managed (Kant 2012). Overall, while biocolours offer an advantage in terms of reduced toxicity and biodegradability, they can also cause problems related to water consumption and potential pollution.

10.3.3 Energy Demand

The production of synthetic dyes is highly energy-intensive and involves several stages, including raw material synthesis, dye formulation, application processes, and effluent treatment. Every step in the process uses a lot of energy, which adds to its overall environmental impact. For example, effluent treatment methods such as the Fenton process demand substantial energy and involve additional chemicals like H_2SO_4 and NaOH, further increasing the ecological footprint of synthetic dye manufacturing (Grisales et al. 2018) (Filho et al. 2024). In contrast, the production of biocolours typically has a lower energy requirement. Derived from natural sources, biocolours are often produced using energy-efficient methods such as fermentation and solvent extraction. For instance, bacterial pigment production via fermentation can be performed with the lowest possible energy input, especially when traditional dyeing techniques are utilized. This makes biocolours a more sustainable option in terms of energy consumption (Tufvesson 2010) (Poo et al. 2024). However, scaling up the industrial production of biocolours presents challenges. Meeting large-scale demand may require the use of modern extraction and dyeing technologies, which could result in increased energy usage. Therefore, to maintain the environmental advantages of biocolours, it is crucial to improve these emerging technologies for energy efficiency.

10.3.4 Waste Generation and Pollution

Synthetic dyeing processes are highly water-intensive, requiring approximately 150–200 litres of water for each kilogram of textile produced (Liu et al. 2024). A sizeable portion of these dyes, estimated at 10–15%, remains unfixed and is discharged into wastewater, leading to severe water pollution (Jorge et al. 2023). The effluents left behind contain a complex mixture of pollutants, including heavy metals such as chromium and lead, salts, and a range of organic compounds, all of which are difficult to treat and pose serious threats to aquatic ecosystems. The production of synthetic dyes takes lots of energy, contributing to air pollution through the emission of greenhouse gases. Exposure to synthetic dyes has been connected to various health problem, including skin allergies, respiratory issues, and even cancer. Workers in dye manufacturing industries are particularly more vulnerable due to prolonged exposure to these toxic chemicals (Kishor et al. 2021). Synthetic dye production creates substantial waste, including chemical byproducts, unreacted dyes, and solvents. Often, these materials enter waterways, worsening pollution and harming aquatic life. Furthermore, synthetic dyes contribute to microplastic pollution, as synthetic fibres shed during washing and accumulate in the environment (Irfan et al. 2023; Li et al. 2024).

In contrast, biocolours offer a more environmentally sustainable and health-conscious alternative. Biocolours produce less waste as it's derived from natural sources, and are biodegradable, allowing for composting or reuse and thus reducing the overall waste footprint. Unlike synthetic dyes, biocolours do not contribute to microplastic pollution and are generally non-toxic and hypoallergenic, making them a safer, more eco-friendly option (Mouro et al. 2023). The production and application of biocolours result in fewer hazardous by-products and help in reducing environmental contamination, as summarized in Table 10.2. Their use decreases the chemical load in wastewater, making treatment easier and less harmful to ecosystems (Tkaczyk et al. 2020). Additionally, producing biocolours typically requires less energy and generates lower greenhouse gas emissions (Fried et al. 2021). Overall, switching from synthetic dyes to biocolours presents significant environmental and health benefits, making them a more sustainable and responsible choice for the textile industry.

Table 10.2 Comparison between synthetic dyes and biocolours

Aspect	Synthetic dyes	Biocolours
Water usage	150–200 litres of water per kg of textile during dyeing (Liu et al. 2024)	Lower water usage because of simpler extraction and application processes
Wastewater load	10–15% of dyes discharged into water; contains salts, metals, and toxic organics (Liu et al. 2024; Kishor et al. 2021)	Minimal hazardous chemicals: effluents are mostly biodegradable (Fried et al. 2021)
Air pollution/emissions	High energy use and GHG emissions from synthesis and processing (Atlas of sustainable colour(S) 2025)	Lower energy demands and minimal GHG emissions (Fried et al. 2021)
Soil pollution	Dye sludge disposal contaminates soil with toxic residues (Atlas of sustainable colour(S) 2025)	Biodegradable residues pose minimal soil risk (Fried et al. 2021)
Toxicity/health risk	Allergies, respiratory issues, carcinogens; hazardous for workers	Generally non-toxic, hypoallergenic; safer for workers and consumers (Fried et al. 2021)
Chemical usage	Heavy metals (e.g., chromium, lead), mordants, and petrochemicals used	Mostly plant- or microbe-derived; fewer or no toxic mordants required
Biodegradability	Poor, synthetic dyes are persistent in the environment	High; natural dyes degrade quickly in soil and water (Fried et al. 2021)
Treatment complexity	Wastewater requires advanced treatment methods	Easier treatment due to lower chemical load

10.4 Regulatory Aspects of Biocolours: Regulatory Frameworks Across the Globe

10.4.1 Standards Set by Organizations Such as the FDA (U.S.), EFSA (EU), and FSSAI (India)

Natural food colourants, often known as biocolours, come from minerals, plants, and animals. To guarantee the safety, quality, and proper use of biocolours in food items, several organizations, including the FDA (U.S.), EFSA (EU), and FSSAI (India), have established particular criteria and regulations. The main rules and regulations established by these organizations are broken down as follows:

10.4.1.1 U.S. Food and Drug Administration (FDA)

The use of colour in food, drugs, and cosmetics is regulated by various laws and regulations, each providing guidelines and information about their components. The regulation began with the 1906 US Food and Drug Act (FDA), which aimed to prevent the use of misbranded colours. This was followed by the 1938 Federal Food, Drug, and Cosmetics Act, which introduced inspections for the use of colours in cosmetics and drugs. In 1960, the FDA approved a specific list of colours permitted for use in the food industry (Chittaragi and Menon 2023). The US Federal Food, Drug, and Cosmetic Act (FFDCA) and its amendments, including the Food Additive and Color Additive Amendments, the Food Safety and Modernization Act (FSMA) of 2012, the Fair Packaging and Labeling Act (FPLA) of 1966, and the Public Health Security and Bioterrorism Preparedness and Response Act (Bioterrorism Act) of 2002, serve as the primary foundation for the federal regulation of colour additives in the United States. Under Title 21 of the Code of Federal Regulations, the USFDA, a division of the Department of Health and Human Services (HHS), is principally responsible for enforcing the laws relating to food colouring (CFR Citation 2025). Parts 70–82 and 101 contain rules on petitions and labelling and list the specifications and rules for the use of approved colour additives, referring also to Part 58 on good laboratory practice for testing (Lehto et al. 2016). The FDA mandates the safety of colour additives in food, specifying their use types, maximum amounts, and declaration on food labels after approval for their intended use.

10.4.1.2 European Food Safety Authority (EFSA) EU

In the European Union, there is a comprehensive framework governing the use of biocolours as food additives to ensure both safety and quality. At its core, Regulation (EC) No. 1331/2008 establishes a standardized procedure for authorizing food additives, including biocolours. This process ensures that each additive must be evaluated rigorously before being allowed on the market. Building on this, Regulation (EC) No. 1333/2008 specifically addresses food colouring. It defines the rules for using colour additives ingredient across different food categories and provides a positive list of colours that have been approved for usage. Additionally, Regulation (EC) No. 1129/2011 amends and refines these guidelines by setting highest usage levels and clarifying how these additives should be safely incorporated into food products. Together, these regulations create an integrated legal structure that protects consumer health and on other hand it supports industry innovation. Detailed annexes within Regulation (EC) No. 1333/2008 specify the exact food categories, the list of approved colours, and the permitted limits for each additive. This fastidious approach ensures that food products containing biocolours not only enhance visual appeal but also meet stringent safety standards, contributing to a trustworthy and sustainable food supply system in the EU (Lehto et al. 2016).

10.4.1.3 Food Safety and Standards Authority of India (FSSAI)

Flavouring compounds and bio-colourants are frequently employed to draw in customers. Due to their critical role in the visual presentation of the food, food colourings have an enormous influence on food. The Food Safety and Standards Authority of India, or FSSAI, has authorized specific food colours and flavours that are out of harms for human consumption in India. Food products may only contain flavours and colours approved by the FSSAI. The Food Safety and Standards (Food Product Standards and Food Additives) Regulations, 2011, include a list of approved food colouring and flavouring compounds together with their standards, which were supplied by the FSSAI. Food products containing biocolours must be safe to eat. By using toxicity studies to evaluate safety, FSSAI makes sure that these colours comply to the necessary safety regulations (Varghese and Ramamoorthy 2023).

10.4.2 Permitted Biocolours and Acceptable Daily Intake (ADI), Compliance with REACH (EU), CPSIA (U.S.), and Other Safety Guidelines Regarding Biocolours

There are numerous rules, safety requirements, and recommendations pertaining to the use of biocolours in food that guarantee their safe application and conformity to national and international legislation. In addition to the requirements established for approved biocolours, the Acceptable Daily Intake (ADI), REACH compliance (EU), and CPSIA compliance (US) are critical regulatory frameworks pertaining to biocolours.

10.4.2.1 Permitted Biocolours and Acceptable Daily Intake (ADI)

The quantity of a food additive consumed daily for the duration of a lifetime without posing a consequential danger to health is known as the Acceptable Daily Intake (ADI), and it is calculated on a bodyweight basis as shown in Table 10.3. "Without appreciable risk" refers to the certainty that, given the state of knowledge, no harm will occur, even after a lifetime of exposure to the additive in question. A range of milligrams per kilogram of body weight per day is typically used to express the ADI.

When colour additives are employed at "Good Manufacturing Practice (GMP)" levels in the United States, it indicates that their usage in food is restricted to the bare minimum required to provide the intended colour effect, rather than having a set upper limit in milligrams per kilogram. This method, which is applied to a variety of colour additives, ensures safety while permitting their usage in food.

Table 10.3 Available ADI (mg/kg) values for certain colours approved both in the EU and the US (Lehto et al. 2016)

Name EU/US	EU	US
Curcumin/turmeric	3	(GMP)
Caramels	100 (E 150c); 300 (caramel group)	(GMP; also, GRAS)
Annatto/annatto extract	6 (bixin); 0.3 (norbixin)	(GMP)
Paprika extract/paprika, paprika oleoresin	24 (paprika extract); 1.7 for carotenoids in paprika extract	(GMP)
Lycopene	0.5	(GMP)

10.4.2.2 REACH Compliance (EU)

REACH, which stands for Registration, Evaluation, Authorization, and Restriction of Chemicals, is an EU regulation that came into effect on June 1, 2007. Its primary goal is to enhance the competitiveness of Europe's chemical industry while safeguarding human health and the environment. A key feature of REACH is its commitment to minimizing animal testing and promoting other toxicity assessment methods. REACH applies to nearly all chemicals, from those found in everyday products like paints and cleaning supplies to those used in industrial processes. Companies must prove the safety of their chemicals sold in the EU, manage any associated risks, and provide detailed safety information to the European Chemicals Agency (ECHA). If risks cannot be controlled, authorities can impose usage restrictions. Ultimately, REACH aims to phase out hazardous chemicals and encourage safer alternatives, protecting consumers and associated workers while promoting sustainability in the industry (Scruggs et al. 2014).

10.4.2.3 CPSIA Compliance (US)

In the United States, consumer product safety is governed by the Consumer Product Safety Improvement Act (CPSIA). While it primarily pertains to products for children, it also affects food products that potentially include biocolours. The CPSIA demands testing of food products to guarantee that they do not contain lead or other harmful metals, which includes biocolours. Manufacturers and importers of biocolours are required to follow stringent lead and phthalate standards and submit to third-party testing in order to guarantee conformity with the CPSIA. With particular restrictions on lead and phthalates, the CPSIA aims to reduce chemical dangers, especially for products intended for children (Leone and Berger 2009).

10.4.2.4 International Codex Alimentarius

The FAO and WHO created a set of worldwide food standards known as the Codex Alimentarius. Guidelines for the safe use of food additives, such as natural colourants, are provided by Codex. It provides scientifically supported assessments of food additives, such as biocolours, to aid the work of different regulatory bodies. The Codex General Standard for Food Additives (GSFA) outlines the permitted use of biocolours across various food categories. The 55th session of the Codex

Committee on Food Additives (CCFA55) reviewed and updated provisions related to food colours (Laganà et al. 2017).

10.4.3 Testing and Certification

10.4.3.1 Toxicological Assessments: Allergenicity, Mutagenicity, and Other Safety Studies

For biocolours to be safe in food and other consumer goods, toxicological evaluations are essential. A variety of safety studies, such as those on allergenicity, mutagenicity, carcinogenicity, etc., are included in these evaluations of the possible health concerns that biocolours represent to people.

10.4.3.1.1 Allergenicity

Testing for allergenicity determines whether a material can cause allergic reactions in those who are vulnerable. This is specifically crucial for biocolours because natural compounds can still trigger allergic reactions in some individuals, regardless of whether they are sourced from plants or animals. A mouse model called the local lymph node assay (LLNA) was created to assess a chemical's capacity to cause cutaneous sensitization. In contrast to conventional guinea pig techniques, the LLNA offers significant advantages for animal welfare. The assay is based on measuring the events that occur during the skin sensitization induction phase, particularly the proliferation of lymphocytes in the draining lymph nodes, which is a characteristic of a skin sensitization response (Gerberick et al. 2006). People who are allergic to certain plants or fruits may occasionally react to biocolours made from those plants or fruits. The reason for this is a phenomenon known as cross-reactivity, in which the immune system of the body incorrectly perceives comparable proteins in other substances as allergens. For instance, a banana-derived colourant might cause a reaction in individuals with a banana allergy (Julanon et al. 2023).

10.4.3.1.2 Mutagenicity

A substance's ability to induce genetic mutations that perhaps result in cancer or other inherited disorders is evaluated by mutagenicity testing. It is a crucial part in determining whether any food additive, including biocolours, is safe. A bacterial test called the Ames test is used to determine whether a chemical can cause mutagenesis. It makes use of bacterial strains (*Salmonella typhimurium*) that need histidine to develop. If a chemical is mutagenic, it can cause these bacteria to undergo mutations that enable them to grow on a histidine-free media without histidine, proving that the chemical is a mutagen (Samiei et al. 2015). Cultured mammalian cells are used in in vitro mammalian cell testing to determine whether a chemical alters genetic material (DNA) (White et al. 2019).

10.4.3.1.3 Carcinogenicity (Cancer Risk)

Studies on carcinogenicity assess the ability of a substance to cause cancer following prolonged exposure. These investigations are necessary to make sure that lifetime consumption of biocolours does not raise the risk of cancer. Assessing the possible carcinogenic consequences of chemicals requires long-term animal studies, especially with rodents like rats and mice. To determine whether the drug causes tumour formation or other types of cancer, researchers expose the animals to large dosages of the material over a considerable period of their lives. This approach assists in identifying possible carcinogens and provides information to regulators regarding the safety of compounds for human exposure. (Human Foods Program 2017).

10.4.3.2 Environmental Safety Testing: Biodegradability, Aquatic Toxicity, etc. Regarding Biocolours

Assessing the effects of biocolours on the environment, specifically in relation to their biodegradability, aquatic toxicity, and general environmental persistence, is the main goal of environmental safety testing. Since biocolours can enter water bodies through runoff or wastewater discharge, these tests are crucial to ensuring that they don't affect ecosystems over the long term, especially aquatic systems.

10.4.3.2.1 Biodegradability Testing

Biodegradability tests examine how quickly a chemical is broken down by natural biological processes, such as the action of bacteria, fungi, or other microorganisms. This is a crucial factor for biocolours since materials that do not biodegrade could remain in the environment and eventually harm it. OECD 301 Tests (Organization for Economic Cooperation and Development): These tests assess a chemical's capacity for biodegradation. The OECD 301A (Closed Bottle Test) and OECD 301B (CO2 Evolution Test) are common tests that assess the level of microbial deterioration over a given time frame (usually 28 days) (OECD 2006). The amount of oxygen that microorganisms need to break down a material is measured by biological oxygen demand, or BOD, tests. Given that microbial breakdown necessitates a substantial amount of oxygen, a high BOD implies that the material is biodegradable.

10.4.3.2.2 Aquatic Toxicity Testing

To find out if biocolours can damage ecosystems when they enter water bodies, aquatic toxicity testing evaluates how they affect aquatic animals (such as fish, invertebrates, and algae). When releasing biocolours into rivers, lakes, or the ocean, these tests are essential for determining how they can impact aquatic life. Aquatic animals are used in acute and chronic toxicity experiments to evaluate the detrimental effects of chemicals on aquatic life over brief and extended periods of time, respectively. Acute tests, which typically last 14 days or fewer, look at the acute lethality brought on by large dosages of a drug. These tests expose aquatic organisms (usually fish, daphnia, or algae) to high concentrations of a substance for a brief period to assess lethal effects (e.g., LC50, the concentration at which 50% of the organisms die). At lower, continuous exposures, chronic testing (lasting more than 14 days) assesses sublethal effects like growth, development, and reproductive capabilities (Bhardwaj et al. 2012).

10.4.3.3 Certification for Natural or Eco-friendly Labelling for Biocolours (e.g., GOTS for Textiles)

Certifications for eco-friendly or natural labelling for biocolours aid in proving the products' environmental responsibility, sustainability, and safety. These certificates guarantee that biocolours fulfil requirements concerning their composition, manufacturing process, and effects on the environment. There are several other significant certifications and labelling schemes that can apply to biocolours, particularly for usage in food, cosmetics, and textiles, even though GOTS (Global Organic Textile Standard) is a well-known certification for organic textiles.

10.4.3.3.1 GOTS (Global Organic Textile Standard)

The goal of GOTS is to establish a global standard for organic textiles that that ensures traceability from the moment raw materials are harvested to the completion of environmentally and socially responsible production and labelling, giving consumers the certainty that the material is GOTS-certified organic. If natural dyes, including biocolours, satisfy certain environmental and toxicological requirements, such as making sure that dyes and colorants don't include dangerous chemicals or heavy metals, GOTS permits their usage. This GOTS standard includes the production, processing, manufacturing, packaging, labelling, exporting, importing, and distributing of fibre goods, yarns, textiles, and clothing (GOTS 2023).

10.4.3.3.2 Ecocert

Ecocert certification is a stringent procedure conducted by an impartial organization to confirm that a system, service, or product satisfies social and environmental standards. It guarantees that procedures and goods are real, open, and compliant with sustainable norms. Making sure that products are created with natural or organic materials is the main goal of Ecocert certification. This would require that the colourants used in biocolours be derived entirely from natural sources, such as fruits, vegetables, or plants, rather than synthetic chemicals. When the certification procedure is complete, Ecocert decides and, if it is favourable, provides a written guarantee known as a certificate (Vermeer et al. 2010).

10.4.4 Challenges in Regulatory Compliance

10.4.4.1 Variation in Definitions and Standards for Natural and Biodegradable Across Countries Regarding Biocolours

Countries may have quite different definitions and standards for natural and biodegradable biocolours because of variations in legal systems, cultural norms, and scientific discoveries.

10.4.4.1.1 Natural

Foods that are unprocessed or just mechanically treated are commonly referred to as "natural" (Stich 2016). But far too frequently, a colour that is referred to as natural is only so because the substance that gives it the colour is present in nature. The qualities of the colouring component can be altered and the natural starting material compromised by processing stages including chemical transformation and any additional manufactured, isolated, or purified chemical ingredients. The FDA regulates colours, which are often referred to as "natural" colours, which are derived from plants, minerals, and animals. However, these colours are not subject to FDA certification and often lack precise definitions and industry-accepted quality standards. This can lead to misbranding, contamination, inappropriate processing techniques, and postproduction adulteration, posing a significant risk. (Simon et al. 2017). Natural colours are also frequently described in the EU as coming from natural sources such as minerals, plants, and animals. These colours' safety is assessed by the European Food Safety Authority (EFSA). Compared to the United States, the European Union has more stringent regulations about what constitutes "natural" colour, and colours that are the result of chemical processing might not be categorized as such (Lehto et al. 2016). In India, the use of natural colours in food and textiles is regulated by the Food Safety and Standards Authority of India (FSSAI), which outlines that natural colours should be extracted from plant or animal sources (Varghese and Ramamoorthy 2023). The definition of "natural" can vary greatly. In some locations, processing methods may allow a greater range of compounds to be labelled "natural," while others implement stricter rules.

10.4.4.1.2 Biodegradable

Biodegradability is a key consideration in environmental protection standards. Although the Environmental Protection Agency (EPA) does not have a specific criterion for biocolours, biodegradability is typically understood to mean that a material may naturally decompose into non-toxic components in a fair amount of time. In the United States, a "biodegradable" biocolour would probably satisfy these requirements, though testing and standards may differ based on the application (e.g., textiles vs. food) (Guin et al. 2017). The EU's biodegradability regulations are more stringent. The biodegradability of materials used in paints, cosmetics, and food packaging is carefully assessed under the REACH (Registration, Evaluation, Authorization, and Restriction of Chemicals) regulation. For instance, a biocolour would have to adhere to the biodegradability requirements specified in EN 13432 for packaging. The term "biodegradable" in the EU usually means that a material can decompose in an environment (soil, water) in a predetermined amount of time (Leone and Berger 2009). The definitions of natural and biodegradable biocolours are determined by national laws, with significant variations in the degree of stringency or relaxation in each nation with respect to the origins of the colours and their effects on the environment. Although standards are trending toward alignment as global awareness of sustainability grows, notable regional variations still exist.

10.4.4.2 Limited Standardized Testing Protocols for Novel Biocolours

Adoption and regulation of innovative biocolours are hampered by the lack of uniform testing procedures. For these biocolours to be safe, high-quality, and environmentally compatible, testing procedures are essential. Here are some of the key difficulties and gaps in standardized testing guidelines for novel biocolours:

Inconsistent Global Standards Biocolour testing systems and benchmarks differ by nation and region. For instance, the approval procedures for biocolours used in food and cosmetics may differ between the US and the EU (Lehto et al. 2016).

Varying Definitions of "Natural" Because "natural" can mean various things to different people throughout the world, new biocolors that are taken from natural sources may not be evaluated or subjected to the same standards depending on the location in which they are utilized. The creation of reliable testing standards is hampered by this lack of agreement on terminology (Stich 2016).

Unclear Standards for Life Cycle Analysis (LCA) Although LCA is a widely used technique to analyze the environmental consequence of various products, there aren't many established procedures for LCAs that are especially applied to biocolours. Because of this, it is challenging to review the actual environmental advantages of biocolours over synthetic colourants. Resource procurement, energy use, waste, and the possibility of bioaccumulation are a few examples of variables that LCA may include (Menoufi 2011).

Advantages of Biocolours in Regulatory Compliance As worldwide consciousness of environmental preservation grows, biocolours have become a promising replacement to synthetic dyes across multiple industries. Their rise is driven by ecological concerns and alignment with regulatory compliance and environmental benchmarks. Here are several important regulatory benefits of biocolours:

10.4.4.3 Alignment with Global Sustainability Goals (UN SDGs)

In 2015, the United Nations established 17 Sustainable Development Goals (SDGs) as part of a 15-year plan to end extreme poverty, combat inequality, and protect the planet. Adopted by all 193 UN member states, these goals aim to create a better future for everyone, encompassing objectives in areas like climate action, education, and economic prosperity, ensuring that no one is left behind as summarised in Table 10.4. They help nations and corporations identify common goals and opportunities for action. (United Nations 2015). Biocolours, sourced from plants, algae, fungi, and microorganisms, provide an eco-friendly alternative to synthetic dyes. Their application across industries benefits environmental conservation, human health, and sustainable practices. Biocolours play a crucial role in advancing the United Nations Sustainable Development Goals (UN SDGs) and enhancing global sustainability.

Table 10.4 Table highlighting the United Nations Sustainable Development Goals and the contributions of biocolours, along with examples

SDG	Goal title	Biocolour contribution	Examples	References
SDG 3	Good Health and Well-being	Lowers health risks by substituting harmful synthetic dyes with safe, biodegradable alternatives	Textile workers using natural dyes experience fewer respiratory and skin issues	Devi et al. (2025)
SDG 6	Clean Water and Sanitation	Generates low-toxicity, biodegradable waste that helps reduce freshwater pollution	Turmeric and madder dyes generate wastewater with lower Chemical Oxygen Demand (COD) and Biochemical Oxygen Demand (BOD) values	Ahmed et al. (2024)
SDG 12	Responsible Consumption and Production	Uses renewable and waste-based resources to promote circular economy practices	Using fruit peels or flower waste to extract natural dyes helps reduce reliance on petrochemicals	Elsahida et al. (2019)
SDG 13	Climate Action	Reduces carbon emissions during production and supports carbon-neutral alternatives	Microbial pigments, such as carotenoids and melanin, have a smaller carbon footprint	Devi et al. (2025)
SDG 15	Life on Land	Encourages sustainable agricultural practices, safeguards biodiversity, and minimizes harmful land-use methods	Cultivating Indigofera or safflower enhances agroforestry practices and supports various ecosystem services	Elsahida et al. (2019)

10.4.4.4 Reduced Scrutiny Compared to Synthetic Dyes Due to Lower Toxicity

Biocolours, being natural and less toxic than synthetic dyes, benefit from regulatory advantages. While synthetic dyes face strict assessments due to potential health risks, biocolours experience less scrutiny, leading to quicker product approvals and increased market acceptance. They often qualify for eco-labelling and sustainability incentives, particularly in countries promoting green chemistry and the circular economy, offering subsidies and tax benefits for using biodegradable materials.

Examples and Benefits

The EU's REACH Regulation classifies synthetic dyes as Substances of Very High Concern (SVHC) due to their hazards, requiring strict compliance and testing (EFSA 2016). Biocolours, in contrast, are usually exempt as they pose minimal risks to health and the environment. In India, biocolours in textiles and cosmetics can earn eco-labels like Eco Mark and Green Pro certification. These certifications reflect environmental responsibility, enabling manufacturers to access sustainable procurement and attract eco-conscious consumers.

10.4.4.5 Opportunities for Regulatory Incentives

Biocolours presents industries with opportunities to leverage regulatory incentives for sustainability, including eco-label certifications and government subsidies. Biocolours are biodegradable and non-toxic, making them eligible for various eco-labelling certifications, such as the EU Ecolabel and GreenPro Certification in India. The EU Ecolabel is awarded to products that meet high environmental standards throughout their lifecycle, while GreenPro certifies sustainability based on multiple environmental parameters, appealing to eco-conscious consumers and manufacturers (EU Ecolabel 2024). Government subsidies and tax benefits support this shift towards greener practices, significantly easing the financial burden for businesses transitioning to sustainable alternatives like biocolours. For example, India's National Bioplastics Policy offers capital subsidies and interest concessions for adopting biodegradable materials, fostering green innovation (EY-India 2024). Additionally, companies producing biocolours may qualify for carbon credits, further lowering their tax liability and enhancing their sustainability leadership. Overall, biocolours align with UN Sustainable Development Goals, offering significant ecological and economic advantages while facilitating compliance with regulatory standards. Their lower toxicity and biodegradability make them a strategically beneficial choice for innovative industries.

10.5 Future Perspectives

Biotechnology research is increasingly focused on genetic engineering to enhance biocolour production. Techniques like CRISPR-Cas9 editing and metabolic pathway optimization enhance pigment biosynthesis by improving yields, stability, and functionality. These advancements may reduce production costs and energy use while ensuring biocolours meet industrial performance standards. Harmonizing international standards is essential for maximizing the global potential of biocolour technologies. A consistent framework for safety, quality, and performance allows products from various regions to meet shared criteria. Increasing coordination among international regulatory bodies streamlines product development, enhances market access, and fosters cross-border collaboration. This unified approach boosts consumer confidence by ensuring compliance with strict environmental and health regulations. Turning waste into valuable resources is revolutionizing the biocolour industry. By repurposing agricultural and industrial byproducts, manufacturers reduce environmental impact while supporting a circular economy. Innovative extraction techniques make it possible to harness these materials for pigment production, turning waste into a sustainable asset for eco-friendly manufacturing. The future of biocolours lies in the convergence of sustainability, innovation, and harmonized regulatory policies. As industries increasingly shift towards eco-friendly alternatives, biocolours are poised to become mainstream in textiles, food, cosmetics, and pharmaceuticals. Key future directions include:

- *Biotechnological Advancements*: Synthetic biology and metabolic engineering can significantly improve the yield, stability, and spectrum of biocolours from microbial and algal sources, enabling cost-effective and scalable production.
- *Sustainable Feedstocks*: Use of agricultural waste, food by-products, and renewable biomass for pigment extraction will enhance the circular bioeconomy and reduce environmental impact.
- *Green Extraction Technologies*: Development of solvent-free or low-impact extraction techniques, such as supercritical fluid extraction or enzyme-assisted processes, will further reduce the carbon footprint of biocolour production.
- *Integrated Life Cycle Assessment (LCA)*: Incorporating LCA tools in biocolour development will provide comprehensive insight into their environmental impact across all stages, from raw material sourcing to disposal.
- *Global Regulatory Harmonization*: Future regulatory frameworks must aim at global alignment to facilitate international trade, reduce ambiguity, and ensure uniform safety and quality standards for biocolours.
- *Public Awareness and Market Acceptance*: Education and transparent labelling will play a pivotal role in consumer acceptance and market penetration of biocolour-based products.
- *Policy Support and Incentives*: Governments and institutions must provide policy incentives, R&D funding, and infrastructure support to accelerate the biocolour industry's transition from being a niche market to becoming the norm.

10.6 Conclusion

The transition from synthetic dyes to biocolours marks a significant step toward sustainable and eco-conscious industrial practices. This chapter emphasizes how biocolours, derived from natural sources such as plants, microbes, algae, and insects, offer numerous environmental benefits, including biodegradability, reduced toxicity, and a lower ecological footprint. Their use in industries like textiles, food, cosmetics, and pharmaceuticals contributes not only to environmental conservation but also to human health and safety. However, transitioning to biocolours on a large scale comes with its own set of challenges. Issues like inconsistent yields, a limited palette of colours, extracting methods that can be resource-heavy, and the seasonal availability of raw materials need careful consideration. Fortunately, sustainable production methods such as microbial fermentation, leveraging agro-industrial waste, and employing green extraction technologies can help alleviate these hurdles. On the regulatory side, biocolours used in consumer products are subject to rigorous safety evaluations by global and national bodies such as the FDA, EFSA, FSSAI, and BIS. Regulatory approval demands comprehensive data on toxicity, stability, purity, and standardization. The absence of globally harmonized regulations, however, continues to be a barrier to international commercialization and widespread use. In conclusion, while biocolours offer a promising alternative to synthetic dyes, their success depends on advancing sustainable technologies, creating supportive policies, and developing globally harmonized regulatory frameworks. With ongoing research, industry collaboration, and public awareness, biocolours can become a potential component of green innovation, supporting a cleaner environment and a healthier future for both producers and consumers.

References

Ahmad MT, Sushil M, Krishna M (2012) Influence of dye industrial effluent on physico chemical characteristics properties of soil at Bhairavgarh, Ujjain, MP, India. Res J Environ Sci 1(1):50–53

Ahmed F, Repon MR, Pranta AD, Yasar MSR, Rakib MF, Ahmed I, Emon TA, Nishat TA (2024) Eco-sustainable printing of cellulosic polymeric material using bio-colorants and bio-cross-linkers. SPE Polymers 6(1). https://doi.org/10.1002/pls2.10159

Atlas of Sustainable Colour(s) (2025) https://aosc.online/colour-journal-environmental-impact-of-dyes.html

Beatley T (2014) Ethical duties to the environment. In: Center for resource economics eBooks. Island Press, pp 131–148. https://doi.org/10.5822/978-1-61091-491-8_14

Bhardwaj S, Gupta D, Bihani SGL (2012) Study of acute, sub acute and chronic toxicity test. Int J Adv Res Pharma Bio Sci 1(3):103–129

Chavan R (2011) Environmentally friendly dyes. Elsevier eBooks, pp 515–561. https://doi.org/10.1533/9780857093974.2.515

Chittaragi, D., & Menon, J. S. (2023). Bio-colours from spices. IntechOpen eBooks. https://doi.org/10.5772/intechopen.109830

Choudhury AKR (2018) Eco-friendly dyes and dyeing. Adv Mater Technol Environ 2:145–176

Chung K (2016) Azo dyes and human health: A review. J Environ Sci Health C 34(4):233–261. https://doi.org/10.1080/10590501.2016.1236602

Code of Federal Regulations(CFR) (2025) Visited on 27th Aug 2025.https://www.ecfr.gov/current/title-21/chapter-I/subchapter-A/part-74

CFR 2016 should be cited as (CFR 2025) in the text and in the reference list as Code of Federal Regulations (CFR) 2025. Visited on 27th August 2025. https://www.ecfr.gov/current/title-21/chapter-I/subchapter-A/part-74

Damant A (2011) Food colourants. In Elsevier eBooks, pp 252–305. https://doi.org/10.1533/9780857094919.2.252

De Oliveira Barreto JV, Casanova LM, Neves A, Junior R-M, Reis-Mansur MCPP, Vermelho AB (2023) Microbial pigments: major groups and industrial applications. Microorganisms 11(12):2920. https://doi.org/10.3390/microorganisms11122920

Devi S, Panghaal D, Kumar P, Malik P, Ravi E, Mittal S (2025) Eco-friendly innovations in textile dyeing: A comprehensive review of natural dyes. Advances in Research 26(1):204–212. https://doi.org/10.9734/air/2025/v26i11247

Durán N, Esposito E (2000) Potential applications of oxidative enzymes and phenoloxidase-like compounds in wastewater and soil treatment: a review. Appl Catal B Environ Energy 28(2):83–99. https://doi.org/10.1016/s0926-3373(00)00168-5

Elsahida K, Fauzi AM, Sailah I, Siregar IZ (2019) Sustainability of the use of natural dyes in the textile industry. IOP Conf Ser Earth Environ Sci 399(1):012065. https://doi.org/10.1088/1755-1315/399/1/012065

Ernst, Young India (2024) Visited on 27th Aug, 2025.https://www.ey.com/en_in/newsroom/2024/09/need-for-national-bioplastic-policy-providingfiscal-benefits-and-formulating-a-regulatory-framework-vital-to-boost-india-s-bioplastics-industry-and-promotesustainability-ey-assocham-report

European Food Safety Authority (EFSA) (2016) Visited on 27th Aug, 2025.https://www.efsa.europa.eu/en/press/news/160914-0

European Union Ecolabel EU Ecolabel (2024) Visted on 27th Aug, 2025. https://environment.ec.europa.eu/news/newlegislation-gives-eu-ecolabel-big-thumbs-2024-06-24_en
EY India (2024) should be cited as (EY-India 2024) and in the reference list as Ernst & Young India (2024). Visited on 27th August, 2025. https://www.ey.com/en_in/newsroom/2024/09/need-for-national-bioplastic-policy-providingfiscal- benefits-and-formulating-a-regulatory-framework-vital-to-boost-india-s-bioplastics-industry-and-promotesustainability-ey-assocham-report
EU Ecolabel – Home (2025) should be cited as (EU-Ecolabel 2024) in the text and in the reference list as European Union Ecolabel EU Ecolabel (2024). Visted on 27th August, 2025. https://environment.ec.europa.eu/news/newlegislation-gives-eu-ecolabel-big-thumbs-2024-06-24_en
European Food Safety Authority (2024) should be cited as (EFSA 2016) in the text and in the reference list as European Food Safety Authority (EFSA) 2016. Visited on 27th August 2025. https://www.efsa.europa.eu/en/press/news/160914-0
Filho WL, Dinis MAP, Liakh O, Paço A, Dennis K, Shollo F, Sidsaph H (2024) Reducing the carbon footprint of the textile sector: an overview of impacts and solutions. Text Res J 94(15–16):1798–1814. https://doi.org/10.1177/00405175241236971
Fried R, Oprea I, Fleck K, Rudroff F (2021) Biogenic colourants in the textile industry – a promising and sustainable alternative to synthetic dyes. Green Chem 24(1):13–35. https://doi.org/10.1039/d1gc02968a
Gerberick GF, Ryan CA, Dearman RJ, Kimber I (2006) Local lymph node assay (LLNA) for detection of sensitization capacity of chemicals. Methods 41(1):54–60. https://doi.org/10.1016/j.ymeth.2006.07.006
Global Organic Textile Standard (GOTS) (2023) Visited on 27, Aug 2025.https://global-standard.org/images/resourcelibrary/documents/standard-and-manual/GOTS_7.0__DE_signed.pdf
Grisales CM, Salazar LM, Garcia DP (2018) Treatment of synthetic dye baths by Fenton processes: evaluation of their environmental footprint through life cycle assessment. Environ Sci Pollut Res 26(5):4300–4311. https://doi.org/10.1007/s11356-018-2757-9
Guin JP, Bhardwaj Y, Varshney L (2017) Mineralization and biodegradability enhancement of methyl Orange dye by an effective advanced oxidation process. Appl Radiat Isot 122:153–157. https://doi.org/10.1016/j.apradiso.2017.01.018
GOTS IWG 2011 should be cited as (GOTS 2023) in the text and in the reference list as Global Organic Textile Standard (GOTS) (2023). Visited on 27, August 2025. https://global-standard.org/images/resourcelibrary/documents/standard-and-manual/GOTS_7.0__DE_signed.pdf
Human Foods Program (2017) Redbook 2000: IV.C.4.a. subchronic toxicity studies with rodents. U.S. Food and Drug Administration
International Working Group on Global Organic Textile Standard (2011) Global Organic Textile Standard (GOTS) Version 3.0. Global Organic Textile Standard
Irfan H, Beg T, Naeem F, Irfan M, Naqvi S, Shengyuan Y (2023) Sustainable technique of dyeing bio-degradable polyester using henna extract. Pigm Resin Technol 53. https://doi.org/10.1108/prt-02-2023-0015
Jayavani L, Shahin Banu Z (2025) A review article on recent production & application of natural dyes or Biocolours from microorganisms. International journal on. Sci Technol 16(1). https://doi.org/10.71097/ijsat.v16.i1.1639
Jorge AM, Athira K, Alves MB, Gardas RL, Pereira JF (2023) Textile dyes effluents: A current scenario and the use of aqueous biphasic systems for the recovery of dyes. J Water Process Eng 55:104125. https://doi.org/10.1016/j.jwpe.2023.104125
Joshi SS, Shinde NH (2017) Study of different Biocolours and its application in cosmetic products. Int J Sci Eng Res (IJSER) 5(10):6–8. https://doi.org/10.70729/ijser171925
Julanon N, Thiravetyan B, Unhapipatpong C, Xanthavanij N, Krikeerati T, Thongngarm T, Wongsa C, Songnuan W, Naiyanetr P, Sompornrattanaphan M (2023) Not just a Banana: the extent of fruit cross-reactivity and reaction severity in adults with Banana allergy. Foods 12(13):2456. https://doi.org/10.3390/foods12132456
Kadolph S (2008) Natural dyes: a traditional craft experiencing new attention. Delta Kappa Gamma Bull 75(1):14

Kant R (2012) Textile dyeing industry an environmental hazard. Nat Sci 04(01):22–26. https://doi.org/10.4236/ns.2012.41004

Keerthana T, Sai Subhashini S (2025) Microbial pigments: types, production methods, and diverse applications in various industries: a review. Int J Sci Technol (IJSAT) 16(1). https://doi.org/10.71097/ijsat.v16.i1.1593

Khoo HE, Azlan A, Tang ST, Lim SM (2017) Anthocyanidins and anthocyanins: colored pigments as food, pharmaceutical ingredients, and the potential health benefits. Food Nutr Res 61(1):1361779. https://doi.org/10.1080/16546628.2017.1361779

Kishor R, Purchase D, Saratale GD, Saratale RG, Ferreira LFR, Bilal M, Chandra R, Bharagava RN (2021) Ecotoxicological and health concerns of persistent coloring pollutants of textile industry wastewater and treatment approaches for environmental safety. J Environ Chem Eng 9(2):105012. https://doi.org/10.1016/j.jece.2020.105012

Klüppel H (2005) The revision of ISO standards 14040-3 – ISO 14040: environmental management life cycle assessment principles and framework – ISO 14044: environmental management life cycle assessment requirements and guidelines. Int J Life Cycle Assess 10(3):165. https://doi.org/10.1065/lca2005.03.001

Kuddus M, Singh P, Raveendran S, Singh R (2024) Microbial pigments. CRC Press eBooks. https://doi.org/10.1201/9781003353980

Kusumlata N, Ambade B, Kumar A, Gautam S (2024) Sustainable solutions: reviewing the future of textile dye contaminant removal with emerging biological treatments. Limnol Rev 24(2):126–149. https://doi.org/10.3390/limnolrev24020007

Laganà P, Avventuroso E, Romano G, Gioffré ME, Patanè P, Parisi S, Moscato U, Delia S (2017) Chemistry and hygiene of food additives. In: Springer briefs in molecular science. https://doi.org/10.1007/978-3-319-57042-6

Lehto S, Buchweitz M, Klimm A, Straßburger R, Bechtold C, Ulberth F (2016) Comparison of food colour regulations in the EU and the US: a review of current provisions. Food Addit Contamin Part A 34(3):335–355. https://doi.org/10.1080/19440049.2016.1274431

Leone F, Berger BJ (2009) The consumer product safety improvements act, its implementation and its liability implications. Def Counsel J 76:300

Li X, Zhu L, Ding X, Wu X, Wang L (2023) Climate change and the textile industry: the carbon footprint of dyes. AATCC J Res 11(2):109–123. https://doi.org/10.1177/24723444231212954

Li C, Zhang T, Zhou X, Cheng Z, Xu T, Li Z, Hong J (2024) Carbon–water–energy footprint impacts of dyed cotton fabric production in China. J Clean Prod 467:142898. https://doi.org/10.1016/j.jclepro.2024.142898

Liu Y, Chen J, Duan D, Zhang Z, Liu C, Cai W, Zhao Z (2024) Environmental impacts and biological technologies toward sustainable treatment of textile dyeing wastewater: a review. Sustainability 16(24):10867. https://doi.org/10.3390/su162410867

Menoufi KAI (2011) Life cycle analysis and life cyle impact assessment methodologies: a state of the art

Mouro C, Gomes AP, Costa RV, Moghtader F, Gouveia IC (2023) The sustainable bioactive dyeing of textiles: a novel strategy using bacterial pigments, natural antibacterial ingredients, and deep eutectic solvents. Gels 9(10):800. https://doi.org/10.3390/gels9100800

OECD (2006) Revised introduction to the OECD guidelines for the testing of chemicals, section 3. OECD Publishing, Paris. https://doi.org/10.1787/9789264030213-en

Parisi ML, Fatarella E, Spinelli D, Pogni R, Basosi R (2015) Environmental impact assessment of an eco-efficient production for coloured textiles. J Clean Prod 108:514–524. https://doi.org/10.1016/j.jclepro.2015.06.032

Parmar M, Phutela UG (2015) Biocolors: the new generation of additives. Int J Curr Microbiol App Sci 4(7):688–694

Pizzicato B, Pacifico S, Cayuela D, Mijas G, Riba-Moliner M (2023) Advancements in sustainable natural dyes for textile applications: A review. Molecules 28(16):5954. https://doi.org/10.3390/molecules28165954

Poo CP, Kamalian L, Yang Z, Lau Y (2024) Bio dyes and bio pigments: the sustainable approach in industrial textile dyeing and printing processes. AHFE Int 158. https://doi.org/10.54941/ahfe1005559
Rathoure AK, Patel H, Patel T, Bagrecha D, Patel U, Modi J (2019) Environmental issues in chemical dyeing and printing industries. Octa J Environ Res 7(2):040–046
Robinson T, McMullan G, Marchant R, Nigam P (2001) Remediation of dyes in textile effluent: a critical review on current treatment technologies with a proposed alternative. Bioresour Technol 77(3):247–255. https://doi.org/10.1016/s0960-8524(00)00080-8
Rodriguez-Amaya DB (2015) Natural food pigments and colorants. Curr Opin Food Sci 7:20–26. https://doi.org/10.1016/j.cofs.2015.08.004
Samiei M, Asgary S, Farajzadeh M, Bargahi N, Abdolrahimi M, Kananizadeh U, Dastmalchi S (2015) Investigating the mutagenic effects of three commonly used pulpotomy agents using the Ames test. DOAJ (Directory Open Access J) 5(1):121–125. https://doi.org/10.5681/apb.2015.017
Scalbert A (1991) Antimicrobial properties of tannins. Phytochemistry 30(12):3875–3883. https://doi.org/10.1016/0031-9422(91)83426-l
Scruggs CE, Ortolano L, Wilson MP, Schwarzman MR (2014) Effect of company size on potential for REACH compliance and selection of safer chemicals. Environ Sci Pol 45:79–91. https://doi.org/10.1016/j.envsci.2014.10.001
Sharma J, Sharma S, Soni V (2021) Classification and impact of synthetic textile dyes on aquatic Flora: a review. Reg Stud Mar Sci 45:101802. https://doi.org/10.1016/j.rsma.2021.101802
Simon JE, Decker EA, Ferruzzi MG, Giusti MM, Mejia CD, Goldschmidt M, Talcott ST (2017) Establishing standards on colors from natural sources. J Food Sci 82(11):2539–2553. https://doi.org/10.1111/1750-3841.13927
Stich E (2016) Food color and coloring food: quality, differentiation and regulatory requirements in the European Union and the United States. In: Handbook on natural pigments in food and beverages, pp 3–27. https://doi.org/10.1016/b978-0-08-100371-8.00001-4
Tilman D, Reich PB, Isbell F (2012) Biodiversity impacts ecosystem productivity as much as resources, disturbance, or herbivory. Proc Natl Acad Sci 109(26):10394–10397. https://doi.org/10.1073/pnas.1208240109
Tkaczyk A, Mitrowska K, Posyniak A (2020) Synthetic organic dyes as contaminants of the aquatic environment and their implications for ecosystems: a review. Sci Total Environ 717:137222. https://doi.org/10.1016/j.scitotenv.2020.137222
Tufvesson L (2010) Environmental assessment of green chemicals-LCA of bio-based chemicals produced using biocatalysis. Doctoral dissertation, Lund University
United Nations (2015) Transforming our world: the 2030 agenda for sustainable development (A/RES/70/1). United Nations General Assembly
Varghese R, Ramamoorthy S (2023) Status of food colorants in India: conflicts and prospects. J Consum Prot Food Saf 18(2):107–118. https://doi.org/10.1007/s00003-023-01427-y
Vermeer D, Clemen B, Michalko A, Nguyen D, Noyes C, Akella A, Bunting J (2010) In: Golden J (ed) An overview of ecolabels and sustainability certifications in the global marketplace. Nicholas Institute for Environmental Policy Solutions Duke University. Interim Report Document, pp 10–11
Warren-Thomas E, Ahrends A, Wang Y, Wang MMH, Jones JPG (2023) Rubber's inclusion in zero-deforestation legislation is necessary but not sufficient to reduce impacts on biodiversity. Conserv Lett 16(5). https://doi.org/10.1111/conl.12967
White PA, Luijten M, Mishima M, Cox JA, Hanna JN, Maertens RM, Zwart EP (2019) In vitro mammalian cell mutation assays based on transgenic reporters: a report of The International Workshop on Genotoxicity Testing (IWGT). Mutat Res Genet Toxicol Environ Mutagen 847:403039. https://doi.org/10.1016/j.mrgentox.2019.04.002

The Future Perspective of Biocolours in a Sustainable World

11

Bharti Minhas, Aayushi Verma, Shivani, Tamanna, Nitesh Kumar, Neelam Kumari, and Neelam Kaushik

Abstract

The growing concern over environmental degradation and the shift toward sustainable practices have brought biocolours to the forefront of eco-friendly solutions in various industries. This chapter explores the future perspectives of biocolours as a sustainable alternative to synthetic dyes, focusing on their potential for reducing environmental impact, improving social equity, and promoting economic viability. Natural colours or biocolours are becoming more and more necessary in our modern world because synthetic colours are carcinogenic, non-biodegradable, and harmful to both human health and the environment. Businesses in the food, pharmaceutical, and cosmetics sectors must use natural-source colours to satisfy customer demand for natural products and preserve economic competitiveness. The manufacture of biopigment can be made more efficient and less expensive by using industrial leftovers, agro-waste, or by-products from many industries. Pigment production difficulties are being addressed by developments in bioinformatics methodologies, omics platforms, and molecular investigations. Biotechnology could be the essential key for industrial optimization, environmental impact reduction, and sustainable development. The pharmacological potential of biopigments, including their potential to fight cancer and other disorders, is another area of research interest. To create a greener future, more experimental use of biocolours as a substitute for synthetic colourants in various areas of our life should be welcomed. However, more

B. Minhas · Shivani · N. Kaushik (✉)
Department of Microbiology, Himachal Pradesh University, Shimla, India

A. Verma · Tamanna
Department of Biotechnology, Himachal Pradesh University, Shimla, India

N. Kumar · N. Kumari
Department of Biosciences, Himachal Pradesh University, Shimla, India

R. K. Bhatia, A. Walia (eds.), *Biocolours*,
https://doi.org/10.1007/978-981-95-1738-1_11

investigation is required to determine the best ways to maximize pigment properties like composition and yield through the use of metabolic engineering tools, growth parameter optimization, the use of low-cost organic substrates to increase value, the existence of numerous elicitors for pigment synthesis, stabilizing methods for higher pigment utilization, and safer extraction methods for large-scale extraction.

Keywords

Biocolours · Extraction · Genetic engineering · Sustainability · Industrial applications · Regulation

11.1 Introduction

Biocolours are pigments that originate from natural, biological sources, including microbes, plants, animals, and insects. Chlorophyll, a natural pigment present in plants, is responsible for the green colour in plants and is the most abundant pigment present in nature. Besides chlorophyll, anthocyanins (found in fruits like blueberries), carotenoids (found in tomatoes or carrots), flavonoids, and betalains are among the common natural plant pigments found in nature (Singh et al. 2023). In addition to plant pigments, insects and animals also provided pigments to a lesser extent, which were widely used in ancient times. But due to the difficulties in extracting pigment, excessive use of animals, and costly insect culture, their usage has reduced (Yusuf et al. 2017; Shamim et al. 2014). Microalgae and microorganisms are novel suppliers of biocolours (Morales-Oyervides et al. 2020). The ability to produce pigment on a large scale due to their rapid development with minimal or no effect on the environment, cost effectiveness, simplified handling and genetic manipulation, and insufficient seasonal effects on pigment quality and quantity are some of the benefits that microorganisms have over animal and plant sources (Rana et al. 2021). Additionally, certain microbes, like bacteria and fungi, produce riboflavin, flavins, and niacin pigments. During cheese and yoghurt production processes, these pigments can be used to give natural hue (Singh et al. 2023). When it comes to food colouring, microbe-derived pigments are more efficient than plants as they are more readily available, non-seasonal, easier to scale up, produce more, and require fewer downstream processing (Dufossé 2018).

Since they often contain less dangerous chemicals and have a less detrimental impact on the environment than artificial colours, biocolours are considered to be more environmentally friendly and durable (Renita et al. 2023). These natural pigments have grown in popularity as a substitute for artificial colours as consumer demand for sustainable and harmless products increases (Ghosh et al. 2022). The surroundings may be at risk from artificial hues, which are often made of compounds obtained from hydrocarbons. Contrarily, biocolours are environmentally friendly and often thought to be better for individuals and surroundings. Therefore, to ensure their continuous use for generations to come and to allow for a feasible growth in their use, biological hues must be produced effectively from natural sources. The fabric, culinary, beauty products, art, medicinal, and polymer sectors are seeing a sharp rise of interest for biologically generated hues. Also, the use of biocolours aligns with a growing need for natural and clean-label products because biological pigments maintain integrity and attract to consumers who are mindful of their health (Singh et al. 2023).

The growing need for hues in the global economy may be fulfilled by biocolours derived from various natural resources, which can also serve as a substantial alternative to manmade pigments. However, the high expense of massive manufacturing, reduced stability, hue changes, absence of guidelines, and technical issues that impede the commercial manufacturing of biocolours produced by microbes are some of the disadvantages. Scientists' initiatives to substitute synthetic colours are discouraged by these constraints. Given the bright prospects of biocolours, research should concentrate on developing simpler methods for producing biocolours from renewable resources in order to expand their commercial uses. The major issues in manufacturing biocolours will be resolved by the invention of various biotechnological techniques including synthetic biology, gene editing, and metabolic engineering. Further, utilizing farm waste and manufacturing by-products enhances and expedites the process of biosynthesis by reducing expenses (Barreto et al. 2023). The affordability and convenience of the manufacture of biocolours will expand potential applications in the near future. However, more funding, rules, and laws are needed for their production, usage, and commercialization.

11.2 Technological Advancements in Biocolour Production

Traditional liquid–liquid techniques using biological solvents to separate colours are the basis of most commercial dye extraction processes. However, insufficient production, extended drying periods, and significant synthetic solvent usage are the method's disadvantages and such chemicals further have negative impacts on the planet and the wellness of people. As an outcome, effective and environmentally responsible harvesting strategies have been developed. Also, the commercialization of natural dyes produced from plants and microbe populations depends on the development of cost-effective manufacturing and removal techniques. In order to obtain biocolours from microbes and plants, methods like enzyme-assisted extraction, pressurized liquid extraction, microwave-assisted extraction, ultrasound-assisted extraction, and ultra-high-pressure-assisted extraction have been established. For large-scale extraction, these techniques are usually more efficient, need fewer solvents (water, aqueous, and non-aqueous), and are both profitable and very effective.

11.2.1 Enzymatic Extraction

Enzymes are crucial for the extraction of pigments and improve the standard and uniformity of the end product (Sharma 2014). In order to enhance both efficiency and productivity, enzymatic extraction utilizes hydrolytic enzymes that break down the cell wall of both plants and microbes, releasing internal components into the solvent. The advantages of employing enzymes for pigment extraction are fast extraction, high yields, moderate reaction conditions, and significant specificity. The choice of right enzyme is an important aspect in enzyme-assisted method. For enzyme-assisted extraction, different enzymes such as cellulase, hemicellulase, ligninase, pectinase, and amylase could be used for the extraction of pigments from plants. In this method, plant material is pretreated with a single enzyme or a combination of multiple enzymes to hydrolyse the cell wall and liberate the phytoconstituents attached to lipid and carbohydrate chains inside the cell (Puri et al. 2012; Das et al. 2021). Further, volatile chemicals, hydrophilic and hydrophobic pigments, phenolic compounds, and other bioactive phytoconstituents are extracted from the pretreated plant materials by solvent extraction or pressurized hot water extraction. The concentration of enzymes, substrate, and the treatment time, temperature, pH, etc., are the main functional parameters involved in the enzyme-based extraction process (Adadi et al. 2018). A wide variety of enzymes has been investigated for bicolour extraction. Lombardelli et al. (2021) used an enzyme cocktail of cellulase, xylanase, and pectinase to extract betaine from beetroots. Recently, Stanek-Wandzel et al. (2024) used cellulase, hemicellulase, and pectinase for the extraction of polyphenolic compounds from grape pomace and reported that enzyme-assisted extraction significantly increased the yield of phenolic compounds and resulted in higher total phenolic content.

11.2.2 Microwave-Assisted Extraction (MAE)

Microwave-assisted extraction is gaining popularity as an alternative for conventional extraction methods because of its speed, efficiency, even heating, and minimal solvent consumption for the extraction of water- and oil-based plant pigments (Dahmoune et al. 2014). This method uses the heat energy of microwaves to extract pigment from plants by heating solvent containing sample. This causes analytes from a sample matrix to separate into the solvent. The extraction rate increases with increasing solvent polarity because it absorbs and heats up microwave energy more quickly (Bashir et al. 2024). The extraction time, solvent, temperature, stirring, and microwave power are the factors that affect the microwave extraction process. Microwave extraction enables the faster extraction of organic compounds with a yield that is comparable to or better than conventional extraction methods. However, prolonged microwave heating can raise the temperature of the product and lower the extraction yield through product degradation or compound breakdown or by changing the properties of the extract (Das et al. 2022). Sharma et al. (2021) extracted betaine and betaxanthin from *Amaranthus tricolour* by MAE and obtained the highest yield. Recently, Chambaud et al. (2023) used water-based microwave-assisted extraction (MAE) to extract glycosylated flavonoids and anthraquinones from *Rubia tinctorum* L. and obtained the highest extraction yield with the most intense colour.

11.2.3 Supercritical Fluid Extraction (SFE)

Supercritical fluid extraction is a process of isolating bioactive components/pigments from organic biomass by using a supercritical fluid as an extractant. The low density and high solvating power are the main advantages of a supercritical fluid. CO_2 is the most commonly used supercritical fluid because it is safe, inexpensive, leaves no solvent residue after extraction, is easily available, and can separate heat-sensitive compounds (Uwineza and Waśkiewicz 2020). Further, it can be used for the extraction of bioactive compounds that are prone to degradation at higher temperatures since it may be utilized at temperatures close to the ambient critical temperature (31 °C). Moreover, the solvent strength of supercritical CO_2 can be altered by changing pressure, temperature, and the use of cosolvents (Majik and Gawas 2023). However, its solvating ability for polar analytes is quite poor. The extraction of polar compounds is more effective when modifiers like acetone, ethanol, or MeOH are used. Moreover, viscosity and the solute diffusivity are superior to conventional extraction solvents. Ludwig et al. (2021) reported the highly efficient β-carotene recovery from *Dunaliella salina* by varying the pressure, temperature, and ethanol concentration. Larocca et al. (2023) used extreme CO_2 (CO_2-SFE) at 400 bar, a temperature of up to 60 °C, a CO_2 flow velocity of 6 L/min, and ethanol as a cosolvent to obtain carotenoids from *Rhodotorula* sp. red yeast. The findings demonstrated that although productivity increased noticeably with the inclusion of the co-solvent, but the extraction of carotenoids was not substantially impacted by rising temperatures.

11.2.4 Ultrasound-Assisted Extraction

Ultrasound-assisted extraction (UAE) employs ultrasonic waves to increase the yield of extraction by inducing cycles of compression and decompression through the production of comprehensive phenomenon including bubbling, vibration, crushing, and blending inside a medium. This extraction process has been acknowledged as both efficient and environment friendly. The cell wall can be broken by ultrasound, which speeds up the solubility of extract. With the impact of an ultrasonic wave, UAE rapidly incorporates the components of extracted material into the solvent, producing a multi-component mixed extract and further separation, refinement, and purification are done using the appropriate methods.

The main benefits of using UAE include reducing extraction period, reducing solvent use, working at lower temperatures with no external heating, increasing the yield of extraction, and being easy to use (Mounika et al. 2023). Şahin et al. (2019)

used UAE for the extraction of lutein and carotenoids from *Chlorella.* Similarly, Savic Gajic et al. (2021) extracted carotenoids from orange peel using the UAE approach with olive oil as a solvent. Recently, ultrasound-enzyme-assisted extraction (U-EAT) of polyphenol and amylase enzyme from ramie leaves was carried out by Loan et al. (2024) and reported that U-EAT could enhance the productivity of extraction and give a noticeably larger amount of bioactive substances than an untreated sample.

11.2.5 Ultra-High Pressure-Assisted Extraction

For efficient extraction of bioactive compounds, the ultra-high-pressure extraction technique uses enormous amounts of pressure to infiltrate the solvent within the raw material and dissolve the desired component in the solvent. Comparing this technique to other techniques, this requires less solvents and time. Based on the extraction solvent used, both water-based and oil-based colours can be extracted using pressure-driven liquid extraction. At high pressure, several events like phase transition, adjustments to reaction kinetics, alterations in molecule form are generated, resulting in a reaction that promotes volume reduction and enhances extraction efficiency (Bashir et al. 2024). Compared to traditional liquid extractions, the UHPE approach produces less energy and causes reducedthermal breakdown of the biological compounds (Khan et al. 2019). Ren et al. (2023) employed UHPE of total phenols from *Eucommia ulmoides* leaves. Additionally, ionic liquids (IL)-assisted extraction, ohmic heating (Bai et al. 2019), pulsed electric field (PEF)-assisted extraction (Martínez et al. 2020; Wang et al. 2023) and ultrasonication (Sundaramahalingam and Sivashanmugam 2023) have been recognized as practical and efficient green techniques to enhance the extraction yield of pigments from plant microbial biomass.

11.2.6 Synthetic Biology for the Production of Biocolours

Synthetic biology is a renowned technique in the biotechnology of colours and their microbial producers that combines engineering ideas with biotechnological techniques such as genome editing and DNA sequencing to enhance the production of a bioactive compound by altering bacterial genes. The most important step in the recombinant production of biocolours is the identification of key pathways and functional genes. Rate-limiting enzymes and feedback inhibition typically result in metabolic constraints in natural synthesis pathways. Upon identifying genes required for colour manufacture pathways by direct cloning or a metagenomics technique, researchers create diverse synthesis pathways, following pathway assembly, which target the recombination of desired genes utilizing plasmid or plasmid-free methods. This approach led to an increased yield of β-carotene after the expression of three lipase and carotene pathway-encoding genes from different microbial genera into *Saccharomyces cerevisiae* (Fathi et al. 2021). Further, the production of lycopene pigment was increased thrice by the expression of a combinatorial multi-gene pathway assembly in *E. coli* (Coussement et al. 2017). In another study, lycopene production was significantly increased (up to 73.3 mg/cell dry weight) by using recombinant genes in engineered *S. cerevisiae* that optimized lipid production to pigment accumulation (Ma et al. 2019).

11.2.7 Metabolic Engineering for Biocolour Production

The primary purpose of metabolic engineering is to optimize metabolic processes to produce desired molecules in higher quantities or to create new pathways for the production of valuable compounds. Recent advances in metabolic engineering and molecular biology have made it possible to manipulate genes to produce more pigments and clone the genes involved in pigment biosynthesis. Pigment biosynthetic pathways have been thoroughly investigated and engineered to overproduce a pigment as well as to change the molecular structure and colour of the pigments. For example, *Streptomyces coelicolour* produces the blue pigment Actinorhodin, which has been genetically modified to yield kalafungin, a similar brilliant yellow polyketide that is utilized to make the reddish-yellow anthraquinone. For the construction of an intracellular biosynthetic pathway, new genes or gene clusters that encode functional proteins must be extracted from microbial sources and expressed heterologously in *Escherichia coli* or other microorganisms in order to create an intracellular biosynthesis pathway. Although significant advancements in the microbial synthesis of high-value compounds have been achieved through this approach, the productivity is typically still insufficient to fulfil the demands of large-scale production (Xu et al. 2023). Metabolic engineering and mutagenesis have been used to increase carotenoid production in *R. mucilaginosa* (Wang et al. 2017). For the first time, Furubayashi et al. (2021) documented the microbial synthesis of capsanthin, a distinctive carotenoid often presents in *Capsicum annuum,* by expressing eight genes in *Escherichia coli* to create a heterologous capsanthin biosynthesis pathway. The production level of 0.5 mg/L capsanthin was obtained following the essential overexpression of carotenogenic genes and the reduction of by-product generation. Xinrui et al. (2023) developed an efficient metabolically engineered *Escherichia coli* strain by co-expression of key enzymes of violaxanthin, which is capable of producing a large amount of violaxanthin and can be used for commercial production. Further, metabolic engineering technologies have been applied for the hyperaccumulation of pigments such as anthocyanin in different plants, including tobacco and *Arabidopsis* (Murthy et al. 2023).

CRISPR/Cas technology is another metabolic engineering strategy that has been efficiently used for the production of pigments. In this approach, CRISPR/Cas9 technology could be used to manufacture and create natural pigments by introducing the pigment-encoding genes into the bacterial genome. This technology has been employed in *Saccharomyces cerevisiae* to produce a wide range of products on an industrial scale. *S. cerevisiae* can also be modified to produce natural colours after introducing a colour-producing gene into its genome using the CRISPR/Cas9 technology (Lian et al. 2018). This gene editing tool has been used to improve biosynthetic pathways for the production of increased pigment yields. Limited reports are available regarding the application of CRISPR-Cas9 in microbes to produce biocolours. To optimize the use of the genome editing tools for this purpose, more research is needed.

11.2.8 Genetically Modified Microorganisms in Biocolour Production

The efficiency, production, and adaptability of microbial production of biocolours have been improved by genetic manipulation (Singh et al. 2023). Because of their quick fermentation phases, easy handling, cheaper expenses for raw materials, ability to manipulate microbial genomes, and lack of geographical or seasonal restrictions, microbes are often regarded as superior for the manufacture of pigments. Both natural and engineered bacterial, fungal, or yeast strains produce economically important pigments such as astaxanthin, lycopene, prodigiosin, canthaxanthin, and β-carotene, which are used in a wide range of sectors, including food, pharmaceuticals, fabrics, cosmetics, and others. Cloning the genes necessary for pigment biosynthesis in yeast or bacterial vectors has evolved as a more useful and cost-effective industrial production process. Some industrially significant microbes that may be used to genetically alter the pigment synthesis and produce customized recombinants include *E. coli*, *Bacillus subtilis*, *S. cerevisiae*, *P. putida*, *Corynebacterium glutamicum*, and *Pichia pastoris* (Sen et al. 2019). For example, the tomato pigment lycopene was produced by genetically modifying *S. cerevisiae* (Chavan et al. 2011). By controlling the expression of heterologous genes encoding CrtY (lycopene cyclase), CrtW (β-carotene ketolase), and CrtZ (hydroxylase), the pathway of carotenoid biosynthesis for astaxanthin production was engineered in recombinant *Corynebacterium glutamicum*, yielding 0.4 mg/L/h astaxanthin (Henke et al. 2016). In a different study, 9.5 + 0.6 mg/L of callistephin, a red pigment found in strawberries, was produced using synthetic consortia of 4-strain *E. coli* polyculture (Jones et al. 2017). Coussement et al. (2017) optimized the lycopene production process in modified *E. coli* and achieved 448 mg of lycopene/g of cell dry weight. The engineered *C. glutamicum* was also tried for anthocyanin production by co-expressing the genes for anthocyanidin synthase (ANS) and 3-O-glucosyltransferase (3GT) from *Petunia hybrida* and *Arabidopsis thaliana*, respectively. By carefully regulating the co-expressed anthocyanin pathway and optimizing fermentation conditions, 40 mg/L of cyanidin 3-O-glucoside (C3G) production from catechin was reported (Zha et al. 2018).

11.3 Sustainability and Environmental Impact

Pigments which are derived from natural sources including plants, insects, animals and microbes are safer option for our environment in comparison to the synthetic ones. Figure 11.1 shows the contribution of biocolours in lowering the carbon footprint in multiple ways in contrast to synthetic dyes.

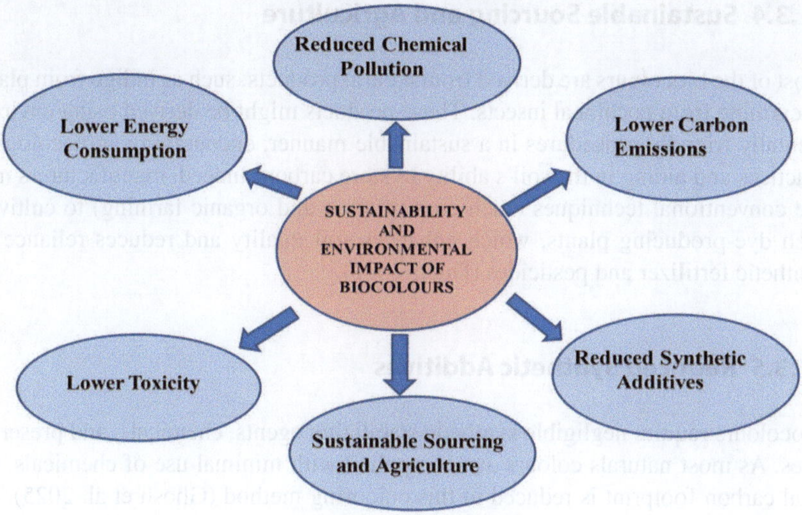

Fig. 11.1 Sustainability and environmental impact of biocolours

11.3.1 Lower Energy Consumption

Biological colours are made using less and simpler resource-intensive techniques; on the other hand, artificial pigments are produced using energy-intensive chemical methods such as petrochemical refining and complex reactions, which significantly raise carbon emissions (Jordan et al. 2024).

11.3.2 Reduced Chemical Pollution

Biocolours bring down the amount of toxic waste in the textile sector, like azo compounds, that are often produced during the manufacturing of artificial colours and have harmful impact on the environment. These colours can be harmful to marine organisms and remain in the environment, which leads to contamination and damage to the environment. Biocolours on the other hand, contribute to lower harmful ecological impact as they are very less toxic and recyclable (Mayegowda et al. 2025).

11.3.3 Lower Carbon Emissions

Biocolours are produced from natural sources and their raw material requirements results in very low carbon footprint. The growing plants helps absorbing CO_2 from the atmosphere, which further lowers the emission. The large production of CO_2 and its release in the environment is caused due to the manufacturing of artificial dyes, specifically when large industrial processes are used (de Oliveira et al. 2024a, b).

11.3.4 Sustainable Sourcing and Agriculture

Most of the biocolours are derived from natural products, such as indigo from plants or carmine from cochineal insects. These products might be derived using environmentally friendly procedures in a sustainable manner, encouraging agroecological practices and aiding in the soil's ability to store carbon. Indeed. manufacturers may use conventional techniques (such crop rotation and organic farming) to cultivate such dye-producing plants, which enhances soil quality and reduces reliance on synthetic fertilizer and pesticides (Duerr 2011).

11.3.5 Reduced Synthetic Additives

Biocolours require negligible synthetic stabilizing agents, chemicals, and preservatives. As most naturals colours can be applied with minimal use of chemicals, the total carbon footprint is reduced in this colouring method (Ghosh et al. 2025). On the other hand, artificial colours usually need extra chemicals to get the desired shade or for colour fastness. These supplemental chemicals usually have a more harmful impact on our ecosystem during manufacturing and disposal. Biocolours use less energy, produce fewer pollutants, release fewer carbon emissions, and use a more environmentally friendly supply system, hence they provide an ecologically beneficial alternative. Despite the fact that the cost and selection of synthetic dyes still dominate several industries, biocolours provide a considerable chance for more ecologically beneficial production in cosmetics, food, fabric, and other sectors (Khattab et al. 2020).

11.3.6 Waste Biomass as Raw Material for Biocolour Production

A sustainable way to reduce waste and carbon footprints and support circular economies is by making use of reusable substances as a primary source to produce biocolour. The utilization of waste biomass, which is abundant in many industries, can have major positive effects on the environment, economy, and society. Agricultural residues have recently been proposed as an alternative raw material to the manufacturing of biocolours to lower overall manufacturing cost and solve environmental concerns. According to Usmani et al. (2020), biomass waste and leftovers from agro-industries can be used as a potential source of microbe nutrition (carbon, nitrogen, and minerals) for the microbial synthesis of biocolours. Biocolours can be produced from pigments present in plant wastes from crops like indigo, turmeric, pomegranate, and beetroot. For example, the outer layer of pomegranates can be utilized to produce natural hues that vary from dark red to orange. For potential usage in fabrics and food colouring, biocolours may be produced from the husks of rice and maize (Samanta et al. 2015). Berry, grape, onion, and carrot leftovers are high in carotenoids and anthocyanins that can be used for biocolour production. It is possible that green pigments such as chlorophyll, present in the stalks of kale or spinach, may be converted into green colour. Hence, by reusing waste, expenses for disposal are decreased and the harmful impacts of organic waste on environment are minimized. In addition, this strategy aid sustainable production process in food and textile industries (Singh et al. 2023). Further, fabric scraps can be used as a substrate for extracting pigments or the microbiological production of biocolours. Using microbes, fungi, or algae grown in waste materials, biocolours can be derived by fermentation techniques. The use of textile scraps to produce biocolours creates a closed-loop procedure where materials are reused rather than thrown away. This reduces the pollution that textile scraps and landfill rubbish cause to the environment. By reusing waste resources in a long-lasting and excellent way and utilizing leftover biomass as a raw material for biocolour manufacture not only reduces detrimental effect like pollution and carbon emissions but also promotes the circular economy as a whole. These opportunities support the production of green technology, resource utilization, and waste reduction in a variety of industries, helping to establish a viable supply chain for the production of biocolour (De Mol and Vandamme 2024).

11.3.7 Circular Economy Approaches in Biocolour Manufacturing

A circular economy approach to biocolour manufacture aims to secure constant reuse of materials all over the manufacturing process, lessen waste, and enhance resource capabilities. The biocolour production sector may increase efficiency and lessen its effect on the ecosystem by integrating the idea of the circular economy. Below are some of the ideas that biocolour production could put in the concept of circular economy:

(a) *Resource Efficiency Using Waste Biomass and Materials*

Rather than depending only on natural sources, biocolour companies may focus on using trash from forestry, agriculture, and food industry as a raw material. This reduces the requirement for natural sources and trash in multiple sectors. For example, utilizing farm waste (like the stems of indigo plants) or fruit and vegetable peelings (like citrus peels and pomegranate husk) for extraction of pigments minimizes the requirement for further resource exploitation. The transformation of waste into advantageous biocolours reduces the necessity for raw materials and conserve ecosystems (Martini et al. 2021). Besides, biocolours may be manufactured by microbe fermentation or biological extraction from recyclable textiles waste (such as cotton and wool scraps), tannin-rich leather scraps, or even leftover paper. This allows companies to turn garbage into valuable products and reduces the burden on waste disposal sites (Seidu et al. 2024).

(b) *Non-toxic and Biodegradable Dyes*

The fact that biocolours are often reusable and made of safe, environment friendly substances hold up the principles of the circular economy. The use of biological pigments lowers the risk of dangerous artificial substances that are commonly used in the colouration industry. For example, biocolours made from plants like indigo, turmeric, or hibiscus can be less detrimental to the surroundings. Consequently, by bringing down the use of hazardous substances, biocolours uplift an additional circular and renewable approach while reducing land, water, and air pollution (Chavan 2011).

(c) *Sustainable Manufacturing and Design*

The use of eco-friendly extraction procedures during biocolour manufacturing is another component of the circular economy. This may incorporate low-energy, harmless, and low-water processes, including liquid-based or enzyme-based processes for biocolour extraction which require a minimum amount of water and energy. By applying harmless techniques like fermentation or solvent-free extraction, biocolours may be made from few available resources. These cost-effective, low-waste techniques foster the use of recycled materials and lessen the overall negative effect of biocolour production (Kaur et al. 2022).

(d) *Biomimicry in the Creation of Biocolour*

For example, nature frequently uses harmless approaches to produce vibrant colours that could conceivably be employed as a model for novel production processes. The use of biological systems lessens the reliance on chemical substances and processes, resulting in less harmful and eco-friendly alternatives (Pranta and Rahaman 2024). In the biocolour business, circular economy approaches encourage sustainability by improving usage of resources, lowering trash, and certifying that materials used in manufacturing are an element of a renewable cycle. By following these directions, biocolour companies may remarkably lower their effect on the biodiversity. By encouraging innovation, lessening the reliance on fossil fuels, and encouraging alternative methods of manufacturing and utilizing pigments, this approach not only brings social and economic benefits but also create a better tomorrow (Mariotti et al. 2020).

11.4 Emerging Markets and Applications

11.4.1 Expansion into New Industries

Because of their extensive viability, non-toxicity, and biodegradability, biocolours are establishing in new industries. Increasing customer awareness for products devoid of synthetic chemicals and environment friendly is expanding their growth.

Biocolours are used extensively because of their possible medical and therapeutic uses. Due to sustainability and health benefits, biocolours are having an impact in the following new industries:

(a) *Pharmaceutical Industry*

Pharmaceutical industries utilize biocolours from microbes, higher algae, and plants since they have many biological properties that are good for human well-being. Carotenoids (Ganguly et al. 2019), lutein, zeaxanthin, xanthophyll (Bora et al. 2019), betacyanin and flavonoids (Parmar and Phutela 2015), anthocyanins (Zha and Koffas 2017), and astaxanthins (Li et al. 2018) are some of the important biocolours used in the pharmaceutical industry. It has been proven that microbial pigments such as violacein, carotenoids, anthocyanins, and naphthoquinone are strong antioxidants and exhibit antimicrobial properties (Sen et al. 2019). These characteristics can improve the safety of pharmaceutical formulations and, in particular, the shelf lives of the most perishable compounds. It was demonstrated that an endophytic fungal pigment was more effective than the commercial antibiotic streptomycin agents (Sen et al. 2019). Numerous pigments that are derived from microorganisms, including astaxanthin, canthaxanthin, lycopene, monascorubrin, riboflavin, rubropunctatin, β-carotene, torularhodin, and others, have antimicrobial and anticancer capabilities and exhibit a wide range of biological qualities that are advantageous to humans. It is well known that prodigiosin possesses antibiotic, antibacterial, anti-malarial, and antineoplastic properties (Shaikh 2016). For example, prodigiosin produced by marine isolates, namely, *Streptomyces* sp. and *Zooshikella* sp., has shown potent antibacterial activity against *S. aureus* and other human pathogenic strains (Ramesh et al. 2020). These pigments are being used not only in the formulation of edible products like pills but also as ingredients in topical creams and lotions. Colourless carotenoids are mostly utilized in wellness nutrition and cosmetics to shield the body from oxidative stress and to prevent internal and external stress. Their antioxidative strength, UV protection, and anti-inflammatory properties make them an essential active ingredient in topical formulations. There are many clinical applications for bacterial pigments in the diagnosis of diseases like diabetes mellitus, leukaemia, and cancer. Melanin, a bacterial pigment, serves as a sunblock and shields the human skin from UV rays (Narsing Rao et al. 2017). *Salinicoccus* sp. pigment was discovered to have an inhibitory effect on human cancer cell line proliferation, which could be used to create anticancer medications. For example, prodigiosin from *Pseudoalteromonas* sp. showed cytotoxicity on U937 leukaemia cells (Srilekha et al. 2018). Betanin, a red pigment in beetroot, may offer a defence against certain diseases linked to stress (Poorniammal et al. 2022). Astaxanthin pigment is a beneficial medicinal substance, a strong antioxidant, and a dietary supplement, and it has anti-inflammatory, anti-apoptotic, and antioxidant qualities that exhibit considerable promise for the treatment or prevention of cancerous cells (Mhalaskar et al. 2018). Astaxanthin pigment isolated from the marine endophytic bacterium *Pontibacter korlensis* AG6 was demonstrated to have a lethal effect on the human breast cancer cell line (MCF-7) and also possessed strong antibacterial and antioxidant qualities (Pachaiyappan et al. 2021).

(b) *Paints, Coatings, and the Electronic Industry*

The paint and coatings industry has undergone an important change towards eco-friendliness due to increasing environmental fears. Traditional paints and coatings lead to environmental pollution by emitting high volatile organic compounds (VOCs) or by solvent release. In contrast, bio-based paints and coatings are VOCs free, offering customers an eco-friendly and safe environment. Bio-based coatings are useful for many uses, like floor coatings, wall paints, furniture coatings, and vehicle coatings. Natural pigments like chlorophyll and carotenoids obtained from plants, and pigments from fungi and algae are being used to obtain non-toxic, biodegradable paints (Malik and Sarkhel 2024). Despite the benefits, there are still issues such as higher production costs and achieving comparable performance to traditional coatings. Future research is expected to concentrate on increasing the performance of bio-based colours, improving the recycling process, and developing multifunctional coatings. The continuous adaptation of the coating sector towards sustainability practices shows the importance of reducing environmental impact and support responsible innovation (Malik and Sarkhel 2024).

Biocolours are also gaining importance as a sustainable alternative to synthetic chemicals used in electronic components. Microbe-produced pigments, like melanin, have been shown to have conductivity, which makes them important component for use in organic electronic devices such as screens and printed circuit boards. Recently, melanin-based thin films for wearable on-skin electronics with multimodal sensing were obtained by Contreras-Pereda et al. (2024). These films showed good conductivities with dual charge carriers (electrons and ions) under different conditions. Additionally, the film's high levels of natural compatibility, surface bonding, and conformability to skin were obtained, allowing its use in wearable electronic on-skin devices on pig.

11.4.2 Customization for Niche Applications

One of the most exciting aspects of biocolours is the ability to customize them for niche applications, providing more tailored solutions across industries. The development of non-toxic, biocompatible, biodegradable, and edible packaging is currently one of the most urgent areas in the research and development of new packaging materials. Natural food colouring is used in smart packaging to identify changes in the physical and chemical characteristics of foods that are packed. Such packaging typically consists of biopolymers, their blends, and biocomposites, which are used for increasing the shelf life of food and drugs. Additionally, these systems could incorporate a variety of edible markers for quality control, customer and manufacturer identification, monitoring storage conditions, detecting package integrity violations, and detecting food spoilage (Kalpana et al. 2019). By adding a holographic marker to the coating surface, biopolymer-based edible films and coatings can be functionalized to mark products or sense storage conditions. A consumable holographic marker creates multicolour reflection on the surface of food products. Using the nanoimprint technique, holographic markers were created from a film-forming solution on the surface of thin natural polymer coatings based on starch, gelatin, agar, and starch/gelatin and starch/agar blends (Podshivalov et al. 2020). Since these tags are sensitive to heat and humidity, this technology can be used to mark food and medical items, decoration of food items, and handle storage conditions (Kalpana et al. 2019). Alizarin, curcumin, and betalains are important natural dyes or colours for food safety markers in smart packaging (Priyadarshi et al. 2021). Chicken has been packaged using hibiscus natural colours based on chitosan starch sheets (Kanha et al. 2022).

11.4.3 Growing Market Trends in Developing Countries

As discussed, biocolours have great potential for different industrial applications and compete with the synthetic colours. There will be a considerable demand in the future for naturally occurring colours with an estimated growth rate of 5–10% every year (Puri et al. 2024). The UK, Germany, France, Italy, and Spain are the EU's top markets for natural colours. In recent years, developing economies like China, India, and South Korea have also witnessed growth in their markets. According to a study by Fact.MR in 2024, the biobased pigments and dyes market is expected to grow to 31.29 billion dollars in 2024 and is projected to reach 48.58 billion by 2034 at a compound annual growth rate (CAGR) of 4.5%. With expanding global market, especially in the South Asia Pacific, the Indian market is projected to grow at a CAGR of 6.7% from 2024 to 2034, reaching a market share of 58.6%. Due to customer preference and steady growth in the market, the pharmaceutical, cosmetic, and food and beverage sectors are probably going to see an increase in anticipated demands. The market is viewing increase in demand for sustainable pigments and dyes due to growing environmental awareness. Due to expanding global population, the textile industry is expected to witness the fastest CAGR in terms of needs for biobased pigments and dyes. Nowadays, many companies are replacing petroleum-based ink in their packaging with bio-based pigments, especially soy-based ink. Eco-friendly pigments have the potential to revolutionize a variety of sectors, such as food and beverage, cosmetics and personal care (Fact.MR 2024).

11.5 Challenges and Barriers

11.5.1 High Production Costs Compared to Synthetic Dyes

The expensive process of extraction and the unpredictability of traditional natural colours led to the creation of synthetic colours in the nineteenth century. The availability of raw materials and susceptibility of these natural pigments to diverse environmental factors are some of the issues with which the natural colour business is dealing. In microbial pigment production, one of the crucial factors to take into account is the cost-effectiveness of the specified media, which is vital for industrial processing when done on a large scale (Kalra et al. 2020). The cost is determined by the pigment, the extraction method involved, and every aspect of the bioprocess, both upstream and downstream. Low production yield and high fermentation expenses are two problems that restrict the extensive utilization of pigments in industrial settings. The process of bioproduct extraction, along with other downstream engineering procedures, often accounts for over half of the overall cost of bioprocesses, which offsets their commercialization (Barreto et al. 2023). Due to their high cost index, limited thermal endurance, potential interactions with additional dietary ingredients, and coproduction-related toxicity, natural food colouring has limitations. In comparison, natural colourants require more energy and water than their synthetic equivalents. Additionally, they require greater additions than synthetic colourants in order to get a satisfactory hue (Renita et al. 2023). In the case of plant pigments, a large number of plants required for the production of natural pigments results in costlier raw materials. Furthermore, it takes a long cycle to obtain new sources of floral pigment through the extraction of seasonal flowers. Growing plants especially for the purpose of dyeing could be costly (Azmin et al. 2022).

11.5.2 Limited Stability and Scalability of Biocolours for Industrial Use

Textile fibres typically do not absorb natural pigments easily. But careful consideration must be given to how well they can be extracted and purified and also how to adhere them to the textiles in a secure and environmentally friendly manner. Additionally, the literature on the more fundamental aspects of organic dyeing, such as dye-fixing procedures, chemical characterization of dyes, and dyeing enhancement parameters, is lacking (Grande et al. 2023). New studies on environmentally friendly natural colourant sources are necessary due to growing demand, limited resources, and a number of drawbacks with the currently approved natural pigments, including their limited availability all year long, poor water solubility, instability in the presence of heat, light, and varying pH, and fluctuation in pigment extraction. The industry faces a number of obstacles that hinder the commercialization process. For example, in case of fungal pigments, three key elements are necessary for the fungal-based pigment business to be sustainable and grow: (a) fungal pigments must not contain mycotoxin; (b) pigment yield; and (c) pigment purity and stability. A shorter shelf life resulting from the molecule's instability under various conditions may restrict the use of natural pigments as commercial pigments (Kalra et al. 2020).

Modern genetic engineering techniques enable the production of various economically significant pigments in microbes that either do not natively synthesize them or, when they do, only produce them in trace amounts, e.g., lutein, carotenoids, and zeaxanthin. Pigments extracted from microbes are less stable in some ecological conditions, which reduce their stability and shelf life resulting them to fade over the period of time when compared to artificial pigments. Renita et al. (2023) reported that the heat stability of betacyanins and anthocyanins is low as compared to carotenoids. Natural colourants lose their original colour upon heating. The application of bacterial pigments in industrial circuits is still not common despite being eco-friendly, non-toxic, and biocompatible for human use (Agarwal et al. 2023).

Plant-based pigments also have some limitations. Plant pigments rapidly denature when pH shift occurs and will harm their reproducibility. Plant pigments also need ample monoculture space, rainfall, and pest control along with other unexpected variables that affect the crop health and harvest loss percentage. Even though the plant pigments are biodegradable, natural, and organic, it has few environmental effects because of high water consumption for irrigation along with excessive use of pesticides. Conventional methods for extracting various pigments are water/solvent infusions, Soxhlet extraction, distillations, and maceration. These techniques are widely utilized but have their own drawbacks, including excessive solvent use, subpar extraction, drawn-out processes, and the chemical degradation of pigment structure at increased temperatures, which ultimately results in bioactivity loss. Organic-phase solvents, e.g., petroleum ether, methanol, acetone, benzene, and hexane are frequently used in these extraction procedures. These solvents are highly flammable, poisonous, and non-biodegradable in nature (Barreto et al. 2023).

11.5.3 Regulatory Hurdles and the Need for Standardized Quality

Colour additives are extensively used to impart colour to medicines, food, cosmetics, and the human body (21 Code of Federal Regulations, CFR). In contrast, commercial adulteration of biocolours, either deliberate or unintentional, has been made feasible. Even though biocolouring is approved by FDA, there are no industry-accepted safety protocols or quality criteria for it. In India, Food Safety and Standards Authority of India (FSSAI), which is subordinate to the Ministry of Health and Family Welfare, have the responsibility of enforcing the food regulations established under the Food Safety and Standards Act (2006). Adulteration of biocolourants is an emerging issue and can be done in many forms, from straightforward misbranding of the products or misuse of the term "natural" for labelling to deliberate postproduction adulteration, contamination from raw material sources, or inadequate processing. In order to guarantee safety and reliability in the supply chain, obligatory industry-wide safety standards are required for the manufacturing, processing, and worldwide trading of naturally sourced colourants. Biocolourants are considered safe to use when procured from authorized and well-researched origins, used within recommended limits, proper dosage, administration, biocolourant's purity as well as the extraction procedures used. However, overconsumption of even natural chemicals may cause harm to health. The presence of residual pollutants or by-products in the final products may pose risks. Regulatory authorities analyse biocolourants for safety and regulate daily consumption rates before permitting them for use, as some biocolourants may cause allergies in certain individuals (Bashir et al. 2024). In 2010, European Union regulators recommended the limited amounts of certain colourings that can be consumed daily and mandated warnings on children's food labels. The World Health Organization (WHO) and the United States Food and Drug Administration (US-FDA) have also prohibited certain colours from being used in food, medications, and cosmetic products. Food-grade pigments must receive clearance from regulatory authorities while maintaining the toxicological safety of the product (Kalra et al. 2020).

The production and utilization of biocolours raise environmental concerns because naturally derived dyes can be sourced from endangered, rare, and threatened species or consist of heavy metal-containing mordant dyes as part of the dye molecule (like zinc phthalocyanine, acid blue, copper (II) phthalocyanine, lead (II) phthalocyanine, and cadmium phthalocyanine). The recent edition of the Global Organic Textile Standard barred the use of dyes that were derived from these species or colours which contain heavy metals. Several initiatives have been launched and implemented to resolve this issue, but no natural colour is available at present which is completely sustainable (Afonso et al. 2023). Natural pigments obtained from minerals may have heavy metal contamination, which could be problematic when used for industrial purposes and also contaminates agricultural land and human food through irrigation water. However, several regulations and laws around the world are designed to restrict the amount of arsenic, lead, and other heavy metals that come from pigments used in cosmetics or similar products. Waste and disposal generated from production units, if they are not properly routed and treated, can pass through the tropic chain and bioaccumulate in aquatic life forms. In food sector, application of microbial pigments requires authorization and alignment with legal and regulatory criteria set up by organizations like the Food and Drug Administration (FDA) and the GRAS (Generally Regarded as Safe) label. Some microbial pigment producers have successfully satisfied these criteria (Barreto et al. 2023).

The United States' Natural Products Association (NPA) has listed guidelines on natural product labelling. Natural pigments are difficult to harvest and need expert workmanship. An appropriate and consistent pigment extraction method must be developed without compromising the necessary quality of the pigment before being utilized in the manufacturing of food products. There are few concerns which should be addressed while using natural ingredients, such as identifying the best supplier who minimizes cross-contamination, has hygienic processing practices, and offers high-quality supplies, processes, and services. Because natural ingredients from plants are used in diverse ways, there were comparatively limited regulatory guidelines addressing safety assessment for the natural food product business (Azmin et al. 2022).

11.5.4 Consumer Perception and Acceptance Challenges

The biggest challenge researchers yet have to overcome is the limited consumer's awareness regarding multidimensional benefits of biocolours. Colours significantly influenced the acceptance of products by consumers (Bashir et al. 2024). The idea of "bio-based" products may result in either positive or negative reactions. Consumers generally are unfamiliar with bio-based products, contributing to uncertainty and mixed feelings. This idea is most commonly linked to the environmental concerns and emphasis less on technical or health aspects. Moreover, interpretation of "bio-based" is often generalized or considered in non-specific manner, while consumer evaluations are primarily based on personal use. Customer's attitude towards bio-based goods is complicated in nature and situation-specific. Perception of bio-based products is influenced by factors like the original characteristics of product, manufacturing techniques, usability, and product composition (Sijtsema et al. 2016). Factors like unpleasant aroma or flavour, limited colour shades, poor colour fastness, and high product cost are hindering the path of acceptance of biocolours in the society.

11.6 Integration with Global Trends

11.6.1 Biocolours and the Green Economy

The United Nations Environment Programme (UNEP) has reported the green economy as "one that results in improved social equity and human well-being, while significantly lowering environmental risks and ecological scarcities." It promotes a low carbon footprint, optimizes resource efficiency, and supports socially inclusiveness. The fundamental roles of green economy include sustainable production and consumption of natural resources, along with increased resource efficacy for sustainable development. Sustainable production and consumption is aiming to upgrade the manufacturing practices and consumption behaviour to minimize the utilization of resources along with waste generation and emissions. Optimization of resource efficiency involves the approaches via which resources are being utilized for the benefits of society and intends to lessen the aggregate need for natural resources, waste production, and emissions produced in each unit of service or product (UNEP 2025). India's global initiative "LiFE," or Lifestyle for Environment, will pave the path for the future generations to adopt an "environmentally conscious lifestyle." In India's G20 presidency, India prioritizes the idea of green growth, validating how sustainable development is the base of Indian policymaking (G20 India). For the sustainable consumption and production of biocolours, administrative focus should be on the technical advancements together with improved regulatory frameworks, institutional setups, and tariff designs. Customer's willingness to support sustainable future and their awareness of eco-conscious lifestyles should both rise (UNEP 2025). The emergence of advanced, greener, and more sustainable biocolour extraction protocols for industrial applications is the highlight of the evolving green economy.

11.6.2 Integration into Global Sustainability Goals

A series of 17 Sustainable Development Goals (SDGs) was proposed by the UN within the framework of its 2030 agenda for sustainable development in association with 191 UN member states. Sustainable production and consumption of biocolours extensively works directly or indirectly on the achievement of following SDGs: 2 (achieve food security and improved nutrition and promote sustainable agriculture), 3 (ensure healthy lives and well-being for all at all ages), 6 (ensure availability and sustainable management of water and sanitation for all), 7 (ensure access to affordable, reliable, sustainable and modern energy for all), 9 (develop robust infrastructure, support inclusive and sustainable industrialization and foster innovation), 12 (ensure sustainable consumption and production patterns), 14 (sustainable utilization of water resources), 15 (protect, restore and promote sustainable utilization of terrestrial ecosystems, sustainably manage forests, combat desertification, and halt and reverse land degradation and halt biodiversity loss) (Department of Economic and Social Affairs, UN). Global partnerships like UN-PAGE, i.e., Partnership for Action on Green Economy, work collaborately with partner countries to remodel their existing economies in alignment with SDGs through an inclusive green economy approach (Partnership for Action on Green Economy 2025). Aligning the circular bioeconomy with UN SDGs can lead to significant long-term impact on earth. For example, utilization of agro-industrial waste as a primary source for the biotechnological production of biocolours. The economically sustainable production of valuable metabolites and efficient management of waste promotes economic growth. Further innovations in sustainable technologies require strategic planning and establishment of a sustainable framework (de Oliveira et al. 2024a, b).

11.6.3 Alignment with Clean Label Trends in the Food and Cosmetics Industries

At present, consumers are searching for healthy, balanced, and minimally processed dietary choices. The "clean label" trend has compelled the food industry to manufacture products to match consumer preferences. Till now, the definition of "clean label" is unclear, as a result manufacturers use different labelling approaches to convey claims of their food products being "natural," "fresh," and free from industrial ingredients. The rise in use of food labels is proportional to the increasing number of consumers reading both front package claims, particularly organic and natural, as well as back package details, including list of ingredients and nutritional information (Rose et al. 2020). Consumer awareness and mindfulness of their health has pushed the demand of "clean-label" products in the market, therefore, insist on transparency in the list of ingredients, assuring sustainable food quality, food safety, and security. Product involvement may be the most important consideration for people who are willing to pay a high price. In the modern era of natural skin care and clean label cosmetics, the entire beauty business should be assessed from the standpoint of safety and sustainability. Customer's knowledge of safe and non-toxic skin care products is linked to their willingness to purchase clean label cosmetics. Clean beauty targets to minimize waste production, water utilization, carbon emissions, and recycling of product containers through the use of sustainable ingredients (Lee and Han 2022). Due to the increasing toxicity and carcinogenicity of synthetic dyes, the cosmetics industry is also shifting to safer pigment substitutes. As a result, businesses in the cosmetics sector are spending more money in investigating novel natural pigment sources and utilizing them efficiently to formulate safe, ecofriendly, cost-effective, and more efficiently functional cosmetic products (Agarwal et al. 2023).

11.7 Future Innovations

11.7.1 Development of Sustainable Multifunctional Biocolours Exhibiting Antimicrobial or UV Protective Properties

Microorganisms synthesize coloured pigments that are made up of diverse chemical components with a range of biological activities. Bacteria, fungi, and microalgae have the potential to meet the increased demand of natural pigments (Sajjad et al. 2020). Filamentous fungi (ascomycetous fungi) synthesize a spectrum of tints or pigments and colours comprising yellow, red, purple, orange, green, brown, and blue. These pigments exhibit diverse pharmacological properties such as antibacterial, antiproliferative, antioxidant, immunomodulatory, and anticancer effects. Such fungal pigments produce minimal or no mycotoxin and have increased water solubility. The investigation is still required for identification of such different biodiversities like fungal biodiversity for biocolours (Toma et al. 2023).

Bacterial pigments with antimicrobial properties can serve as a replacement for traditional antibiotics. Infectious microbial strains such as methicillin-resistant *Staphylococcus* strains are susceptible to pigments like undecylprodigiosin extracted from *Streptomyces* species. Moreover, some pigments have antiviral, antifungal, and antiparasitic activities. For example, *Chromobacterium violaceum* synthesize violacein pigment, which inhibits viral replication and shows antifungal activity too. Bacterial pigments such as prodigiosin and violacein possess antiparasitic properties making them effective against *Plasmodium* and *Trypanosoma* species. These microbial bioactive pigments offer novel direction in the production of safer and more sustainable antimicrobial and antiparasitic drugs to address emerging drug resistance (Bashir et al. 2024). In addition, pyocyanin pigment is a stable biocolourant agent synthesized by *P. aeruginosa* in 90–95% of cases, have distinct physical and chemical characteristics contributing in its broad spectrum of biological activities, including antimicrobial, antifungal, antibiofilm, anticancer, anti-inflammatory, antioxidant, and anti-neurodegenerative effects (Abdelaziz et al. 2023).

Plant pigments have been utilized by humans as biocolourants since early life. We need to prioritize the future innovations in the pigment technology to enhance the quality and quantity of the productivity along with strengthening the characteristics of plant pigments. Skin of red onions (*Allium cepa*) biosynthesize anthocyanins and polyphenolic flavonoids which can be utilized as biocolourants. They exhibit notable UV-shielding efficacy because of their complicated composition but at low concentrations. To meet the elevated demands of the textile industry, we can use combination of natural dyes produced from plants with mordants like chitosan made from biomass by-products (Grande et al. 2023). Synthetic UV absorbers are commonly utilized in textiles, but with increasing customer awareness, the demand of sustainable alternatives is also increasing. Natural dyes extracted from *Solanum nigrum*, *Amaranthus viridis*, and orange peel are rich in polyphenols, which can significantly improve the UV protection of fabrics by increasing their Ultraviolet Protection Factor (UPF) to excellent levels (50+). Natural dyes, particularly those containing flavonoids and tannins, have the potential to absorb UV radiations and provide protection by scavenging free radicals. Repeated or prolonged exposure to UV and washing however can reduce this protection. Mordants, such as iron and chitosan, can help to maintain the UV protection after multiple washes. Some natural dyes harvested from grape seed extracts can ensure sustained protection without mordants even after numerous washing cycles (Afonso et al. 2023). In cosmetic industry also, natural pigments can be of great use. Long-term exposure to UV radiation can trigger dermatoheliosis, commonly referred to as photoaging and has the potential to harm DNA which intensifies the risk of developing skin cancer. As a result, sunscreens and moisturizers are now incorporated with pigments that prevent UV damage. The majority of extremophiles synthesize pigments that help them to protect their DNA from damage. Cyanobacteria biosynthesize the UV-A-inducible pigment scytonemin, has the ability to absorb both UV-A and UV-B light, and protect us against harmful UV radiations (Agarwal et al. 2023).

11.7.2 Enhancing Biocolour Longevity and Vibrancy for Industrial-Scale Applications

In the cryosphere, pigments are charismatic traits of the cold-adapted fungi and bacteria that synthesize a variety of pigments as a part of its defence system to protect themselves against environmental stressors like low temperatures, oxidative stressors, and UV radiations. They could serve as a source for natural pigment production, along with their metabolites, on large scale for industrial and biotechnological uses (Sajjad et al. 2020). The *Monascus* genus dominates the industrial production of natural-based pigment in the food sector in Asian nations and has received worldwide attention, as its pigments have multiple health advantages, including cholesterol-lowering, anti-cancer, anti-mutagenic, anti-obesity, anti-diabetic, and anti-inflammatory properties. The possibility of using the innovative approach of using waste produced in large amounts by breweries, such as banana peel, could be an inexpensive fermentable feedstock for the pigment production and is an eco-friendly method for industrial pigment and enzyme application along with sustainable waste management (Silbir and Goksungur 2019; Louhasakul et al. 2023).

Compared to chemical dyeing, the enzymatic method of hair colouring offers a number of significant benefits. Enzyme-based reactions produce non-toxic waste and are less polluting, biodegradable, user-friendly, and simple to handle. The thermostable bacterial laccase-catalysed dyes have recently gained attention. The dye penetrated the hair well without affecting its surface morphology. It had a pleasing colour spectrum that satisfied consumer needs and demonstrated a high level of resistance to fading from shampooing and pH variations (Panwar et al. 2022). In order to prevent biocolours from deteriorating during the extraction techniques, new, moderate, or non-thermal extraction techniques are emerging. In order to elevate efficiency, produce more colour, and utilize fewer resources, modern procedures are replacing or supplementing traditional methods like boiling, solvent extraction, and ultrasonic-assisted extraction. In line with the global sustainable development goals of 2030, these more recent approaches seek to be more sustainable. To increase biocolour stability, shield pigments from heat and enzymatic destruction, and increase bioavailability, methods like high pressure, enzymes, ultrasound, microwave-assisted extraction, pulsed electric fields, ohmic heating, and supercritical fluid extraction are under investigation. Biocolours are also protected from processing conditions via nanoencapsulation. These advanced techniques are said to be better than traditional ones since they are quick, energy-efficient, and have no effect on the sensory qualities of food (Renita et al. 2023).

11.7.3 Potential of Genetically Engineered Microorganisms for Customized Pigment Production

In order to eliminate toxic synthetic dyes, efforts should be made to optimize these bacteria for the synthesis of colours at an affordable cost and to improve strains through genetic engineering. Combinatorial genetic engineering is currently under investigation, in accordance with the growing number of carotenogenic gene sequences. Through gene combination or the creation of novel carotenoids, including ones that have never been reported previously, like multi-hydroxylated carotenoids, which could have strong antioxidant properties, researchers were able to produce more efficient biosynthesis. Utilizing conventional strain enhancement procedures, advanced genetic engineering methods for strain development, genome shuffling, and fermentation techniques to scale up production to industry levels can also result in the sustainable production of high-use microbial pigments. Large-scale biopigment manufacturing will require modern biotechnological interventions in biorefineries to ensure successful commercialization and, eventually, to strengthen the bioeconomy (Toma et al. 2023). Genetic modifications, such as over-expressing key enzymes or inserting/deleting specific genes, help to improve pigment yield and overcome the limitations of naturally occurring production. Engineered strains of *Corynebacterium glutamicum* and *Escherichia coli* have been developed to produce carotenoids like β-carotene and zeaxanthin, while *Rhodobacter sphaeroides* has been modified for lycopene production (Bashir et al. 2024). Canthaxanthin, an orange-red keto-carotenoid, is found in animals including crustaceans, some fish, and birds, as well as microorganisms like various bacterial species, fungi, and algae. Canthaxanthin is widely used as a colouring agent in a variety of food products, in the pharmaceutical industry, as a supplement to animal feed, and in cosmetics with powerful antioxidant properties. Novel strategies that combine genetic engineering and molecular biology have been investigated to make new transgenic strains with high canthaxanthin yields that can be produced from genetically modified organisms with improved environmental tolerance, bringing a new era and revolution in the pigment business (Gaur and Bera 2023).

Metabolic engineering, combined with cost-reducing strategies using industry by-products, has significantly advanced the production of microbial pigments. Genetic engineering allows the optimization of pigment production by modifying genes or introducing new biosynthetic pathway to enhance yield. Techniques like multi-omics and AI-driven approaches are used to analyse and predict microbial growth and optimize production processes. Examples include improving lutein carotenoid production in *Chlorella saccharophila* and carotenoid production in *Isochrysis galbana*. Synthetic biology also plays a crucial role by enabling the control and improvement of pigment biosynthesis, such as increased carotenoid production in yeast (*Saccharomyces cerevisiae*) and bacteria. CRISPR/Cas9 is another promising tool to genetically engineer microbes for natural pigment production. The CRISPR-Cas9 system is reliable for the reconstruction and optimization of biosynthetic pathways. Marine environments are emerging as rich sources of novel pigments with unique properties, including antimicrobial activity, which could be useful in diverse industries. The study of marine microbes offers the possibility of discovering more stable and durable natural pigments for industrial applications (Barreto et al. 2023).

11.8 Conclusion

The modern inclination in society for "natural" pressured the manufacturers to find new sources of biocolours. Currently, microbial pigments, plant pigments, and fungal pigments are becoming the centre of attraction in the pigment industry. Their role as colourants and additives with enhanced formulations comprising bioactive features such as antioxidants, antimicrobials, antivirals, anti-inflammatories, UV protectives, flame-retardants, insect-repellents, and many more is highly appreciated. The production, utilization, and commercialization of biocolours require additional investments, laws, and regulations. The marketing approach used by the industry put more emphasis on the presence of natural pigments and colour in products than on their acknowledged health benefits. However, while adhering to existing regulations, the food colouring sector offers an untapped opportunity to improve consumer education and elucidate the biological properties and potential health advantages of natural colourants. With advantages for the economy, ecology, and industry, microbial pigments are an underappreciated substitute for conventional mineral and synthetic colours. Despite ongoing difficulties, large-scale production has a lot of potential for the food, cosmetics, textile, and pharmaceutical industries. The public's perception of natural colours is being positively influenced by their safety, non-toxicity, and environmental friendliness. As public awareness of environmentally friendly products that come from renewable resources evolved, biocolours attracted researchers from traditional and diversified operations to develop effective eco-friendly and cleaner process approaches. Understanding the practical and financial feasibility of these procedures should be the aim of scale-ups and their use in industrial settings.

References

Abdelaziz AA, Abo Kamer AM, Al-Monofy KB, Al-Madboly LA (2023) *Pseudomonas aeruginosa's* greenish-blue pigment pyocyanin: its production and biological activities. Microb Cell Factories 22:110

Adadi P, Barakova NV, Krivoshapkina EF (2018) Selected methods of extracting carotenoids, characterization, and health concerns: a review. J Agric Food Chem 66(24):5925–5947

Afonso TB, Bonifacio-Lopes T, Costa EM, Pintado ME (2023) Phenolic compounds from by-products for functional textiles. Materials 16:7248

Agarwal H, Bajpai S, Mishra A, Kohli I, Varma A, Fouillaud M, Dufossé L, Joshi NC (2023) Bacterial pigments and their multifaceted roles in contemporary biotechnology and pharmacological applications. Microorganisms 11:614

Azmin SNHM, Sulaiman NS, Nor MSM, Abdullah PS, Kari ZA, Pati S (2022) A review on recent advances on natural plant pigments in foods: functions, extraction, importance and challenges. Appl Biochem Biotechnol 194:4655–4672

Bai F, Bian K, Huang X, Wang Z, Fan H (2019) Pressure induced nanoparticle phase behavior, property, and applications. Chem Rev 119(12):7673–7717

Barreto JVDO, Casanova LM, Junior AN, Reis-Mansur MCPP, Vermelho AB (2023) Microbial pigments: major groups and industrial applications. Microorganisms 11(12):2920

Bashir I, Pandey VK, Dar AH, Dash KK, Shams R, Mir SA et al (2024) Exploring sources, extraction techniques and food applications: a review on biocolours as next-generation colourants. Phytochem Rev 23(4):1–26

Bora P, Das P, Bhattacharyya R, Barooah MS, Saikia Barooah M (2019) Biocolour: the natural way of colouring food. J Pharma Phytochem 8(3):3663–3668

Chambaud M, Colas C, Destandau E (2023) Water-based microwave-assisted extraction of pigments from madder optimized by a box–behnken design. Separations 10(8):433

Chavan RB (2011) Environmentally friendly dyes. In: Handbook of textile and industrial dyeing. Woodhead Pub, Cambridge, pp 515–561

Contreras-Pereda N, Suárez-García S, Pfattner R, Ruiz-Molina D (2024) Melanin-inspired conductive thin films for multimodal-sensing wearable on-skin electronics. Mater Today Chem 35:101855

Coussement P, Bauwens D, Maertens J, DeMey M (2017) Direct combinatorial pathway optimization. ACS Synth Biol 6(2):224–232

Dahmoune F, Spigno G, Moussi K, Remini H, Cherbal A, Madani K (2014) *Pistacia lentiscus* leaves as a source of phenolic compounds: microwave-assisted extraction optimized and compared with ultrasound-assisted and conventional solvent extraction. Ind Crop Prod 61:31–40

Das S, Nadar SS, Rathod VK (2021) Integrated strategies for enzyme assisted extraction of bioactive molecules: a review. Int J Biol Macromol 191:899–917

Das A, Basak S, Chakraborty S, Dhibar M (2022) Microwave: an ecologically innovative, green extraction technology. Curr Anal Chem 18(8):858–866

De Mol ML, Vandamme EJ (2024) Arts, cultural heritage, sciences and micro–/bio–/techno–/logy: impact of biomaterials and biocolourants from antiquity till today. J Ind Microbiol Biotechnol:kuae049

de Oliveira CRS, de Oliveira PV, Pellenz L, de Aguiar CRL, da Silva Júnior AH (2024a) Supercritical fluid technology as a sustainable alternative method for textile dyeing: an approach on waste, energy, and CO_2 emission reduction. J Environ Sci 140:123–145

de Oliveira F, Dias AC, Sanchez-Munoz S, Balbino TR, Santos-Ebinuma V, da Silva SS (2024b) Sustainability feasibility of fungi-based biocolourants by biotechnological routes. Chem Eng J 494:152942

Department of Economic and Social Affairs, UN, https://sdgs.un.org/

Duerr S (2011) The handbook of natural plant dyes: personalize your craft with organic colours from acorns, blackberries, coffee, and other everyday ingredients. Timber Press

Dufossé L (2018) Microbial pigments from bacteria, yeasts, fungi, and microalgae for the food and feed industries. In: Grumezescu AM, Holban AM (eds) Handbook of food bioengineering, natural and artificial flavoring agents and food dyes. Academic, pp 113–132

Fact.MR (2024). https://www.factmr.com/report/1050/bio-based-pigments-and-dyes-Market

Fathi Z, Tramontin LRR, Ebrahimipour G, Borodina I, Darvishi F (2021) Metabolic engineering of *Saccharomyces cerevisiae* for production of β-carotene from hydrophobic substrates. FEMS Yeast Res 21(1):foaa068

Furubayashi M, Kubo A, Takemura M, Otani Y, Maoka T, Terada Y et al (2021) Capsanthin production in *Escherichia coli* by overexpression of capsanthin/capsorubin synthase from *Capsicum annuum*. J Agric Food Chem 69(17):5076–5085

G20 India, https://www.g20.in/

Ganguly S, Bandyopadhyay K, Dutta P, Pramanik A (2019) A review on microbial-pigment: a good source of biocolour. Pharma Innov J 8:416–420

Gaur V, Bera S (2023) Microbial canthaxanthin: an orange-red keto carotenoid with potential pharmaceutical applications. Biotechnologia 104(3):315–328

Ghosh S, Sarkar T, Das A, Chakraborty R (2022) Natural colourants from plant pigments and their encapsulation: an emerging window for the food industry. LWT 153:112527

Ghosh J, Repon MR, Pranta AD, Rupanty NS, Khan F, Noor T (2025) Bioactive component integrated textiles: a promising source of medicine and healthcare. J Eng Fibers Fabrics 20:15589250241308561

Grande R, Raisanen R, Dou J, Rajala S, Malinen K, Nousiainen PA, Osterberg M (2023) In situ adsorption of red onion (*Allium cepa*) natural dye on cellulose model films and fabrics exploiting chitosan as a natural mordant. ACS Omega 8:5451–5463

Henke NA, Heider SA, Peters-Wendisch P, Wendisch VF (2016) Production of the marine carotenoid astaxanthin by metabolically engineered *Corynebacterium glutamicum*. Mar Drugs 14(7):124

Jones JA, Vernacchio VR, Collins SM, Shirke AN, Xiu Y, Englaender JA et al (2017) Complete biosynthesis of anthocyanins using *E. coli* polycultures. mBio 8(3):e00621–e00617

Jordan J, Lima TDMD, Nugent M, Laaksonen P (2024) Effect of organic biocolourants on lightfastness and mechanical properties of polylactic acid. Colour Technol:1–11

Kalpana S, Priyadarshini SR, Leena MM, Moses JA, Anandharamakrishnan C (2019) Intelligent packaging: trends and applications in food systems. Trends Food Sci Technol 93:145–157

Kalra R, Conlan XA, Goel M (2020) Fungi as a potential source of pigments: harnessing filamentous fungi. Front Chem 8:369

Kanha N, Osiriphun S, Rakariyatham K, Klangpetch W, Laokuldilok T (2022) On-package indicator films based on natural pigments and polysaccharides for monitoring food quality: a review. J Sci Food Agric 102(15):6804–6823

Kaur P, Singh S, Ghoshal G, Ramamurthy PC, Parihar P, Singh J, Singh A (2022) Valorization of agri-food industry waste for the production of microbial pigments: an eco-friendly approach. In: Advances in agricultural and industrial microbiology: volume 1: microbial diversity and application in agroindustry. Springer, Singapore, pp 137–167

Khan SA, Aslam R, Makroo HA (2019) High pressure extraction and its application in the extraction of bio-active compounds: a review. J Food Process Eng 42(1):e12896
Khattab TA, Abdelrahman MS, Rehan M (2020) Textile dyeing industry: environmental impacts and remediation. Environ Sci Pollut Res 27(4):3803–3818
Larocca V, Martino M, Trupo M, Magarelli RA, Spagnoletta A, Ambrico A (2023) Evaluation of carbon dioxide supercritical fluid extraction (CO_2-SFE) on carotenoids recovery from red yeast cells. Biomass Convers Biorefinery:1–10
Lee J, Han KK (2022) Sustainable changes in beauty market trends focused on the perspective of safety in the post-COVID-19 period: literature review. J Cosmet Dermatol 21:2700–2707
Li D, Liu J, Wang X, Kong D, Du W, Li H et al (2018) Biological potential and mechanism of prodigiosin from Serratia marcescens subsp. lawsoniana in human choriocarcinoma and prostate cancer cell lines. Int J Mol Sci 19(11):3465
Lian J, HamediRad M, Zhao H (2018) Advancing metabolic engineering of *Saccharomyces cerevisiae* using the CRISPR/Cas system. Biotechnol J 13(9):1700601
Loan LTK, Vinh BT, Tai NV (2024) Ultrasound-enzyme assisted extraction of polyphenol and amylase enzyme from *Vietnamese ramie* leaves: predictive modeling development and optimization study. J Microbiol Biotechnol Food Sci 13(5):e10700
Lombardelli C, Benucci I, Mazzocchi C, Esti M (2021) A novel process for the recovery of betalains from unsold red beets by low-temperature enzyme-assisted extraction. Foods 10(2):236
Louhasakul Y, Wado H, Lateh R, Cheirsilp B (2023) Solid-state fermentation of Saba banana peel for pigment production by *Monascus purpureus*. Braz J Microbiol 54:93–102
Ludwig K, Rihko-Struckmann L, Brinitzer G, Unkelbach G, Sundmacher K (2021) β-carotene extraction from *Dunaliella salina* by supercritical CO_2. J Appl Phycol 33:1435–1445
Ma T, Shi B, Ye Z, Li X, Liu M, Chen Y et al (2019) Lipid engineering combined with systematic metabolic engineering of *Saccharomyces cerevisiae* for high-yield. Metab Eng 52:134–142
Majik MS, Gawas UB (2023) Recent advances in extraction of natural compounds. In: Meena SN, Nandre V, Kodam K, Meena RS (eds) New horizons in natural compound research. Academic, pp 17–33
Malik P, Sarkhel D (2024) Advancements in bio-based materials and low-voc formulations: paving the way for sustainable innovation in the coatings industry. J Coat Technol Innov 2(2):34–38
Mariotti N, Bonomo M, Fagiolari L, Barbero N, Gerbaldi C, Bella F, Barolo C (2020) Recent advances in eco-friendly and cost-effective materials towards sustainable dye-sensitized solar cells. Green Chem 22(21):7168–7218
Martínez JM, Delso C, Álvarez I, Raso J (2020) Pulsed electric field-assisted extraction of valuable compounds from microorganisms. Compr Rev Food Sci Food Saf 19(2):530–552
Martini S, Afroze S, Roni KA, Setiawati M, Kharismadewi D (2021) A review of fruit waste-derived sorbents for dyes and metals removal from contaminated water and wastewater. Desalin Water Treat 235:300–323
Mayegowda SB, Chikkud V, Barua S, Manjula NG (2025) Plant, animal, and microbial sources of dyes and mordants. In: Advances in natural dyes for environmental protection. Apple Academic Press, pp 61–95
Mhalaskar SR, Thorat SS, Kulthe AA (2018) Stability analysis of food bio-colour extracts from broken rice through solid state fermentation. Int J Pure Appl Biosci 6(6):304–313
Morales-Oyervides L, Ruiz-Sánchez JP, Oliveira JC, Sousa-Gallagher MJ, Méndez-Zavala A et al (2020) Biotechnological approaches for the production of natural colourants by talaromyces/penicillium: a review. Biotechnol Adv 43:107601
Mounika D, Rao BE, Babji Y, Ambedkar YR, Rao TS (2023) Impact of different concentrations of cinnamon essential oil on the physico-chemical, microbiological, and sensory characteristics of refrigerated pork sausages. J Meat Sci 18(2):31
Murthy HN, Joseph KS, Paek KY, Park SY (2023) Anthocyanin production from plant cell and organ cultures in vitro. Plants 13:117
Narsing Rao MP, Xiao M, Li WJ (2017) Fungal and bacterial pigments: secondary metabolites with wide applications. Front Microbiol 8:1113

Pachaiyappan A, Sadhasivam G, Kumar M, Muthuvel A (2021) Biomedical potential of astaxanthin from novel endophytic pigment producing bacteria Pontibacter korlensis AG6. Waste Biomass Valoriz 12:2119–2129

Panwar P, Dey B, Sheikh JN, Dutta T (2022) Thermostable bacterial laccase for sustainable dyeing using plant phenols. RSC Adv 12:18168

Parmar M, Phutela UG (2015) Biocolours: the new generation additives. Int J Curr Microbiol App Sci 4(7):688–694

Partnership for Action on Green Economy (2025) https://www.un-page.org/

Podshivalov A, Toropova A, Fokina M, Uspenskaya M (2020) Surface morphology formation of edible holographic marker on potato starch with gelatin or agar thin coatings. Polymers 12(5):1123

Poorniammal R, Prabhu S, Sakthi AR (2022) Biocolourants and their applications in food industry. Biotica Res Today 4(7):524–528

Pranta AD, Rahaman MT (2024) Extraction of eco-friendly natural dyes and biomordants for textile colouration: a critical review. Nano-Struct Nano-Objects 39:101243

Priyadarshi R, Ezati P, Rhim JW (2021) Recent advances in intelligent food packaging applications using natural food colourants. ACS Food Sci Technol 1(2):124–138

Puri M, Sharma D, Barrow CJ (2012) Enzyme-assisted extraction of bioactives from plants. Trends Biotechnol 30(1):37–44

Puri RG, Dalal KS, Dhuldhaj U, Chaudhari BL (2024) Biocides for coating formulations: current status and future challenges. In: Industrial microbiology and biotechnology: a new horizon of the microbial world. Springer Nature, Singapore, pp 605–633

Ramesh C, Vinithkumar NV, Kirubagaran R, Venil CK, Dufossé L (2020) Applications of Prodigiosin extracted from marine red pigmented bacteria *Zooshikella* sp. and actinomycete *Streptomyces* sp. Microorganisms 8:556

Rana B, Bhattacharyya M, Patni B, Arya M, Joshi GK (2021) The realm of microbial pigments in the food colour market. Front Sustain Food Syst 5:603892

Ren XY, Jin XY, Zong W (2023) Optimization of ultra-high pressure-assisted extraction of total phenols from *Eucommia ulmoides* leaves by response surface methodology. Open Chem 21(1):20220309

Renita AA, Gajaria TK, Sathish S, Kumar JA, Lakshmi DS, Kujawa J, Kujawski W (2023) Progress and prospective of the industrial development and applications of eco-friendly colourants: an insight into environmental impact and sustainability issues. Foods 12(7):1521

Rose N, Reynolds T, Kolodinsky J (2020) P90 consumer use of food labels increases as "clean label" trend continues. J Nutr Educ Behav 52(7):S58–S59

Şahin S, Nasir NTBM, Erken İ, Çakmak ZE, Çakmak T (2019) Antioxidant composite films with chitosan and carotenoid extract from *Chlorella vulgaris*: optimization of ultrasonic- assisted extraction of carotenoids and surface characterization of chitosan films. Mater Res Express 6(9):095404

Sajjad W, Din G, Rafiq M, Iqbal A, Khan S, Zada S, Ali B, Kang S (2020) Pigment production by cold-adapted bacteria and fungi: colourful tale of cryosphere with wide range applications. Extremophiles 24:447–473

Samanta KK, Basak S, Chattopadhyay SK (2015) Recycled fibrous and nonfibrous biomass for value-added textile and nontextile applications. In: Environmental implications of recycling and recycled products. Springer, Singapore, pp 167–212

Savic Gajic IM, Savic IM, Gajic DG, Dosic A (2021) Ultrasound-assisted extraction of carotenoids from orange peel using olive oil and its encapsulation in ca-alginate beads. Biomolecules 11(2):225

Seidu RK, Eghan B, Acquaye R (2024) A review of circular fashion and bio-based materials in the fashion industry. Circ Econ Sustain 4(1):693–715

Sen T, Barrow CJ, Deshmukh SK (2019) Microbial pigments in the food industry-challenges and the way forward. Front Nutr 6:7

Shaikh Z (2016) Biosynthesis of prodigiosin and its applications. IOSR J Pharm Biol Sci 11(6):1–28

Shamim G, Ranjan SK, Pandey DM, Ramani R (2014) Biochemistry and biosynthesis of insect pigments. Eur J Entomol 111(2):149–164
Sharma D (2014) Understanding biocolour-a review. Int J Sci Technol Res 3(1):294–299
Sharma A, Mazumdar B, Keshav A (2021) Valorization of unsalable *amaranthus tricolour* leaves by microwave-assisted extraction of betacyanin and betaxanthin. Biomass Convers Biorefinery 13:1251–1267
Sijtsema SJ, Onwezen MC, Reinders MJ, Dagevos H, Partanen A, Meeusen M (2016) Consumer perception of bio-based products: an exploratory study in 5 European countries. NJAS Wageningen J Life Sci 77:61–69
Silbir S, Goksungur Y (2019) Natural red pigment production by *Monascus purpureus* in submerged fermentation systems using a food industry waste: Brewer's spent grain. Foods 8:161
Singh T, Pandey VK, Dash KK, Zanwar S, Singh R (2023) Natural bio-colourant and pigments: sources and applications in food processing. J Agric Food Res 12:100628
Srilekha V, Krishna G, Mahender P, Charya MS (2018) Investigation of in vitro cytotoxic activity of pigment extracted from *Salinicoccus* sp. isolated from Nellore Sea coast. J Mar Med Soc 20(1):31–33
Stanek-Wandzel N, Krzyszowska A, Zarębska M, Gębura K, Wasilewski T, Hordyjewicz-Baran Z, Tomaka M (2024) Evaluation of cellulase, pectinase, and hemicellulase effectiveness in extraction of phenolic compounds from grape pomace. Int J Mol Sci 25(24):13538
Sundaramahalingam MA, Sivashanmugam P (2023) Production of microbial carotenoid using innate inherent of the food industry wastewater. J Mol Liquid 388:122772
Toma MA, Rahman MH, Rahman MS, Arif M, Nazir KHMNH, Dufossé L (2023) Fungal pigments: carotenoids, riboflavin, and polyketides with diverse applications. J Fungi 9:454
UNEP (2025) https://www.unep.org/
Usmani Z, Sharma M, Sudheer S, Gupta VK, Bhat R (2020) Engineered microbes for pigment production using waste biomass. Curr Genomics 21(2):80–95
Uwineza PA, Waskiewicz A (2020) Recent advances in supercritical fluid extraction of natural bioactive compounds from natural plant materials. Molecules 25:3847
Wang Q, Liu D, Yang Q, Wang P (2017) Enhancing carotenoid production in *Rhodotorula mucilaginosa* KC8 by combining mutation and metabolic engineering. Ann Microbiol 67:425–431
Wang M, Zhou J, Castagnini JM, Berrada H, Barba FJ (2023) Pulsed electric field (PEF) recovery of biomolecules from *chlorella*: extract efficiency, nutrient relative value, and algae morphology analysis. Food Chem 404:134615
Xinrui D, Bo L, Yihong B, Weifeng L, Yong T (2023) Metabolic engineering of *Escherichia coli* for high-level production of violaxanthin. Microb Cell Factories 22(1):115
Xu S, Gao S, An Y (2023) Research progress of engineering microbial cell factories for pigment production. Biotechnol Adv 65:108150
Yusuf M, Shabbir M, Mohammad F (2017) Natural colourants: historical, processing and sustainable prospects. Nat Prod Bioprospect 7:123–145
Zha J, Koffas MA (2017) Production of anthocyanins in metabolically engineered microorganisms: current status and perspectives. Synth Syst Biotechnol 2(4):259–266
Zha J, Zang Y, Mattozzi M, Plassmeier J, Gupta M, Wu X et al (2018) Metabolic engineering of *Corynebacterium glutamicum* for anthocyanin production. Microb Cell Factories 17:1–13

MIX
Papier aus verantwortungsvollen Quellen
Paper from responsible sources
FSC® C105338

If you have any concerns about our products,
you can contact us on
ProductSafety@springernature.com

In case Publisher is established outside the EU,
the EU authorized representative is:
Springer Nature Customer Service Center GmbH
Europaplatz 3, 69115 Heidelberg, Germany

Printed by Libri Plureos GmbH
in Hamburg, Germany